NEW CELL ADHESION RESEARCH

NEW CELL ADHESION RESEARCH

PATRICK NOTT
AND
MATTHEW TEMPLE
EDITORS

Nova Biomedical Books
New York

For permission to use material from this book please contact us:
Telephone 631-231-7269; Fax 631-231-8175
Web Site: http://www.novapublishers.com

LIBRARY OF CONGRESS CATALOGING-IN-PUBLICATION DATA

New cell adhesion research / editors, Patrick Nott & Matthew Temple.
p. ; cm.
Includes bibliographical references and index.
ISBN 978-1-60692-378-8 (hardcover)
1. Cell adhesion. 2. Cell adhesion molecules. I. Nott, Patrick. II. Temple, Matthew.
[DNLM: 1. Cell Adhesion. 2. Cell Adhesion Molecules. 3. Signal Transduction. QU 55.7 N532 2009]
QH623.N49 2009
611'.0181--dc22

2008047205

Published by Nova Science Publishers, Inc. ✦ New York

CONTENTS

PREFACE

Cell adhesion is the binding of a cell to another cell or to a surface or matrix. This book discusses the mitochondrial retrograde signaling, a recently discovered cellular signaling pathway, that may work in irradiated mammalian cells. The question whether radiation of visible and near IR radiation can activate this cellular signaling pathway is also reviewed.

Novel experimental techniques that probe cellular forces exerted at the sites of adhesive contacts are examined, as well as how these forces can explain the roles of cell adhesion in orchestrating multi-cellular behavior. Furthermore, the cell adhesion molecule trophinin is discussed, which is uniquely involved in the process of human embryo implantation. The roles of nano and microtopography in the substratum upon cell adhesion are addressed, as well as the effects of substratum and near-substratum mechanical properties on cell adhesion.

CADM1, a new mast-cell adhesion molecule that mediates interaction with fibroblasts, nerves, and smooth muscles is discussed in this book, as well as carbon nanoparticles that are used as substrates for cell adhesion and growth. In addition, by reducing cell adhesion and inducing cell proliferation, diabetes increases the risk for oral carcinogenesis. The results of this animal model study are examined as well. Finally, the epithelial cell adhesion molecule, EpCAM, has been studied and subsequently discussed in this book.

Chapter I - The mitochondrial retrograde signaling is an information channel between mitochondrial respiratory chain and the nucleus for the transduction signals regarding the functional state of the mitochondria. The present review examines the question whether radiation of visible and near IR (IR-A) radiation can activate this cellular signaling pathway. The reactions with the primary photoacceptor, cytochrome c oxidase, as well as signaling pathways between mitochondria, plasma membrane, and nucleus were explored on the cell adhesion model using various chemicals that modify different enzyme systems. Functions of cytochrome c oxidase as a signal generator as well as a signal transducer in irradiated cells are outlined. It is concluded that mitochondrial retrograde signaling, a recently discovered cellular signaling pathway, may also be at work in irradiated mammalian cells.

Chapter II - The regulation of cell-extracellular matrix and cell-cell adhesion is essential for tissue development and homeostasis. Stable cell junctions maintain normal tissue architecture, but, during embryogenesis, tissue regeneration and cancer cell migration, adhesive contacts loosen and become traction sites for cellular rearrangement. Previous studies have identified essential molecules for the assembly and maintenance of cell adhesion. Yet, we know very little about the mechanical role of cell adhesion as a site of force transmission. Cellular traction forces are generated by the actin cytoskeleton, and the remodeling of the actin network is thought to regulate the strength and dynamics of cell adhesion. The precise molecular linkage between the adhesion complex and the actin cytoskeleton, however, remains elusive. This review discusses novel experimental techniques that probe cellular forces exerted at the sites of adhesive contacts, and how these forces can explain the roles of cell adhesion in orchestrating multi-cellular behavior.

Chapter III - Morphological observations of human and monkey embryo implantation sites indicate that blastocyst trophectoderm cells that have adhered to the uterus proliferate and invade, whereas trophectoderm not in contact with the uterine epithelium remains a monolayer. This finding suggests that the initial adhesion triggers activation of cells in trophectoderm. Adhesion molecules which are specifically up-regulated in the plasma membrane during early pregnancy are the subject of much interest because they may function as receptors for the blastocyst. This review described the cell adhesion molecule trophinin uniquely involved in the process of human embryo implantation. This review also covers other cell adhesion molecules, integrins, cadherins, selectins, basigin, galectins and tissue transglutaminase.

Chapter IV - The ability of cancer cells to interact with their surrounding environment directly affects many aspects of their behaviour. In addition to interactions with extracellular matrix constituents that can promote cell survival, changes in intercellular adhesive contacts can facilitate cell dissociation and migration. The activity of Src, a nonreceptor protein tyrosine kinase, and associated Src family members, can regulate the signalling and function of cell-cell adhesion proteins; these kinases are frequently elevated in tumour tissue and cells where they are associated with a gain in aggressive cell features including increased migration and invasion. Subsequently, the ability of inhibitors of Src family kinases to positively regulate cell-cell adhesion and suppress cell migration and invasion has been recently demonstrated. Here, the review the current understanding of the interplay between Src and receptors that mediate cell-cell adhesion and the implications for tumour development and spread.

Chapter V - Recent publications and the theories arising from their results are surveyed for three areas of work on cell-substratum and cell-cell adhesion. These areas are 1. The roles of nano and microtopography in the substratum upon cell adhesion. 2. The effects of substratum and near-substratum mechanical properties on cell adhesion. 3. Mechanotransduction effects and cell adhesion. It is concluded that all three are important and that the patterns of force interaction with the cell may pay important roles not only in adhesion but also in cell spreading and gene expression activation.

Chapter VI - Mast cells are a native composer of connective tissue of the skin dermis and intestinal and respiratory mucosa, and their interaction with fibroblasts is important for mast cell development and survival, because mast cells express c-kit receptor tyrosine kinase and fibroblasts express c-kit ligand. Part of mast cells are in close approximation with nerves in the skin and intestinal and respiratory mucosa. Independent lines of accumulated evidence indicate the existence of an intensive bidirectional crosstalk between mast cells and sensory nerves and suggest that mast cells and sensory nerves may be viewed as a functional unit, which could be of crucial importance in neuroimmunological pathways. Mast cells appear to have a property of influencing smooth muscle function via not only such nerve-mast cell effects, but also direct pathways. In bronchial asthma, mast cells infiltrate the airway smooth muscle layer, and interact directly with smooth muscle cells, suggesting pathogenic roles for mast cells in airway obstruction. Current studies on mast cell biology identified a novel adhesion molecule of mast cells, namely cell adhesion molecule-1, CADM1 [formerly referred to as Immunoglobulin Superfamily member 4 (IGSF4), Spermatogenic Immunoglobulin Superfamily (SgIGSF), Synaptic Cell Adhesion Molecule (SynCAM), Tumor Suppressor in Lung Cancer-1 (TSLC1), or Nectin-like Molecule-2 (Necl-2)]. This molecule is unique, because it serves as not only simple glue but also appears to promote functional communication between mast cells and fibroblasts, nerve or smooth muscle cells.

Chapter VII - Carbon nanoparticles, such as fullerenes, nanotubes and nanodiamonds, have been considered as promising building blocks in nanotechnology for broadscale nanostuctured devices and materials, e.g., microchips, nanorobots and biosensors, carriers for controlled drug and gene delivery or tracers for novel imaging technologies. Soon after macroscopic production was established, fullerenes, as donors and acceptors of electrons, started to be targeted for use in photodynamic therapy against tumor and microbial cells and also for quenching oxygen radicals. It has been discovered that nanodiamonds can also act as antioxidant and anti-inflammatory agents.

The interaction of carbon nanoparticles with cells and tissues has been investigated mainly using suspensions of these particles in cell culture media or other fluids. Relatively little is known about the influence of layers consisting of carbon nanoparticles on cell-substrate adhesion. Carbon nanoparticles can be advantageously used for surface modification of various artificial materials developed for the construction of tissue replacements and other body implants. In the form of films deposited on the surface of implants, carbon nanoparticles can not only improve the mechanical and other physical properties of a body implant, but can also enhance the attractiveness of the implant for cell colonization. The latter effect is probably mediated by the surface nanostructure of the films, which, at least to a certain degree, mimics the architecture of physiological cell adhesion substrates, such as extracellular matrix molecules. On nanostructured substrates, the cell adhesion-mediating molecules are adsorbed in the appropriate spectrum, amount and spatial conformation that make specific sites on these molecules (e.g. amino acid sequences like RGD) accessible to cell adhesion receptors. In addition, the nanostructured surfaces have been reported to enhance the adsorption of vitronectin, preferred by osteoblasts for their adhesion. The electrical conductivity of carbon nanotube films has also had beneficial effects on the growth and maturation of osteoblasts. Thus, the nanoparticles deposited on the bone-anchoring parts of bone, joint or dental replacements can improve the integration of these devices with the surrounding bone tissue. In addition, carbon nanoparticles admixed into polymers, designed for the fabrication of three-dimensional scaffolds for bone tissue engineering, could decorate the walls of the pores in these materials, and thus promote the ingrowth of bone cells. The interaction of cells with substrates modified with carbon nanoparticles could be further intensified by functionalizing these particles with various chemical functional groups or biomolecules, including KRSR-containing adhesion oligopeptides, recognized preferentially by osteoblasts.

This interdisciplinary review, based mainly on the authors'results, deals with the interaction of bone-derived cells with glass, polymeric and silicon substrates modified with fullerenes C_{60}, nanotubes, nanocrystalline diamond as well as other carbon allotropes, such as amorphous hydrogenated carbon or pyrolytic graphite. In general, these modifications have resulted in improved adhesion, growth, viability and osteogenic differentiation of these cells. The use of carbon and/or metal-carbon composite nanoparticles for the construction of patterned surfaces for regionally-selective adhesion and directed growth of cells is also discussed in this work.

Chapter VIII - Oral cancer is the seventh most common malignancy, worldwide. It constitutes a major health problem because it is still associated with poor prognosis and about 50% five-year survival, despite advances in surgery, radiation and chemotherapy treatment over the past two decades. Oral squamous cell carcinoma (OSCC) accounts for more than 90% of oral cancer cases. In the development of OSCC have been implicated environmental factors such as tobacco smoking and heavy alcohol consumption, infectious agents such as some types of human papilloma virus, genetic alterations involving oncogenes and tumor suppressor genes, as well as epigenetic changes. Recently, some epidemiological studies have also incriminated diabetes mellitus as a risk factor for the development of oral premalignant lesions and OSCC.

In order to investigate the possible association of diabetes with increased risk for oral cancer, as well as the underlying molecular mechanism of this association an experimental model of OSCC induced in diabetic and normal Sprauge–Dawley rats was established. Diabetes type I was induced by a single intraperitoneal injection of streptozotocin in 20 rats and verified by measuring glucose levels in blood samples obtained by tail prick in a strip-operated sensor. In 14 of the diabetic rats, as well as in 12 rats without diabetes, chemical carcinogenesis was induced by a five-month application of the carcinogen 4-nitroquinoline N-oxide on the hard palate of the animals. Six diabetic rats as well as 6 non-diabetic rats were used as controls of normal oral tissue without carcinogenesis. The four groups of animals were used in order to study the putative effect of diabetes on signal transduction pathways involved in the sequential stages of oral carcinogenesis: normal mucosa, hyperplasia, dysplasia, early invasion, well differentiated OSCC and moderately differentiated OSCC.

The diabetic rat model was used to investigate whether the molecular basis for the putative association of diabetes with oral oncogenesis involves the insulin receptor and its substrate-1 (IR and IRS-1, respectively), as well as the influence of diabetes on signal transduction pathways involving oncogene proteins EGFR, erbB2, erbB3, FGFR-2, FGFR-3, FAK, c-myc, N-ras, ets-1, H-ras, c-fos and c-jun, tumor suppressor gene proteins p53 and p16, apoptosis-related markers Bax and Bcl-2, and cell proliferation-related marker Ki-67. The obtained findings suggest that the insulin and insulin-like growth factor-1/IR/IRS-1 signalling pathway is altered by diabetes, leading to reduced cell adhesion and possibly increased risk for oral cancer. In addition, diabetes seems to promote the activation of Ras/Raf/MAPK signal transduction pathway, mainly by induction of receptors erbB2 and erbB3, resulting in increased cell proliferation. Finally, diabetes does not seem to influence apoptotic mechanisms during oral carcinogenesis. An interesting extrapolation of these findings is that the type I diabetic patients who are not insulin-regulated may have a greater risk for developing oral cancer than the general population does. This hypothesis remains to be examined by large epidemiological studies.

Chapter IX - Liver fibrosis is a pathological feature induced by the progression of chronic liver injuries and is characterized by the accumulation of an excess amount of extracellular matrices in the damaged lesions, as an abnormal wound healing response. Activated hepatic stellate cells have been postulated as the cells responsible for the advancement of liver fibrogenesis. Therefore, the molecular function of activated hepatic stellate cells has been investigated vigorously in considering the cells potential therapeutic targets for preventing and curing liver fibrosis. Hepatic stellate cells have been shown to express various molecules, such as cytokines, cytokine receptors, extracellular matrices and adhesion molecules. Here, the authors review two molecules in the liver with the potential of cell-cell and/or cell-matrix interaction: neural cell adhesion molecules (NCAM) and osteonectin. NCAM is a glycoprotein mediating cell-cell and cell-matrix interaction in a homophilic and heterophilic manner, respectively. In the human liver, NCAM expression was found in the stellate cells of undamaged or slightly damaged livers in addition to nerve fibers and natural killer cells. In the rat liver, NCAM was also detected in the activated stellate cells, but not in the quiescent stellate cells. However, the quiescent stellate cells came to express NCAM following treatment with several cytokines, such as transforming growth factor beta-1, platelet-derived growth factor, epidermal growth factor and insulin-like growth factor. The second molecule described here, osteonectin, is a calcium-binding matricellular glycoprotein and has varied functional potential for modulating cell-matrix interaction, for example its counteradhesive effects. Osteonectin has been demonstrated to be expressed in the various interstitial cells of injured and fibrotic tissues. It has also been found in the myofibroblastic cells located in chronically injured liver tissues, however only scantly in intact stellate cells. In addition, the number of osteonectin-positive human stellate cells in livers with chronic hepatitis increased in comparison with the number found in undamaged or slightly damaged livers and in livers with cirrhosis. Osteonectin has also been found in the myofibroblastic cells of the tumor capsule and along the capillaries within hepatocellular carcinoma. Further investigation of the function of these molecules in stellate cells is imperative to elucidate the molecular mechanisms of the communication between stellate cells and matrices.

Chapter X - This chapter will focus on current findings on the epithelial cell adhesion molecule EpCAM, recapitulate past achievements, and eventually look into the future of research on EpCAM and therapeutic potentialities

Short Communication - Recently, the authors showed that surface coatings of mucins, i.e. large-sized amphiphilic glycoproteins found at the mucosal surfaces of all vertebrates, are capable of reducing the uptake of human neutrophils to a polymeric model biomaterial. In the present investigation, human-derived epithelial HEK293, fibroblastic MRC-5 and osteoblastic MG63 cells were studied microscopically during contact with tissue culture-treated polystyrene (TCPS) substrates coated with bovine (BSM), porcine (PGM) and human (MG1) mucins, respectively. The authors found that, for an initial period of two days, all mucin-coated substrates exhibited strongly reduced cell uptakes compared to the non-coated controls. Over time, successively more cells established on the surfaces. However, excluding the BSM-coated substrate which probably was insufficiently coated, the in-growth of cells was significantly slower to the mucin-coated than to the non-coated substrates. Particularly, the PGM- and MG1 coated substrates showed a very limited in-growth of fibroblastic cells over a period of one week. Importantly, the cell viabilities remained at a high level (≥95 %) during all time of cell-material contact, indicating that the mucin-coated substrates were essentially non-toxic. A screening experiment suggested that mucin coatings also could be used to prevent the surface adhesion of neuronal cells. Altogether, the authors results demonstrate that mucin coatings could be used to control the in-growth of cells to an artificial material. Specifically, the strong inhibitory effect on fibroblastic cell adhesion suggested that mucin coatings could be used to reduce fibrous encapsulation of implanted materials.

In: New Cell Adhesion Research
Editors: Patrick Nott and Matthew Temple
ISBN 978-1-60692-378-8

Chapter I

Attachment of Cells Is Regulated by Monochromatic Radiation in Red and Near Infrared Optical Region via a Novel Retrograde Mitochondrial Signaling Pathway

Tiina I. Karu
Institute of Laser and Information Technologies of Russian Academy of Science,
Troitsk 142190, Moscow Region, Russian Federation.

Abstract

The mitochondrial retrograde signaling is an information channel between mitochondrial respiratory chain and the nucleus for the transduction signals regarding the functional state of the mitochondria. The present review examines the question whether radiation of visible and near IR (IR-A) radiation can activate this cellular signaling pathway. The reactions with the primary photoacceptor, cytochrome c oxidase, as well as signaling pathways between mitochondria, plasma membrane, and nucleus were explored on the cell adhesion model using various chemicals that modify different enzyme systems. Functions of cytochrome c oxidase as a signal generator as well as a signal transducer in irradiated cells are outlined. It is concluded that mitochondrial retrograde signaling, a recently discovered cellular signaling pathway, may also be in work in irradiated mammalian cells.

1. Introduction

Adhesion is a primordial property of cells basic for tissue integrity. Cell adhesion to an extracellular matrix or artificial substrata (like glass in our experiments) is mediated by integrins [1,2]. Integrin receptors initiate numerous signal transduction pathways that control cell shape, motility, proliferation, and death [3,4]. The integrins and focal adhesion molecules belong to the large family of glycoproteins [5] that do not absorb light in the red-to-near-IR spectral region. Apparently, the modulative effects of irradiation with wavelengths of this spectral region on cell adhesion are not connected with the direct action of light on these molecules.

The first experimental investigations of the modulation of cell adhesive properties by low-power laser light provided evidence that the plating efficiency [6,7] and cell motility [8], as well as the adherence of *Salmonella* to lymphocytes [9] increased after irradiation. The cell-glass attachment was found to be dependent on the dose and wavelength of the continuous wave (CW) 600–860 nm light [10] and on the pulse parameters when the cells were irradiated with 820 nm pulse-modulated light [11]. Low doses of ruby (694 nm), alexandrite (755 nm), and Nd:YAG laser (1064 nm [12] and 532 nm [13]) irradiation induced changes in cell adhesion by modulating the integrin expression pattern [12] and the focal adhesion kinase activity [13]. However, the irradiation at 810 nm did not reveal a significant increase in adhesion of periodontal ligament cells *in vitro* [14].

The adhesion of cells to extracellular matrices is the initial event of their growth *in vitro*. The adhesion of HeLa cells can be increased by irradiation with low-intensity monochromatic visible light. This phenomenon had a well-structured action spectrum [10] and cytochrome c oxidase was considered as the photoacceptor responsible for this spectrum [15]. the adhesion model was chosen here to investigate the possible cellular signaling pathways between mitochondria, the plasma membrane, and the nucleus.

As the first step in these studies, a large-scale experiment was performed with the aim of inhibiting various cellular signaling reactions leading to changes in adhesion reactions with chemicals of known action mechanisms (Figure 1). There are both pros and cons regarding this approach. Pros: fast screening of many chemicals with known action mechanisms that could be involved into light-activated cellular signaling. It is also known that many chemicals that affect certain metabolic pathways inhibit cell adhesion as well. Cons: chemicals usually affect cell metabolism not by a single mechanism but in multiple pathways.

In our experiments [16-22], we first investigated whether the irradiation of a suspension of HeLa cells at λ = 820 nm (f=10 Hz, η=20%, D=60 J/m^2) prior to or after the treatment with different chemicals could modify the effect of the latters on cell adhesion to a glass matrix. In some cases, an action spectroscopy experiment was performed as well. The chemicals tested included inhibitors of the respiratory chain, antioxidants (free-radical scavengers), NO donors, inhibitors of phospholipase A_2 (PLA_2), inhibitors of the flows of monovalent ions through the plasma membrane, oxidants, and thiol-reactive compounds (Figure 1).

Under our experimental conditions, 42.5±2.5% of the total number of cells in a vial (85 000) attached themselves to the vial bottom [10]. Most of the chemicals under study inhibited the attachment of HeLa cells to the glass substrate (Table 1, column 2). fifteen of them inhibited cell attachment significantly; four stimulated cell attachment, while another six had no effect as compared with the control. Some chemicals (melatonin, SNP, H_2O_2, $NaNO_2$, and GTN) were used in different concentrations.

1. Respiratory chain inhibitors
Rotenone (inhibitor at NADH- dehydrogenase level)
Sodium azide (inhibitor at cyt c oxidase level)
Dinitrophenol (uncoupler)
2. Scavengers of ROS (antioxidants)
Mannitol
Melatonin (pharmacological concentrations)
Ethanol
Ascorbic acid
3. Nonpermeable antioxidative enzymes
Catalase
Superoxide dismutase
4. Chemicals that inhibit ion fluxes through plasma membrane
Ouabain (inhibitor of Na^+, K^+ -ATPase)
Amiloride (inhibitor of Na^+/H^+ -antiporter)
5. Chemicals connected with phospholipase A_2 activation
Arachidonic acid (released by PLA_2 activation)
Quinacrine (inhibitor of PLA_2)
6. Thiol-reactive agents
Cysteine
Mercaptoethanol
Hydroquinone
Copper sulphate
Glutathione
Glutathione disulphide
7. Oxidants
Hydrogen peroxide
Methylene blue
8. NO donors
Sodium nitroprusside
Sodium nitrite
Glyceryl trinitrate (nitroglyceryl)

Figure 1. Chemicals used to modify the attachment of HeLa cells to the glass matrix.

2. Prevention and Elimination of Detrimental Effects of Chemicals by Irradiation

The irradiation increased the number of the cells attached to the glass in a dose-dependent manner. The dose dependence of cell adhesion for irradiation at 820 nm (f=10 Hz, η= 20%, I= 3.54 W/m^2) is presented below in Figures 3-4 and marked by *hv*. The wavelength 820 nm was chosen since it was close to one maximum in the action spectrum of cell attachment [10]. The dose-dependence curve was bell-shaped, with a maximum at 60 J/m^2; the percentage of the adhered cells at this point was 64.5±3.1%. It means that in our experimental conditions a new subpopulation (22%) was attached due to the irradiation.

Table 1. The action of chemicals on cell attachment and modification of cell attachment with radiation at 820 nm (f= 10 Hz, η= 20%, τ= 85 s, 60 J/m^2) and chemicals (data from Refs. [16-23]

Chemical	% of attached cells		
	Chemical	Radiation + Chemical	Chemical + Radiation
1	2	3	4
- (Control)	42.5±2.5	-	-
- (Irradiated sample)	64.5±3.1	-	-
1. Arachidonic acid, 1x10^{-5} M	25.5±2.1*	68.0±3.2§	27.4±6.2
2. SOD, 25 U/ml	30.0±4.1*	73.1±3.1§	31.6±2.2
3. Ouabain, 7x10^{-4} M	16.8±3.3*	36.6±2.1§	17.8±1.4
4. Catalase, 2600 U/ml	27.2±5.1*	57.2±2.1§	40.2±4.3§
5. SOD, 250 U/ml	26.7±5.1*	55.9±3.2§	23.9±1.3
6. Ouabain, 7x10^{-5} M	27.1±2.2*	51.6±1.0§	27.1±2.1
7. Sodium azide, 1x10^{-4} M	23.6±2.1*	53.6±1.1§	38.7±3.3§
8. ATP, 5x10^{-5} M	21.0±2.1*	38.0±3.0§	21.0±1.4
9. Rotenone, 1x10^{-5} M	28.0±4.3*	43.5±3.3§	26.3±5.2
10. DNP, 2x10^{-5} M	28.0±7.0*	34.4±3.1	30.5±2.1
11. Quinacrine, 6x10^{-4} M	17.2±5.0*	25.6±4.1	63.8±2.2§
12. Melatonin, 4x10^{-5} M	26.5±5.2*	36.1±7.2	26.1±5.1
13. Mannitol, 2x10^{-3} M	34.4±7.4*	51.6±3.0§	37.1±4.1
14. GSH, 1x10^{-5} M	29.2±5.0*	35.7±3.0	23.6±4.2
15. Hydrogen peroxide, 1x10^{-3} M	16.3±6.2*	19.8±2.1	29.1±3.1§
16. Amiloride, 5x10^{-4} M	69.6±5.9*	91.3±4.0§	69.2±3.1
17. Methylene blue, 1x10^{-3} M	59.6±6.0*	75.6±3.0§	64.0±5.2
18. GTN, 4x10^{-4} M	66.8±4.1*	60.0±2.6	47.2±5.2§
19. Hydroquinone, 1x10^{-5} M	62.3±5.6*	26.4±2.2§	40.2±2.2§
20. GSSG, 1x10^{-4} M	40.0±4.1	70.1±3.0§	25.4±1.3§
21. Sodium nitrite, 4x10^{-4} M	42.1±2.5	58.3±2.1§	43.1±2.2
22. SNP, 5x10^{-4} M	36.1±3.3	32.3±1.0	15.7±4.3§
23. 2-ME, 2x10^{-4} M	42.5±5.0	36.6±5.0	29.8±2.1§
24. Cysteine, 1x10^{-4} M	42.0±2.0	33.0±8.0	6.5±2.2§
25. $CuSO_4$, 2x10^{-5} M	40.7±2.1	39.0±3.1	41.8±6.6

*- significantly different from the control;

§ - significantly different from the action of chemical.

Abbreviations: SOD-superoxide dismutase; DNP- dinitrophenol; ATP- adenosine triphosphate; GSH- glutathione; GSSG- glutathione disulphide; SNP- sodium nitoprusside; 2-ME- 2-mercaptoethanol; GTN- glycerol trinitrate (nitoglycerin)

Let us examine the situation when cells were irradiated with a dose of 60 J/m^2 at 820 nm and chemicals were added before or after the irradiation (Table 1, columns 3 and 4). One can see that preirradiation eliminated the inhibiting effects of chemicals on cell adhesion.

The column 3 in Table 1 shows the percentage of adhered cells when the suspension was irradiated immediately before the addition of a chemical. In these experiments, the radiation dose amounted to 60 J/m^2 (at this dose the increase of cell attachment reached its maximum). A comparison between the data listed in columns 2 and 3 of Table 1 shows that the percentage of attached cell was higher in pre-irradiated samples in most cases. There are four possibilities for this.

Firstly, the pre-irradiation reduced the inhibitive effect of chemicals, but the percentage of attached cells was still lower than that in the control. such chemicals included GSH, H_2O_2, melatonin, quinacrine, DNP, ATP, and ouabain. Secondly, the inhibitive action of the chemical was fully eliminated by pre-irradiation. For example, rotenone strongly inhibited cell attachment (28.0±4.3%). The percentage of attached pre-irradiated cells (43.5±3.3 %) was close to that in the control (42.5±2.5 %). Thirdly, the percentage of attached cells in the pre-irradiated samples was significantly higher than that in the control sample (42.5±2.5 %). This was true for arachidonic acid, SOD, catalase, SOD, ouabain, azide, mannitol, GSSG, and $NaNO_2$. Fourthly, chemicals increased cell attachment. This was true for amiloride and methylene blue. For GTN, the percentages were practically equal. Fifthly, one chemical – hydroquinone - increased cell attachment, but in case of pre-irradiation it inhibited attachment significantly.

Comparison between the data listed in columns 2 and 3 of Table 1 shows that there were four chemicals: SNP, 2-mercaptoethanol, cysteine, and CuSO4, for which the percentage of attached cells in pre-irradiated samples was very close to that in their non-irradiated counterparts. This means that pre-irradiation had no effect on cell attachment with these chemicals present.

The linear regression analysis of the data listed in columns 2 and 3 of Table 1 pointed to the existence of a correlation between them. The correlation coefficient 0.4332 was statistically significant ($p < 0.01$). This analysis involved all 25 pairs of samples. When the subjects for the analysis were only those pairs of samples where the pre-irradiation increased the percentage of attached cells (items 1 through 17 and 20, 21 in Table 1), the correlation coefficient proved higher, namely, 0.6207 ($p < 0.01$) [23]. These correlation coefficients did not show a strong correlation between these two parameters. However, the number of chemicals tested was not large, and they were chosen to act on different metabolic pathways.

A comparison between the data listed in columns 2 and 4 of Table 1 indicated that the difference in the percentage of attached cells between the post-irradiated samples (column 4) and simply chemical-treated samples (column 2) was in most cases not statistically significant. However, there were four chemicals (catalase, sodium azide, quinacrine, and H_2O_2), which increased the percentage of attached cells in the post-irradiated samples (column 4) significantly compared to the simply chemical-treated samples (column 2). In the case of quinacrine, post-irradiation strongly increased cell attachment (from 17.2±5% to 63.8 ±7.2%) unlike the other chemicals. pre-irradiation reduced the cell-attachment inhibition caused by quinacrine by merely a few per cent (from 17.2 ±5% to 25.6±4.1%).

There were also six chemicals (GTN, hydroquinone, GSSG, SNP, 2-ME, and cysteine), which decreased the percentage of attached cells in the post-irradiated samples (compare columns 4 and 2 in Table 1).

The linear regression analysis of the data listed in columns 2 and 4 of Table 1 indicated that there was no correlation between them, the correlation coefficient being equal to 0.1147 ($p = 0.1140$).

So, it was found that irradiation of HeLa cell suspensions at $\lambda = 820$ nm prior to chemical treatment reduced (or even completely eliminated in some cases) the inhibition of cell-to-glass adhesion caused by these chemicals. the existence of a positive and statistically significant correlation between the results of these types of treatment suggests a more general phenomenon. No correlation has been found to exist in the case where the cells were irradiated after chemical treatment.

The irradiation with visible-to-near IR radiation can protect cells against the harmful effect of ionizing [24-26] or UV radiation [27-29]. Recent experiments have provided evidence that the pre-irradiation benefits primary neurons functionally inactivated by toxins [30,31] and decreases methanol-induced retinal toxicity [32,33]. The authors believed that the mechanism behind these events was an upgrading of cytochrome c oxidase by irradiation.

3. A Novel Mitochondrial Signaling Pathway Activated by Radiation in the Visible-to-Near IR Region

The purpose of the experiments, the results of which are presented below was to demonstrate that a signaling pathway exists between the mitochondria (where the suggested photoacceptors are located) and the cellular membrane. In this cycle of experiments we studied a modification of the action spectrum associated with the increase of adhesive properties of the cell membrane (600-860 nm) as well as the modifying of the dose dependence of cell attachment at 820 nm. It is suggested that the nucleus is included in cell signaling pathways mitochondria → cytoplasm → nucleus → plasma membrane [15]. The part of this signaling pathway from mitochondria to called mitochondrial retrograde signaling (paragraph 4 of the present Chapter).

3.1. Modification of Light Action with Chemicals that Act on the Respiratory Chain

The respiratory chains can be manipulated pharmacologically. The chemicals used for that and their action are well known and have been studied for decades. The usual approach is to treat separated mitochondria or submitochondrial particles, but a whole cell as a more complicated system can be used as well [34]. Figure 2 shows the mitochondrial electron transport chain and indicates the points where chemicals used in our experiments are supposed to act. We used the following chemicals from this group: dinitrophenol (DNP), rotenone, sodium azide, methylene blue, and NO donors: sodium nitroprusside (SNP), sodium nitrite ($NaNO_2$), and glyceryl trinitrate (GTN).

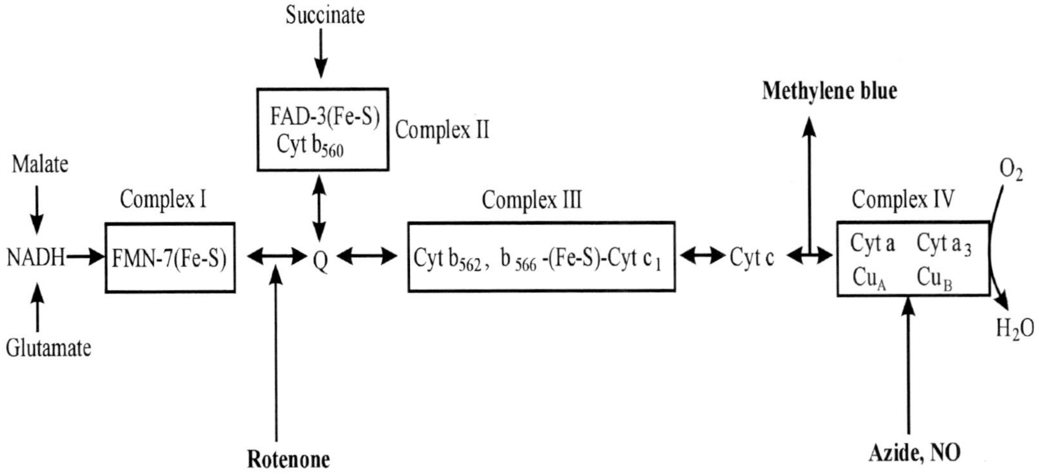

Figure 2. The sites in the mitochondrial electron transport chain where the chemicals used in our experiments act (modified from [36]).

DNP is an uncoupling agent that abolishes the obligatory linkage between the electron transport chain and the phosphorylation system. As a result, ATP synthesis does not occur and the respiration is stimulated. DNP increases cellular respiration by decreasing the mitochondrial membrane potential ($\Delta\psi$). The uncoupling effect of DNP at a cellular level depends on the supply of reducing equivalents [35].

Rotenone is the respiration inhibitor that blocks NADH dehydrogenase (complex I, Figure 2) in the respiratory chain but has no effect on the oxidation of succinate (complex II, Figure 2). azide and nitric oxide block the cytochrome c oxidase catalytic center. In our experiments we used three donors of NO: SNP, GTN, and $NaNO_2$. In addition to multiple regulatory functions at an organism level [37], NO has been recognized as a potential signaling molecule controlling cell respiration [38-44]. It is known that NO in nanomolecular and low micromolecular concentration range reversibly inhibits cytochrome c oxidase in a competitive manner with oxygen [38,40,43,44]. It is known that the NO flux, but not the total NO released, is important in intracellular reactions [45]. However, the concentration of NO as well as the NO flux, which achieves the respiratory chain, are not known *in situ* when NO donors are added to the cell suspension.

Methylene blue acts as an electron shuttle to oxygen that bypasses cytochrome c oxidase (Figure 2). In this way methylene blue increases oxygen consumption and restores mitochondrial electron transport. It was found earlier that methylene blue added to HeLa cells in the darkness stimulated the DNA synthesis rate like He-Ne laser irradiation [46].

Figure 3 presents the data on the modulation of the irradiation effect (λ=820 nm) by DNP, sodium azide, and rotenone. The treatment of cells with DNP before or after irradiation inhibited their attachment independently of the light dose (Figure 3A). The percentages of attached cells were practically the same as in the case of DNP action without irradiation [16]. Comparison between the data in Table 1 and Figure 3A allows one to conclude that the curves of the combined action of irradiation and DNP (Figure 3A) reflect the effects of DNP only. In other words, the irradiation-induced adhesion enhancement was prevented by DNP.

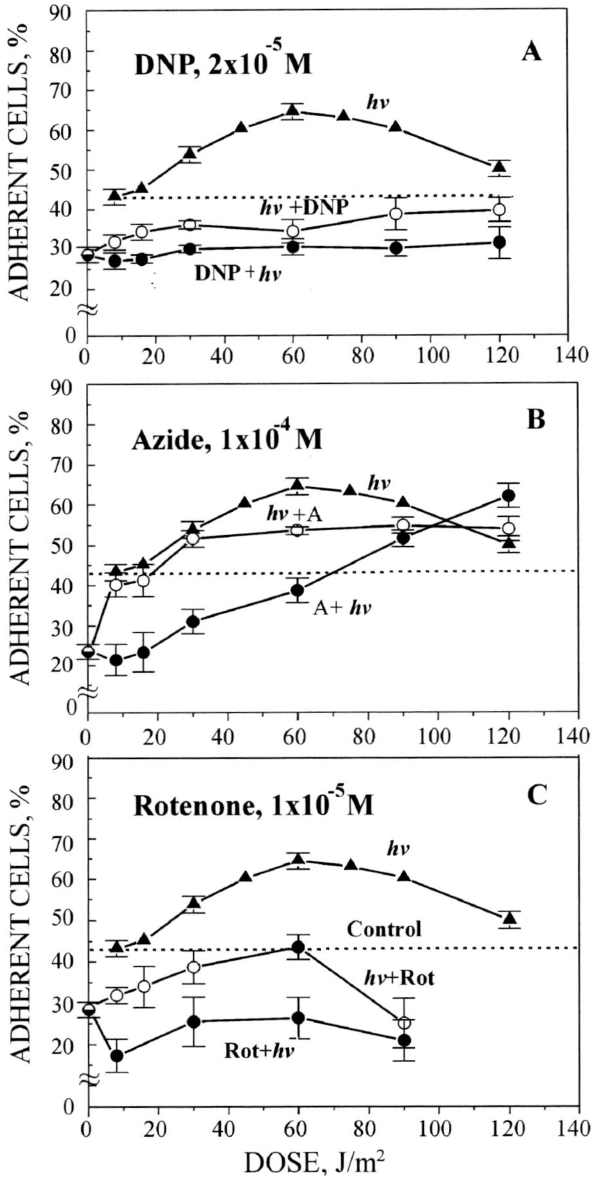

Figure 3. Effects of dinitrophenole (A), azide (B), and rotenone (C) on attachment of HeLa cells. The chemicals were added immediately before or following the irradiation (λ=820 nm, f= 10 Hz, η= 20%) of cells with the dose shown on the abscissa. The dashed line indicates the attachment of intact cells, the curve denoted with hν is the dose dependence of light action without chemicals (modified from [16]).

Azide altered the attachment of cells and the shape of the respective dose-effect curve depending on the treatment sequence (Figure 3B). irradiation before the azide treatment stimulated the attachment of cells as compared with the non-irradiated control. the effect was only slightly dependent on the light dose. The corresponding curve with the threshold near 30

J/m^2 was characterized by a plateau (Figure 3B). The treatment of cells with azide before the irradiation inhibited adhesion at low doses. the inhibitory effect was reversed at doses above 60 J/m^2, when irradiation stimulated cell attachment as compared with the intact control (Figure 3B). This particular curve clearly showed that irradiation eliminated the attachment suppression caused by azide (23.6%, Figure 3B, Table 1).

The treatment of the cells with rotenone before irradiation (Figure 3C) inhibited the adhesion of cells, the percentage of the attached cells at all light doses was less than that when the cells were treated with rotenone alone (28.0%, Figure 3C, Table 1). The treatment of cells with rotenone after irradiation increased their attachment practically to the control level, e.g., irradiation prevented the inhibition effect caused by rotenone (Figure 3C).

Taking together, all these chemicals that act at different points of the respiratory chain, inhibited cell attachment at the concentrations used and prevented the attachment stimulation induced by irradiation. The exception was azide when the light dose was higher than 60 J/m^2 (Figure 3B).

Figure 4 presents the data on the modulation of the irradiation effect (λ=820 nm) by methylene blue. The redox active agent methylene blue (Table 1) stimulated cell adhesion in the darkness (59.6%, Figure 4, Table 1). The magnitude of this effect was comparable to that of the effect of light alone at an optimal dose of 60 J/m^2 (cell attachment 59.6 ± 6% and 64.5 ± 2%, respectively). The data in Figure 4 indicated that the pre- or post-treatment of the cells with methylene blue enhanced the irradiation effect on cell attachment. The stimulating effect was especially apparent at lower light doses (Figure 4).

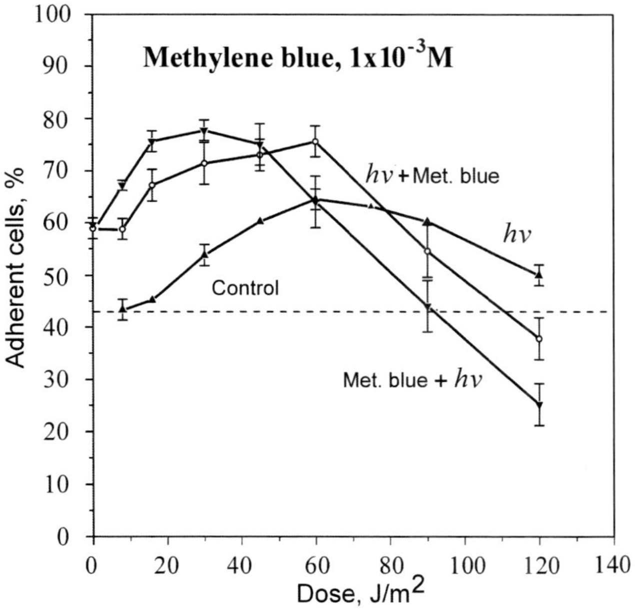

Figure 4. Effects of methylene blue on the attachment of HeLa cells. methylene blue was added immediately before or following the irradiation (λ=820 nm, I=3.54 W/m^2, f=10 Hz, η=20%). The dashed line shows the attachment of the control cells, the curve denoted with hv presents the dose dependence of light action (adapted from [16]).

Modulation of cell attachment by NO depends both on irradiation dose and concentration of NO donors. The exact dose-dependence curves can by found in publications [18,22]. Let us here present the dependences of cell attachment on concentration of three NO donors, SNP, GTN, and $NaNo_2$ (items 22, 18, and 21 in Table 1) added to cell suspension before or after the irradiation at λ=820 nm, 60 J/m² (Figure 5).

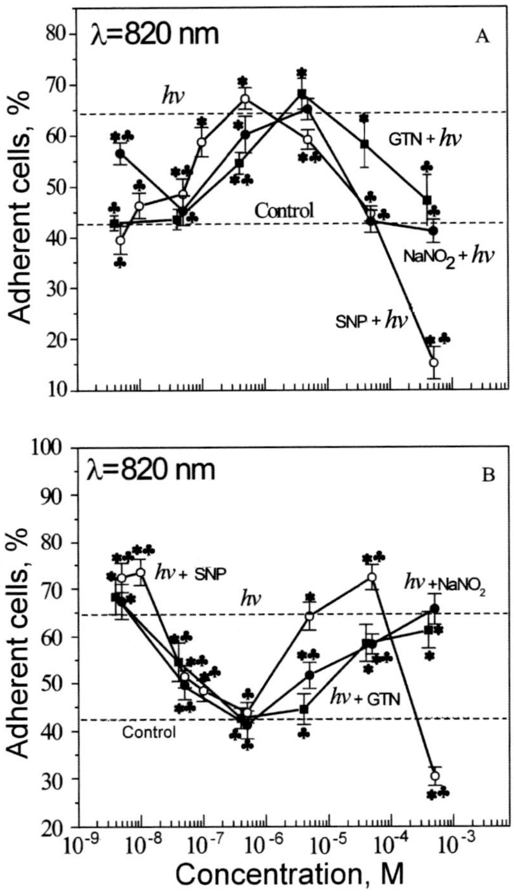

Figure 5. The dependence of cell attachment on the concentration of SNP, GTN, and $NaNO_2$ added to the HeLa cell suspension (A) before or (B) after irradiation at λ=820 nm (I=3.54 W/m², f=10 Hz, η=20%, D=60 J/m²). The dashed line labeled *hv* shows the attachment of cells irradiated at λ=820 nm, 60 J/m² without chemicals added. The other dotted line indicates the attachment of control cells. Asterisks indicate a statistical difference ($p<0.05$) from the control (*) or from the curve labeled *hv* (♣). Adapted from [20].

Figure 5 presents the dependences of the percentage of adhered cells on concentrations of SNP, $NaNO_2$, or GTN added before (Figure 5A) or after (Figure 5B) irradiation (820 nm and 60 J/m²). In the control suspension, 42.5%± 2.5% of cells was attached. Irradiation increased the percent of attached cells to 64.5%±2.0%, meaning that a new subpopulation (22%) was attached due to the irradiation. just this subpopulation is shown between two dashed lines in Figure 5A, B.

Treatment of the cell suspension with NO donors before the irradiation influenced the percentage of attached cells (Figure 5A). It was possible to distinguish three concentration ranges with different types of action. At low concentrations of NO donors (from $4x10^{-9}$ to approximately $5x10^{-8}$ M), the percentages of attached cells were comparable to that in the control, meaning that NO did not allow a subpopulation (~20%) of cells to adhere, and radiation did not stimulate cell attachment. At medium concentrations (approximately between 10^{-7} and 10^{-5} M), NO donors did not influence cell attachment elevated by irradiation. In this case, the percentage of attached cells was close that occurred under irradiation without chemicals. At concentrations higher than $5x10^{-5}$ M, NO donors gradually decreased the percentage of cells stimulated by light to attach. SNP inhibited cell attachment even below the control level. The three curves in Figure 5A did not coincide exactly and the concentration regions were defined rather roughly. This was probably due to differences in NO release mechanisms and thereby the NO fluxes provided by different donors [45,47,48].

The situation was opposite when NO donors were added after irradiation (820 nm; 60 J/m^2) of the cell suspensions (Figure 5B). In this case NO donors did not obstruct the stimulation of cell attachment at low (10^{-9}–10^{-8} M) and high concentrations (above $5x10^{-5}$ M). However, at concentrations from approximately $5x10^{-8}$ to $5x10^{-6}$ M, NO donors eliminated light-induced stimulation of cell attachment. The percentage of attached cells was comparable to that in the control when the concentrations of NO donors were near $5x10^{-7}$ M. It means that NO delivered at this point fully obstructed the light-induced adherence of a subpopulation of cells (~20%).

The complicated dependences of cell attachment on irradiation doses at different concentrations of NO donors [22] as well as dependences on the concentration of NO donors when irradiating at a constant dose (Figure 5A, B) indicated that both factors (irradiation and NO) had a subtle influence on a reaction channel (cellular signaling cascade) between the photoacceptor and cell response to irradiation (attachment).

The maximal amount of cells induced to adhere by 820 nm irradiation in our experimental conditions was ~22% (Figures 3-5). NO in combination with irradiation also influenced attachment of approximately 20% of a cell population (Figure 5). There is no reason to exclude the possibility that NO affected the same subpopulation that was stimulated to adhere by the irradiation. Recall that NO donors did not influence attachment of non-irradiated cells (Figure 5 [22]). So, both agents, photons and NO, quite probably affected the same subpopulation of cells and the final result (change in percentage of attached cells) depended both on the radiation dose [18,20,21,22] and on the concentration of NO donors and the application order of irradiation and NO donors (Figure 5).

Three NO donors used in our study differed from each other by their rates and mechanisms of NO release [47-50]. The concentrations of NO donors added to cells were apparently higher than NO concentrations within cells and mitochondria (their exact values were unknown). It is difficult to discuss why different concentrations of NO donors provided different effects when applied together with the irradiation (Figure 5). One reason was that at high concentrations, NO inhibited many components of the respiratory chain including the oxygen binding site of cytochrome c oxidase [39]. Also, NO could be a substrate, inhibitor or effector of cytochrome c oxidase, depending on cellular conditions [40].

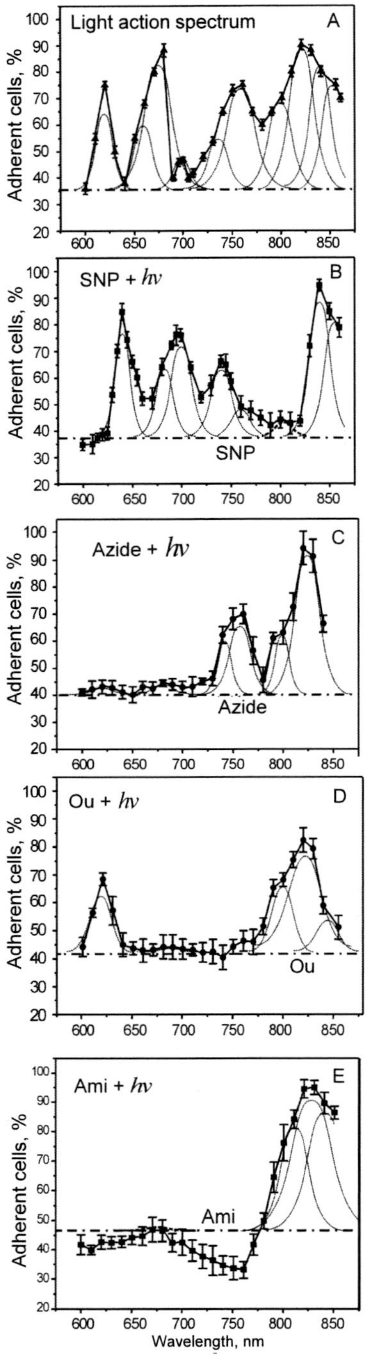

Figure 6. (A). The dependence of cell attachment on wavelength (light action spectrum) when HeLa cell suspensions were irradiated (CW light, D= 52 J/m^2, I= 1.3 W/m^2, τ=40 s, adhesion measurements performed 30 min after irradiation). The action spectra when the following chemicals were added before the irradiation to cell suspension: (B) sodium azide (2x10^{-5} M), (C) SNP (5x10^{-5} M), (D) ouabain (1x10^{-6} M), or (E) amiloride (1.7x10^{-5} M). The dashed lines in Figs. B, C, D, E indicate the attachment of cells in the presence of the respective chemical without irradiation. The fine lines indicate Lorentzian fitting [20-22].

Two chemicals from this set of experiments were used in the action spectroscopy experiments. Figure 6A presents the action spectrum of cell adhesion, whereas Figures 6B, C show the action spectra when azide or SNP were added before the irradiation [15]. The used azide concentration, $2x10^{-5}$ M, did not influence cell attachment without irradiation (40.8 ± 2.5%, marked by dashed line in Figure 6B, vs 42.5 ± 2.5% in control). The doublet bands at 745 and 760 nm and the doublet bands at 800 and 825 nm (Table 2) were present in the action spectrum when azide was added before irradiation (Figure 6B). Comparison of this action spectrum with the control spectrum (Figure 6A) showed the most dramatic changes in the red and far-red regions: complete elimination of the 619 nm peak and the triplet band with the main peak at 675 nm and shoulders at 657 and 700 nm present in the control spectrum (Table 2).

This finding may mean that the putative charge transfer complexes to $Cu_{A_{red}}$ and $Cu_{B_{oxid}}$ (see [15] for explanation) were closed for electron transport in the presence of azide. There were practically no changes in electron transport connected with the suggested d–d transitions in $Cu_{B_{red}}$ chromophores (that are characterized by the doublet bands at 745 and 760 nm), and only a few changes in electron transport occurred, connected with the suggested d–d transitions in $Cu_{A_{oxid}}$ chromophores in the near IR region (the loss of the shoulder at 840 nm, Table 2).

Figure 6C shows the measured action spectrum as well as its deconvolution by the Lorentzian curve fitting when SNP (5×10^{-5} M) was added to cell suspension samples before irradiation. At this concentration, SNP did not influence attachment of non-irradiated cells (37.9 ± 2.1% vs 42.5 ± 2.5% in control). The new action spectrum was characterized by a single band at 642 nm, doublet bands at 685 and 700 nm (main peak), single band at 742 nm (with a weak shoulder at 760 nm) and doublet bands at 842 nm (main peak) and 856 nm (Table 2). On the whole, the comparison of the data in Table 2 allows one to state that two charge transfer channels putatively to $Cu_{A_{red}}$ and $Cu_{B_{oxid}}$ as well as two reaction channels putatively connected with d–d transitions in $Cu_{B_{red}}$ and $Cu_{A_{oxid}}$ were reorganized in the presence of NO (Table 2). The action of NO appeared to be quite different from that of azide, which also reacts directly with the binuclear catalytic center of cytochrome c oxidase. Azide bridges the heme of cytochrome a_3 and Cu_B permanently [51], but NO binds to the catalytic center of cytochrome c oxidase reversibly [40-44].

3.2. Modification of the Light Action Spectrum with Molecules Acting via Interaction with the Cell Membrane

The action spectra were also recorded in the presence of two chemicals, amiloride or ouabain (Figures 6D, E), which cannot cross the plasma membrane but react with it. Ouabain, an inhibitor of Na^+,K^+-adenosine triphosphatase [Na^+,K^+-ATPase] and amiloride, an inhibitor of N^+/H^+ exchanger [NHE], both high molecular weight substances, cannot directly react with cytochrome c oxidase (and the a_3–Cu_B center in particular) like the small ligands: $-N_3$

and NO radicals. Na^+,K^+-ATPase and NHE are enzymes located in the plasma membrane. Na^+,K^+-ATPase moves Na^+ and K^+ against their concentration gradients (active transport); its activity being dependent on the availability of ATP. Ouabain works by attaching to the K^+ channels and blocking its action. Amiloride inhibits the passive transport of Na^+ by nhe by binding to the sodium channels in the extracellular enzyme domain.

Table 2. Comparison of peak positions in control light action spectrum with peak positions in light action spectra when azide, SNP, ouabain or amiloride were added prior to irradiation of cell suspension. Peak positions were calculated from Lorentz fit data; changes < 5 nm were considered to be non-significant. Main peaks are marked bold. Adapted from [20-22]

Control spectrum		Azide, $2x10^{-5}$ M		SNP, $5x10^{-5}$ M	
Band structure. Chromophore putatively responsible for absorption in main band [27]	Peak positions, nm	Peak positions, nm	Changes in comparison with Control	Peak positions, nm	Changes in comparison with Control
Single $Cu_{A_{red}}$	**619**	—	Disappeared	— **642**	Disappeared New peak with the same intensity
Triplet $Cu_{B_{oxid}}$	657 **675** 699 w. sh*	— — —	Disappeared Disappeared Disappeared	— 685 **700**	Disappeared Shift +10 nm and decreased intensity New peak
Doublet $Cu_{B_{red}}$	740 **760**	745 **760**	No changes in peak position and intensity No changes in peak position and intensity	**742** (~770 w. sh.)	New distinct band with the same intensity Disappeared
Quartet $Cu_{A_{oxid}}$	800 s. sh. **820** 840 s. sh. 860 s. sh.	800 w. sh. **825** — Not measured	No changes in peak position and intensity No changes in peak position and intensity Disappeared	(800 v. w. sh.) — **842** 856 s. sh.	Disappeared Disappeared No changes in peak position and intensity No changes in peak position and intensity

Table 2. (Continued)

Control spectrum		Ouabain, $1x10^{-6}$ M		Amiloride, $1.7x10^{-5}$ M	
Band structure. Chromophore putatively responsible for absorption in main band [27]	Peak positions, nm	Peak positions, nm	Changes in comparison with Control	Peak positions, nm	Changes in comparison with Control
Single $Cu_{A_{red}}$	**619**	**620**	No changes in peak position and intensity	—	Disappeared
Triplet $Cu_{B_{oxid}}$	657	—	Disappeared	(~670 v. w. sh.)	Disappeared
	675	—	Disappeared	—	Disappeared
	699 w. sh*	—	Disappeared	—	Disappeared
Doublet $Cu_{B_{red}}$	740	—	Disappeared	—	Disappeared
	760	(~760 v. w. sh.)	Disappeared		Disappeared
Quartet $Cu_{A_{oxid}}$	800 s. sh.	800 s. sh.	No changes in peak position and intensity	811 s. sh.	Shift +11 nm and increase in intensity
	820	**820**	No changes in peak position and intensity	**831**	Shift +11 nm but no change in intensity
	840 s. sh.	840 w. sh.	No changes in peak position but decrease in intensity	842	
	860 s. sh.	—	Disappeared	—	

*w. sh. – weak shoulder; s. sh. – strong shoulder; v. w. sh. – very weak shoulder.

Our interest in these enzymes was due to three circumstances. First, Na^+,K^+-ATPase is the most powerful ATP consumer in a cell, so a connection should exist between light–activated ATP synthesis and the consumption of ATP by this enzyme. Second, NHE contributes substantially to the cellular pH homeostasis (pH_i). Changes in the cellular redox state and pH_i are supposed to take part in the photosignal transduction chain from the mitochondria to the cell nucleus [15]. Third, the studies not related to light action have shown that treatment of cells with ouabain and amiloride not only inhibited the monovalent ion fluxes through the plasma membrane, but also induced various signaling pathways [52,53], e.g. mediated by protein kinases and transcription factors NF-κB and AP-1 [52]. It was suggested that signals produced by binding of ouabain to the cell membrane were transduced to the mitochondria via protein–protein interactions [54]. It is known that an interaction exists between amiloride, amiloride-sensitive channels and NF-κB [55]. Data from the literature provides evidence that ouabain [56,57] and amiloride [58] can modulate cell attachment, and ouabain can also regulate the activity of cellular respiration [59]. The dose dependences of cell attachment when ouabain or amiloride in various concentration are added to HeLa cell suspension, can be found in [19].

Our action spectroscopy results showed that ouabain (Figure 6D) and amiloride (Figure 6E) significantly modify the light action spectrum of the photoinduced increase in the percentage of attached cells (Figure 6A). The light action spectrum in the presence of ouabain was characterized by a single band at 620 nm and a triplet band in the near IR region (main peak at 820 nm with shoulders at 800 and 840 nm, Table 2). Other bands in the red-to-far red region characteristic of the control spectrum (Figure 6A) fully disappeared in the presence of ouabain (Figure 6D). This means that a putative charge transfer channel to $Cu_{A_{red}}$ (619 nm band at) and a channel suggested to be connected with d–d transition in $Cu_{A_{oxid}}$ (820 nm band) worked like in the control cells, but both channels to Cu_B (the charge transfer channel characterized by the band at 680 nm and a channel due to d–d transition characterized by absorption at 760 nm) were closed in the presence of ouabain.

The light action spectrum in the presence of amiloride (Figure 6E) had only the near IR bands, which were characterized by +11 nm shifts of a strong shoulder at 800 nm and a main peak at 820 nm as compared with those in the control spectrum (Table 2). Noteworthy is the fact that the band at 760 nm in the control spectrum was not only eliminated but amiloride also caused a slight inhibition of cell attachment below the control level. Therefore, only one channel, namely that putatively connected with d–d transition in $Cu_{A_{oxid}}$ chromophore, was working in the presence of amiloride. Recall that both ouabain and amiloride in the concentrations used ($1x10^{-6}$ M for ouabain and $1.7x10^{-5}$ M for amiloride) did not statistically significantly influence cell attachment without irradiation (42.1 ± 3.2% and 47.1 ± 2.1%, respectively, vs that in control, 42.5 ± 2.5%). Therefore, despite not binding directly to the catalytic center of cytochrome c oxidase, ouabain and amiloride actively influence cellular respiration in the irradiated cells via inhibition of NHE and Na^+, K^+-ATPase.

3.3. How Does the Mitochondrial Signaling Pathway Work?

The number of HeLa cells attached to the glass substrate increased after irradiation of the cell suspension with certain wavelengths in the visible-to-NIR optical region (Figure 6A). This finding provides evidence that first, a new subpopulation of cells enters an adhesive state because of the irradiation and second, a photoacceptor governs this process. In other words, the irradiation opens some new reaction channel(s) for attachment increase via activation of the photoacceptor. The result that the chemicals added before irradiation strongly changed the light action spectrum shows that reaction channels activated by light are either closed, as evidenced by the full disappearance of bands in the action spectra in Figures 6B-E, or are modified, as evidenced by appearance of new peaks (Figure 6C) or shifted peak positions (Figure 6E). Recall that the chemicals at the concentrations used do not significantly influence cell attachment without irradiation.

The bands in the light action spectrum (Figure 6A) are believed to belong to a not yet identified redox intermediate of cytochrome c oxidase turnover [15]. Redox absorbance changes in living cells recorded in the spectral range 530–890 nm [60] as well as changes in the optical activity of absorbing chromophores at 250–780 nm [61] occurring under laser irradiation support this suggestion.

The result that azide and nitric oxide, which are known as inhibitors of cytochrome c oxidase (Figure 6B, C), modified the action spectrum of cell attachment increase supports the suggestion [15] that cytochrome c oxidase could be a photoacceptor. This result also corroborates a recent hypothesis [62] that the mitochondrial $NO^{\bullet}$-signaling pathway could be an alternative function of cytochrome c oxidase. An unexpected result is that ouabain and amiloride, which do not react directly with cytochrome c oxidase, significantly modified the absorption of the photoacceptor (mirrored by the action spectrum in our case). This finding is consistent with a view [54] that the signal-transducing role played by Na^+, K^+-ATPase from the plasma membrane to mitochondria via protein–protein interactions could be one of its standard functions. The inhibition of cellular respiration by ouabain has been measured experimentally [59]. It is not clear now how the signal from NHE, which is inhibited by amiloride, is transduced to the mitochondria. It is known that amiloride induces fast intracellular acidification [63]. An intracellular signaling pathway is believed to mediate the effects of cell adhesion and receptors of pH_i [58].

The suggested participation of the copper chromophores of cytochrome c oxidase in various redox states ($Cu_{A_{red}}$ $Cu_{A_{oxid}}$, $Cu_{B_{oxid}}$, $Cu_{B_{red}}$) [15] allows us to propose that at least four reaction channels from cytochrome c oxidase, which are regulated by light, may control cell attachment as well. These chromophores are indicated in the first row of Table 2. The percentage of the newly attached subpopulation under CW light radiation was from around 47% at an irradiation wavelength of 820 nm, around 40% at 680 nm to around 33% at 760 nm and around 26% at 620 nm (Figure 6A). The structure of the main bands (Figure 6) resolved by Lorentzian fitting (Table 2) shows not only the complex nature of the reaction channels but also a feature of these channels for fine adjustment. Changes in the peak positions as well as disappearance or appearance of the bands in the light action spectrum in the presence of chemicals show that we are dealing with a dynamic system of photoacceptor(s) fast reacting to changing conditions both in the vicinity of absorbing chromophores as well as in the plasma membrane.

The data shown in Table 2 demonstrated that the only band fully disappearing (in the presence of azide, ouabain, and or amiloride) or strongly modified (by $NO^{\bullet}$) in the light action spectrum was the triplet with the main peak at 675 nm and shoulders at 657 and 700 nm. This finding may point to an increase in reduction of the Cu_B chromophore in our experimental conditions (insofar as these peaks are thought to belong to $Cu_{B_{oxid}}$). A 655-nm band has long been considered to be characteristic of fully oxidized cytochrome c oxidase and disappears by reduction [64]. This peak at 655 nm (putatively at 657 nm in our spectrum, Figure 6A) is believed to belong to the $a_3^{2+} - Cu_B^{2+}$ binuclear center and disappears by the formation of peroxy and ferryl forms of the enzyme in turnover [65]. There is some evidence that the properties of this transition are affected by the oxidation state of Cu_B [66]. Beinert *et al.* [65] also recorded absorption near 680 nm in mixed-valence CO-bound cytochrome c oxidase.

In the study of Szundi et al. (2001), a shoulder at around 700 nm in the ferryl difference spectrum was observed [67]. A similar weak shoulder resolved in our light action spectrum

(Figure 6A) which disappeared in the presence of all chemicals (Figure 6C-E) except SNP (Figure 6B). The absorption near 710 nm increased concomitantly with the reduction of cytochrome c oxidase [67]. The appearance of a new peak at 700 together with a red shift and decrease in the intensity of the peak at 675 nm (Table 2) might indicate that a new charge transfer channel was open in the presence of NO. It has to be emphasized that a very weak absorption of Cu_B chromophore [64] can be strongly amplified in the action spectra.

Comparison of the spectral peak positions in the presence of NO and azide showed that in the first case the distance between the peaks putatively belonging to $Cu_{B_{red}}$ (760 nm in control) and $Cu_{A_{oxid}}$ (820 nm in control) increased, which may be connected with the opening of a new channel at 642 nm. At the same time, there was a shortening of the distance between the peaks putatively characteristic of $Cu_{A_{red}}$ (619 nm in control) and $Cu_{B_{oxid}}$ (680 nm in control, Table 2). In the presence of azide, the situation was quite the opposite. The distance between the peaks at 760 and 820 nm shortened, and two peaks at 619 and 680 nm disappeared completely (Table 2). This finding could be consistent with differences in the binding of these two ligands to the catalytic center of cytochrome c oxidase [51,40,42].

In the presence of chemicals, which do not bind to cytochrome c oxidase (amiloride, ouabain), the doublet bands at 740 and 760 nm putatively characteristic of $Cu_{B_{oxid}}$ (Table 2) fully disappeared. The chemicals, which directly bind to the catalytic center of cytochrome c oxidase, did not influence this doublet (azide) or only modify it (nitric oxide). An absorption at 740 nm, which appears at low temperature measurements by the formation of one of the intermediates (called by authors as compound C), and an absorption at 760 nm (together with absorption at 796 nm), which appears by partial oxidation of cytochrome c oxidase (compound B), have been recorded [68]. Compound B is a peroxy intermediate with a ferric peroxide–cupric copper complex. Compound C is also a peroxy intermediate of the turnover but is different from compound B [69]. The presence of 740 and 760 nm peaks in our control action spectrum might provide some evidence that the photosensitive intermediate was a peroxy compound. The appearance and disappearance of a photosensitive intermediate, possibly an $a_3^{2+} - Cu_B^{1+}$ intermediate, support this suggestion. Recent studies have demonstrated that the peroxy intermediate is really a mixture of several peroxy intermediates [71,72].

Two bands in our control action spectrum are believed to be connected with the Cu_A chromophore. These were the single band at 619 nm and a compound band in the near IR region with a main peak at 820 nm and shoulders at 800, 840 and 860 nm (Table 2). There is a consensus in the literature that a large absorption band centered at 820–830 nm is characteristic of the $Cu_{A_{oxid}}$ chromophore, which gives around 80% of absorption in this region [67]. Cu_A has also been suggested to make a small contribution around 615 nm, which is obscured in part by intense and overlapping transitions from cytochromes a and a_3 [73]. The red absorption band of isolated mitochondria in steady state–coupled turnover is a broad band from around 590 nm to ~630 nm centered near 610 nm [74]. Clore [75] trapped an intermediate absorbing at 610 nm in low-temperature measurements and noticed that for the membrane-bound oxidase there was a red shift of this band as compared with soluble oxidase and also the FWHM of this band was broader in this case. They suggested that this was a charge transfer transition. The suggestion that the band at 619 nm in our light action spectrum (Table 2, Figure 6A) belongs to $Cu_{A_{red}}$ was also supported by the opposite time course of absorption changes in this region ($Cu_{A_{red}}$) and at 830 nm ($Cu_{A_{oxid}}$) [66].

So, treating cell suspensions with sodium azide, SNP, ouabain or amiloride before irradiation significantly modified the action spectrum of cell attachment enhancement. A suggested light-induced mitochondrial signaling pathway can be regulated by small ligands directly binding to the catalytic center of cytochrome c oxidase (N_3, $NO^{\bullet}$) as well as by chemicals specifically binding to plasma membrane enzymes (ouabain, amiloride).

4. Retrograde Mitochondrial Signaling

Mitochondrial retrograde signaling is a pathway of communication in cells from mitochondria to the nucleus that influences many cellular activities under both normal and pathophysiological conditions (reviews: [76,77]). This recently discovered signaling is a opposite signaling pathway to a common and well defined pathway transforming information from the nucleus and cytoplasm to the mitochondria. The retrograde signaling sends information back to the nucleus about changes in the functional state of the mitochondria. The most investigated mitochondrial retrograde signaling pathways so far are those in the budding yeasts *Saccharomyces cerevisiae* (review: [78]) and plant cells (review: [79]). mechanisms of mitochondrial-nuclear crosstalk are described also in mammalian cells: myocytes [80] and cancer cells (review: [81]). It appeared that mitochondria have, alongside their well-known functions, also a function of the retrograde regulation. Recently, it has been suggested that retrograde mitochondrial signaling occurs in irradiated mammalian cells ([82] review: [83]).

One widely used model system for exploring the mechanisms of mitochondrial retrograde signaling is proliferation. It is experimentally proved on this model that mitochondria have the capacity to communicate with the rest of the cell and send signals to the nucleus (review: [77]). The experimental data evidence that this signaling can be mediated via $\Delta\Psi_m$, generation of ROS, changes in Ca^{2+} flow, $NO^{\bullet}$ binding to cytochrome c oxidase, to name the most important characteristics (review: [77]).

Changes in the proliferation of mammalian cell cultures are rather well studied after irradiation with light of different wavelengths, doses, and intensities (reviews: [84-87]). The initial phase of the proliferation, adhesion of cells to a matrix, as well as DNA and RNA synthesis rate are cellular responses studied most systematically. The irradiation of mammalian cells causes an upregulation of various genes. The cDNA microarray technique was used to human fibroblasts irradiated at 628 nm. The gene expression profiles studied (9982 in summary) revealed that 111 genes of 10 categories were upregulated [88]. The analysis of the action spectra of DNA and RNA synthesis rate as well as cell adhesion in the wavelength range 330-860 nm allowed to conclude that cytochrome c oxidase, the terminal enzyme of the mammalian respiratory chain is the responsible photoacceptor [15]. This suggestion was later confirmed in works [31,89,90]. The action spectra with exact peak positions can be found in [150].

Mitochondrial retrograde signaling has been considered as a cellular stress response (reviews: [77,78]) as well as an adaptive response (review: [76]). It has been suggested also that the action of visible and near IR radiation on mammalian cells and tissues can be described within the framework of the adaptation syndrome, which means that responses have much in common, regardless of the type of irritant [92]. When the action of a physical factor is increased (e.g. its intensity, dose), cyclic changes occur in the metabolic activity of mitochondria [93]. First, a stimulation characterized by a threshold and phase of increase occurs. After a strict maximum and a phase of decrease, the control level is reached and then an inhibition could be recorded. Specific characteristics exist relating to the photobiological character of the light action [92]. These are the absorption of light quanta by the photoacceptor molecules and the presence of cellular signaling pathways, mitochondrial retrograde signaling included. In other aspects like manifestation of bell-shaped dose and intensity dependences, the light action characteristics respond to general concepts of adaptation syndrome [94] and are generalized in pharmacology as Arndt-Schulz law.

4.1. $\Delta\Psi_m$ Signaling

The mitochondrial respiratory chain produces energy, which is stored as an electrochemical gradient called mitochondrial membrane potential ($\Delta\Psi_m$). This electrochemical gradient consists of an electrical transmembrane potential and a proton gradient (ΔpH) and drives the synthesis of ATP. Mitochondrial retrograde signaling was initially defined by altered $\Delta\Psi_m$ (review: [76]). Later, other characteristics like changes in concentration of mitochondrial ROS and Ca^{2+} were introduced. Below, we will consider every parameter separately. One has to emphasize that there is a delicate balance between positive and negative influence of every mitochondrial signaling element ($\Delta\Psi_m$, $[Ca^{2+}]_i$, ROS) to signaling pathways (reviews: [77,78]). A detailed analysis of this balance goes beyond of frames of this review.

It is well documented on the models of isolated mitochondria and whole cells that monochromatic radiation in visible and IR-A region can affect $\Delta\Psi_m$ and stimulate ATP synthesis. The illumination of isolated rat liver mitochondria increased ATP synthesis and the consumption of O_2 [95,96,97]. Irradiation with light at wavelengths of 415 nm [96], 602 nm [99], 632.8 nm [97], 650 nm and 725 [95] enhanced ATP synthesis. Light at wavelengths of 477 and 554 nm [96] did not influence the rate of this process. Oxygen consumption was activated by illuminating with light at 365 and 436 nm, but not at 313, 546 and 577 nm [99,99]. Irradiation with light at 632.8 nm increased $\Delta\psi_m$ and ΔpH [100,101], caused changes in mitochondrial optical properties, modified some NADH-linked dehydrogenase reactions [100] and increased the rate of ADP/ATP exchange [102]. Mitochondrial RNA and protein synthesis [103], replication of mitDNA [104] as well as both cytosolic and mitochondrial protein syntheses [105,106]. were stimulated by He-Ne laser irradiation. He-Ne laser irradiation increased not only O_2 uptake by cytochrome c oxidase but also electron transfer and proton pumping activity of this enzyme [107]. In the case of state 4 respiration, the 351-nm and 458-nm laser irradiation accelerated the oxygen consumption by rat liver mitochondria; such acceleration was not observed with 514.5 nm irradiation. On the contrary, in the case of state 3 respiration, the 514.5-nm laser irradiation activated the oxygen consumption by mitochondria. Activation did not occur with 458-nm irradiation but 351-nm irradiation reduced the oxygen consumption in respiration state 3 [108]. Irradiation at 660 nm increased state 3 oxygen consumption at coupling II and III sites, as well as increasing the respiratory control ratio [109].

On the level of a single cell and during the real-time recording, a maximal increase in mitochondrial membrane potential 30% of its basal value was observed at 2 min after a 15-s irradiation at 647 nm (about 165 mV as compared with control value of 140 mV). Then $\Delta\Psi_m$ decreased gradually being back on the basal level 4 min later [110]. $\Delta\Psi_m$ fluctuations have been associated with the opening of the mitochondria permeability transition pore (MPTP). It was suggested in work [110] that the MPTP operates in the low conductance state in the case of low power laser irradiation.

It is known that the treatment of cells with mitochondria-specific ionophores (uncoupling agents) that abolish the obligatory linkage between the electron transport chain and the oxidative phosphorylation induces alterations in $\Delta\Psi_m$ followed by upregulation of a number of genes in the nucleus (review: [76]). On the model of the increase in the number of attached cells, the treatment of cell suspension before irradiation at 820 nm with ionophore dinitrophenol (DNP) as well as respiratory chain inhibitors rotenone (the inhibitor of complex I in the respiratory chain) and azide (inhibitor of complex IV) strongly modified the light effect. These chemicals that acted on the respiratory chain but at different points did not allow an appearance of the attachment stimulation induced by the irradiation [20,21]. Treating of isolated mitochondria with rotenone and complex III inhibitor antimycin A as well as with oligomycin (inhibitor of ATP synthase) or with various uncoupling agents abolished any increase of $\Delta\psi_m$ induced by He-Ne laser irradiation [97].

Increase and decay of $\Delta\Psi_m$ as well as changes in cytokine gene expression and subcellular localization of promyelocytic leukemia protein were measured after irradiation of human keratinocytes with light at 780 nm [111]. The $\Delta\Psi_m$ was increased immediately after the end of irradiation and then subsequently decayed within 200 minutes by an exponential curve. Following irradiation, the expression of interleukin-1L, interleukin-6, and keratinocyte growth factor genes were transiently upregulated. The expression of the proinflammatory gene interleukin-1β was suppressed. The subnuclear distribution of promyelocytic leukemia protein (a cell-cycle checkpoint protein) was altered from discrete domains to its dispersed form within less than 1 hour after the irradiation. The results of work [111] show the connection between a short-time activation of the mitochondrial photoacceptor and secondary cellular responses after mitochondrial retrograde signaling.

A connection between primary events in mitochondria and following retrograde signaling is also seen in data published in work [101]. He-Ne laser irradiation of melanoma cells induced immediately an increase in $\Delta\Psi_m$ and ATP concentration via enhanced cytochrome c oxidase activity and promoted later phosphorylation of JNK/activator protein-1 expressions (JNK is Jun N-terminal kinase) as well as induced an increase in cell proliferation. Irradiation-induced cell proliferation was significantly abrogated by the addition of chemicals that decreased $\Delta\Psi_m$ or inhibited JNK [101]. This investigation binds together light action upon the photoacceptor (cytochrome c oxidase), a modulation of mitochondrial respiratory chain activity, and cell proliferation.

4.2. ROS Signaling

Recent evidence highlights a specific role of mitochondrial ROS in redox cellular signaling and in mitochondrial retrograde signaling in particular. It appeared that mitochondrial ROS are not just damaging by-products of the respiration, but an important factor in cellular signaling [112,113], including retrograde mitochondrial signaling [79,114]. Primary ROS made by mitochondria is superoxide anion ($O_2^{\bullet-}$), which is converted to H_2O_2 either by spontaneous dismutation or by the enzyme superoxide dismutase (SOD). The main source of $O_2^{\bullet-}$ in mitochondria is the ubisemiquinone radical intermediate ($QH^{\bullet}$), formed during the Q cycle at the Q_0 site of complex III of the respiratory chain [113,114]. Flavin mononucleotide group of complex I of the respiratory chain is also a source of $O_2^{\bullet-}$. It is known that ROS generation site in complex I is liberating $O_2^{\bullet-}$ predominantly to the mitochondrial matrix space and that in complex III liberates it with preference to the cytosolic space [115].

Broad-band radiation 760-1440 nm induced the formation of mitochondria-derived ROS in cultured human dermal fibroblasts. It was suggested on the basis of these measurements that mitochondrial retrograde signaling pathways are in work [82]. The formation of mitochondrial $O_2^{\bullet-}$ was found to be relevant for MMP-1 (matrix metalloproteinase-1) upregulation that needed a functional mitochondrial respiratory chain. In summary, the experiments performed in work [82] evidenced that the respiratory chain activity was a determinant of the light response.

Mitochondrial ROS generation has been measured earlier in connection of investigating the mechanisms of low power laser therapy (review: [87]). One has to underline the complex work of Alexandratou et al. [110]. In their experiments, ROS generation was monitored alongside modulation of $\Delta\Psi_m$, intracellular pH (pH_i), and ($[Ca^{2+}]_i$) after irradiation at 647 nm on single-cell level in real time scale. A 15-s irradiation evoked a progressive accumulation of H_2O_2 during 8 min after the irradiation [110]. One has to note that both stimulation and inhibition of the respiratory chain can result in enhanced ROS generation [116].

A series of experiments was performed to investigate possible signaling pathways between mitochondria and membrane. Number of cells attached to a glass matrix was increased and then decreased by a bell-shaped curve when the cells were irradiated at 820 nm (a band in the action spectrum of cell adhesion increase [10,11], see also Figures 3, 4). The treatment of the cells with ROS scavengers (antioxidants) before the irradiation eliminated the attachment stimulation induced by the irradiation, but did not influence the basal level ([16] review: [87]. Mannitol, melatonin, ethanol, and ascorbic acid were used in these experiments. Let us look closer at the action of melatonin, because its action was studied most extensively: firstly in dependence on its concentration (both in physiological and pharmacological concentration ranges) and secondly, dependent on the wavelength of radiation used [117]. Pineal gland hormone melatonin is a significant scavenger of free radicals. The antioxidative action mechanism of melatonin is believed to be different from the action of antioxidants such as vitamin C and E or glutathione [118]. Melatonin as a highly lipophilic small molecule quickly reaches mitochondria and can have a direct action on respiratory chain carriers [119]. Melatonin regulates the mitochondrial redox state, as it was shown experimentally for brain and liver mitochondria [120].

In the model system of stimulation of number of cells to attach (the action spectrum in Figure 6A), melatonin modified the action spectrum in the IR-A region only. In this particular region, the peak at 831 nm, which is characteristic for the light action spectrum, was absent in the spectrum recorded in the presence of melatonin. The band in the action spectra at 820-830 nm is thought to belong mainly to oxidized Cu_A chromophore of cytochrome c oxidase [15]. Other peak positions in the light action spectrum at 618, 668, and 751 nm (see Figure 6A) were not altered in the presence of melatonin [117]. The elimination of the peak with maximum near 830 nm evidences that Cu_A chromophore of cytochrome c oxidase (this chromophore gives around 85% of absorption in this region [65]) becomes more reduced in the presence of melatonin. It is known that the peak at 820-830 nm disappears with the reduction of cytochrome c oxidase [65]. Action spectroscopy experiments with other chemicals are reviewed in the last part of the present review.

In another series of experiments, the oxidative agents H_2O_2 (in a low concentration range) and methylene blue increased the cell adhesion in the dark. The magnitude of this effect was comparable to the magnitude of the effect of radiation at λ=820 nm alone at an optimal dose [16,19]. Methylene blue added to the cells in dark caused also stimulation of DNA synthesis in a comparable per cent with the stimulation caused by He-Ne laser radiation [46]. Methylene blue acts as electron shuttle to oxygen that bypasses cytochrome c oxidase (see Figure 2).

An involvement of plasma membrane into signal transduction between mitochondria and nucleus is evidenced by patch-clamp studies of ionic currents through the plasma membrane under He-Ne laser radiation [121]. No laser light effects were found in patch configurations where the disruption of the plasma membrane and cellular homeostasis occurred (voltage-activated whole-cell, inside-out, and double whole-cell recordings). Only in the case of cell-attached recordings, i.e. in conditions of cellular integrity, the background single-channel currents through the plasma membrane were sensitive to He-Ne laser radiation. This finding proved to be valid in the case of both excitable (neurons, cardiomyocytes) and nonexitable (glia) cells. Although the nature of the light-sensitive background channels was not studied in detail, these channels were thought to be ATP-dependent K^+ channels and Ca^{2+}-dependent K^+ channels [121]. The ATP-dependence of light-sensitive currents through the plasma membrane supports the suggestion about mitochondrial origin of the photoacceptor. this study evidenced that cellular signaling pathways beginning from mitochondria should be in work for recording currents through the plasma membrane under the irradiation. Otherwise, the light-sensitive currents could be recorded in all patch configurations.

Two circumstances considering complex changes in the mitochondrial characteristics of retrograde signaling should be underlined. First, there exists a unique mitochondrial oscillator that depends on oxidative phosphorylation, ROS, and permeability of mitochondrial inner membrane ion channels [122]. It was hypothesized in [123] that the balance between $O_2^{\bullet -}$ efflux through inner membrane anion channels and the intracellular ROS scavenging capacity plays a key role in the oscillation mechanism. Synchronized whole cell oscillations in mitochondrial metabolism which were triggered by a local release of ROS by a single local laser flash after a delay of ~40 s in more than 70% of the mitochondrial population, are thought to be a mechanism working in all types of cells that contain mitochondria [123]. Oscillations in $[Ca^{2+}]_i$ after the irradiation at 647 nm [110] are described in the next part of this review.

Secondly, spatiotemporally synchronized whole cell oscillations in mitochondrial metabolism were initiated only after a specific threshold level of mitochondrially produced ROS was exceeded, and they did not involve the MTPT or $[Ca^{2+}]_i$ overload. A local release of ROS by ~1% of mitochondria triggered the oscillations throughout the entire volume of the cell [123]. One has to note that the strict threshold characterizes dose and intensity dependences of various cellular bell-shaped responses to the irradiation (reviews: [85,86]).

4.3. Ca^{2+} Signaling

Mitochondria play an important role in calcium storage and Ca^{2+} homeostasis, sequestering these ions upon their release from endoplasmic reticulum or following increased Ca^{2+} uptake across the plasma membrane. These organelles can release their Ca^{2+} to increase its local concentrations in subcellular regions to activate different processes, including mitochondrial proliferation and mitochondrial retrograde regulation in response to accumulation of ROS [113]. There appears to be a multifactor cross-talk among Ca^{2+}, $\Delta\Psi_m$, and ROS, centered on the mitochondrion. Remarkable that this "love-hate triangle" [116] is in work both in excitable and nonexcitable cells [123]. Various agents that uncouple respiration and oxidative phosphorylation cause elevated concentrations of cytosolic free Ca^{2+} and upregulation of a number of genes (review: 1). It is known that the uncoupling of respiration and ATP synthesis not only alters $\Delta\Psi_m$ but also affects mitochondrial Ca^{2+} accumulation and ROS generation [116]. A sustained 3- to 8-fold increase of steady-state Ca^{2+} concentration as a part of retrograde signaling leads to the activation of calcineurin, which in turn activates NF-κB (nuclear factor kappa B) and other signaling pathways (review: [76]).

An increase in $[Ca^{2+}]_i$ has been for long time one of the early events measured after the irradiation of various types of cells (latest review: [87]. Insofar as we a dealing with a photobiological phenomenon, every modulation in $[Ca^{2+}]_i$ or $[Ca^{2+}]_m$ can occur only after absorbing the light by a photoacceptor. This circumstance is often forgotten when discussing Ca^{2+} effects. $[Ca^{2+}]_i$ was first measured after irradiation of human lymphocytes with a He-Ne laser [124] and later in sperm cells [125,126] and isolated mitochondria [127,128], hepatocytes [129], nerves [130] and mast cells [131] after irradiation with various wavelengths. After laser irradiation of mast cells with light at 405 nm, $[Ca^{2+}]_i$ was increased, followed by histamine release [131]. The increase in mitochondrial membrane potential $\Delta\Psi_m$ caused a stimulation of c-fos gene expression in a Ca^{2+}-dependent manner [103].

The irradiation of hepatocytes with a He-Ne laser caused an increase in $[Ca^{2+}]_i$ and the increase in cell membrane potential correlated with it. The increases of both parameters took place in an oscillatory manner [129]. Laser irradiation at 647 nm triggered recurrent spikes in $[Ca^{2+}]_i$, measured at the single-cell level. An immediate increase in $[Ca^{2+}]_i$ was observed reaching a peak at about 60s after laser stimulation for 15 s. Then the $[Ca^{2+}]_i$ level returned to the basal value. After 240 s, without any further stimulation, a second global Ca^{2+} oscillation occurred and after 320 s, the cell returned to its initial resting $[Ca^{2+}]_i$ level [110].

4.4. NO• Signaling

Nitric oxide (NO•) is considered as an important intramitochondrial signaling molecule, which modulates mitochondrial respiration by direct binding to cytochrome c oxidase [131,132]. This cGMP – independent (cGMP is cyclic quanosine monophosphate) action of NO• is known to induce the production of ROS and trigger redox signaling [132]. It is supposed that NO• produced in mitochondria under physiological conditions may induce mitochondrial retrograde signaling through altered $\Delta\psi_m$ (review: [76].

NO• is a free radical generated in biological systems by nitric oxide synthases (NOS). Mitochondria have their own NOS (mtNOS), enzyme activity of which is localized in the mitochondrial inner membrane [134]. It was hypothesized that the inhibition (or modulation) of mitochondrial respiration by NO• may represent a novel biochemical pathway regulating the supply of O_2 and energy to tissues under dynamic conditions [133].

Cytochrome c oxidase, which is considered as the photoacceptor in photobiomodulation, shares control over ATP synthesis with several other components of the oxidative phosphorylation machinery. Even under metabolic demand, cytochrome c oxidase rarely has more than 20% of the total control over ATP synthesis [135]. This means that NO• -mediated changes in O_2 consumption can occur without significant effects on ATP synthesis rate. It was proposed in work [135] that this property of oxidative phosphorylation has an important implication for mitochondrial ROS generation and cellular redox signaling. This is because the inhibitor of cytochrome c oxidase NO• not only affects O_2 consumption, but also impacts on the redox status of the respiratory chain an thus its $O_2^{\bullet-}$-generation. Therefore, a window exists in which low level of NO• can regulate mitochondrial ROS generation without affecting ATP synthesis.

Donors of NO• added to the cellular suspension before irradiation at 820 nm eliminated in concentration-dependent way the radiation-induced increase in the number of cells attached to the glass matrix [22]. Dependences of NO•-mediated effects on the wavelength used for irradiation (action spectra) and possible underlying mechanisms of initial reactions of mitochondrial retrograde signaling are presented in the part 3.1 of this review.

4.5. Transient Alkalization of Intracellular pH (pH_i)

It was suggested after the first recordings of light action spectra that the photoexcitation of a photoacceptor in mitochondria is the event that initiates a cascade of redox changes and modulations of biochemical reactions in a cell [136]. This photosignal transduction chain (or cellular signaling in contemporary language) was supposed to lead to a proliferation increase. In this cascade, a transient modulation of cellular redox state was thought to be the crucial element of cellular signaling [136,85]. It was suggested that the irradiation causes also a transient increase in pH_i [136,137]. pH_i is the parameter tightly connected with cellular redox potential. Indeed, the pH jump was soon measured after He-Ne laser irradiation in prokaryotic cells of *E. coli* [138] and later, in the human fibroblasts [110]. The measurements on the single fibroblast level evidenced that the alkalization of 0.21±0.08 units occurred 6 min. after a 15-s irradiation at 647 nm. After 15 min., the pH_i returned to its resting level [110]. This result evidences that a transient alkalization occurs as a part of retrograde cellular signaling from mitochondria to the nucleus. It was found in the same work that Na^+/H^+ antiporter did not participate as a regulator mechanism in the alkalization of pH_i after laser stimulation. Also, no dissipation of cellular pH gradient occurred [110].

Thiol-reactive chemicals (glutathione, 2-mercaptoethanol, cysteine) eliminated the increase in the number of attached cells induced by irradiation at 820 nm. Treatment of the cells with glutathione disulphide, hydroquinone, and copper sulphate eliminated light-induced attachment to the glass matrix only at certain light doses ([17], review: [18]). These experiments evidenced that fluctuations in cellular redox balance were involved in the enhancement of cell attachment induced by the irradiation. It also means that mitochondrial retrograde signaling pathways are influenced by intracellular redox balance. One has not to forget complex nature of cellular signaling pathways. For example, adding of mercaptans and cupric ions to cells trigger large and rapid Ca^{2+} release from sarcoplasmic reticulum vesicles [139].

A redox sensitive transcription factor NF-κB is considered to be one of the key regulatory molecules in oxidative stress-induced cell activation [140] including retrograde signaling [141]. ROS, in particular H_2O_2 are implicated in the activation of NF-κB and the degradation of its inhibitor IκB. Inhibitors of the respiratory chain rotenone and antimycin inhibited pharmacologically induced IκB degradation indicating that the ROS had mitochondrial origin [142]. It is believed that an activation of NF-κB by retrograde signaling involves a novel mechanism, distinct from the known activation mechanisms by cytokines and chemokines (review: [76]).

4.6. Changes in Fission-Fusion Homeostasis of Mitochondria

Mitochondria have the capacity to communicate with the nucleus also by the structural changes of the organelle itself, e.g. by changes of fission-fusion homeostasis in a dynamic mitochondrial network (review: [77]). It appeared that the morphology of the mitochondria is an element in mitochondria-nuclear communications. Mitochondria are not static; they are in constant movement within cells, and numerous fusion and/or fission events take place. These fusion and fission events are accompanied by variations in mitochondrial size, number and mass, which are triggered by a variety of physiological stimuli. In metabolically active cells, mitochondria make up ~40% of the cytoplasm. Mitochondria replicate during cell proliferation by a mechanism of recruitment of new proteins, which are added to preexisting subcompartments [143].

Coordinated changes in fission-fusion homeostasis of mitochondria are a part of complex of communications during retrograde signaling (review: [77]). The dynamic nature of mitochondria might protect them by ensuring that regional changes in membrane potential $\Delta\Psi_m$ are always transient. A key messenger to activate the mitochondrial biogenesis program in various cell types is $NO^{\bullet}$ via $NO^{\bullet}$-cytochrome c oxidase system. $NO^{\bullet}$- induced mitochondrial biogenesis plays an important role in regulation of cellular metabolism operating a sensitive feedback system, retrograde response what enables the cell to compensate for defects in electron transport chains [143].

Light effects on mitochondrial ultrastructure are investigated only in some extent. Mitochondrial ultrastructure was changed after the irradiation of isolated rat liver organelles with a He-Ne laser under experimental conditions in which ATP extrasynthesis took place. In addition to well known orthodox, condensed, and swollen mitochondrial conformations, previously unknown bipartite mitochondria with different degrees of attenuation in their mid regions, and small atypical mitochondria were observed [144]. Small atypical mitochondria were characterized biochemically [145].

The experiments performed with yeast cells [146,147]. allowed to conclude that a short-time He-Ne laser irradiation (λ=632.8 nm) influenced in dose-dependent manner the mitochondrial ultrastructure along the changes in cell proliferation rate at least during 7 following generations. A short-time irradiation of cells with doses that stimulated the respiratory activity caused changes in ultrastructure of the giant mitochondrion and mitochondria-endoplasmic reticulum associations in progeny cells. Irradiation in doses suppressing the respiratory activity caused damaging effects on mitochondria of the cells of successive generations [146,147]. The mechanisms of mitochondrial biogenesis in irradiated cells as well as the ways in which progeny cells respond to maintain mitochondrial functions is a research field of the future. This problem is especially interesting taking into account that the mitochondrion has its own DNA that is inherited maternally. It is also worthy to mention that yeasts, which do not depend on their mitochondrially produced ATP to survive, actually live longer when retrograde signaling is active [114]. Actually, the retrograde response links metabolism with chromatin-dependent gene activation and genome stability in yeast aging [148].

5. Conclusion

Direct targeting of drugs to mitochondria has long been a dream of pharmacologists. New therapeutic strategies in connection with mitochondrial signaling elements $\Delta\psi_m$, ATP, and ROS have been highlighted recently [116]. These strategies concern cardio- and neuroprotection as well as ischemic preconditioning. One has to note that laser irradiation has successfully used in all these cases not only on animal models but also clinically (latest review: [87]). Laser irradiation may have some limitations in clinical practice due to limited penetration depth into tissues. On the other side, by this method, possible deleterious side effects of mitochondrially-targeted drugs can be avoided. Last experimental data [31,32,33,149] evidence that laser phototherapy may facilitate recovery from retinal injuries and other ocular diseases wherein mitochondrial dysfunction is postulated to play a role. A more detailed understanding of the role of mitochondria and retrograde mitochondrial signaling mechanisms in irradiated cells is needed for further development of laser phototherapy. It is every reason to believe on the basis of experimental data gathered so far that mitochondrial retrograde signaling, a recently discovered cellular signaling pathway, is in work also in the irradiated cells.

Acknowledgment

This work was partly supported by Russian Foundation for Basic Research (Grant №07-02-00205a).

References

[1] Gumbiner, B.M. (1996). Cell adhesion: the molecular basis of tissue architecture and morphogenesis. *Cell 84*, 345–357.

[2] Zamir, E. and Geiger, B. (2001). Molecular complexity and dynamics of cell-matrix adhesions. *J. Cell Sci. 114*, 3583–3590.

[3] Watt, F.M. (2002). Role of integrins in regulating epidermal adhesion, growth and differentiation. *EMBO J. 21*, 3919–3926.

[4] Aplin, A.E., Howe, A.K. and Juliano, R.L. (1999). Cell adhesion molecules, signal transduction and cell growth. *Curr. Opin. Cell Biol. 11*, 737–744.

[5] Horton, M.A., Ed. (1996). *Molecular Biology of Cell Adhesion Molecules*. Glasgow: John Wiley.

[6] Boulton, M. and Marshall, J. (1986). He-Ne laser stimulation of human fibroblast proliferation and attachment *in vitro*. *Lasers Life Sci. 1*, 125–134.

[7] Karu, T.I., Pyatibrat, L.V. and Kalendo, G.S. (1987). Biostimulation of HeLa cells by low-intensity visible light. V.Stimulation of cell proliferation *in vitro* by He-Ne laser radiation. *Nuovo Cimento D 9*, 1485–1494.

[8] Haas, A.F., Isseroff, R. Wheeland, R.G. Rood, P.M. and Graves, P.J. (1990). Low-energy He-Ne laser irradiation increases the motility of cultured human karatinocytes. *J. Invest. Dermatol. 94*, 822–826.

[9] Passarella, S., Casamassima, E., Quagliariello, E., Caretto, E.G. and Jirillo, E. (1985). Quantitative analysis of lymphocyte-salmonella interaction and effect of lymphocyte irradiation by He-Ne laser. *Biochem. Biophys. Res. Comm. 130*, 546–552.

[10] Karu, T.I., Pyatibrat, L.V., Kalendo, G.S. and Esenaliev, R.O. (1996). Effects of monochromatic low-intensity light and laser irradiation on adhesion of HeLa cells *in vitro*. *Lasers Surg. Med. 18*, 171–177.

[11] Karu, T.I., Pyatibrat, L.V. and Kalendo G.S. (2001). Studies into the action specifics of a pulsed GaAlAs laser (λ=820 nm) on a cell culture. II. Enhancement of the adhesive properties of cellular membranes: Dependence on the dark period between pulses. *Lasers Life Sci. 9*, 211–217. [also: Dokl. Akad. Nauk. (Moscow) *369*, 116-118, 1999].

[12] van Leeuwen, R.L., dekker, S.K., Byers, H.R., vermeer, B.J. and Grevelink, J.M. (1996). Modulation of $\alpha 4\beta 1$ and $25\beta 1$ integrin expression: heterogenous effect of Q-switched ruby Nd:YAG and Alexandrite lasers on melanoma cells *in vitro*. *Lasers Surg. Med. 18*, 63–71.

[13] Zhu, N.W., Perks, C.R., burd, A.R. and Holly, J.M. (1999). Changes in the levels of integrin and focal adhesion kinase (FAK) in human melanoma cells following 532 nm laser tratment. *Int. J. cancer 82*, 353–358.

[14] Kreisler, M., Meyer, C., Stender, E., Daublander, M., Willershausen-Zonnchen, B. and d'Hoedt, B. (2001). Effects of diode laser irradiation on the attachment rate of peritondal ligament cells: an *in vitro* study. *J. Peridontol. 72*, 1312–1317.

[15] Karu, T. (1999).·Primary and secondary mechanisms of action of visible-to-near IR radiation on cells. *J. Photochem. Photobiol. B: Biol. 49*, 1–17.

[16] Karu, T.I., Pyatibrat, L.V. and Kalendo G.S. (2001). Cell attachment modulation by radiation from a pulsed semiconductor light diode (λ=820 nm) and various chemicals. *Lasers Surg. Med. 28*, 227-236.

[17] Karu, T.I., Pyatibrat, L.V. and Kalendo G.S. (2001). Thiol reactive agents eliminate stimulation of cell attachment to extracellular matrices induced by irradiation at λ =820 nm: possible involvement of cellular redox status into low power laser effects. *Laser Therapy 11*, 177–187.

[18] Karu, T.I., Pyatibrat, L.V. and Kalendo G.S. (2001). Donors of NO and pulsed radiation at λ=820 nm exert effects on cells attachment to extracellular matrices. *Toxicol. Lett. 121*, 57–61.

[19] Karu, T.I., Pyatibrat, L.V. and Kalendo G.S. (2001). Cell attachment to extracellular matrices is modulated by pulsed radiation at 820 nm and chemicals that modify the activity of enzymes in the plasma membrane. *Lasers Surg. Med. 29*, 274–281.

[20] Karu, T.I., Pyatibrat, L.V. and Kalendo G.S. (2004). Photobiological modulation of cell attachment via cytochrome c oxidase. *Photochem. Photobiol. Sci. 3*, 211–216.

[21] Karu, T., Pyatibrat, L. and Afanasyeva, N. (2004). A novel mitochondrial signaling pathway activated by visible-to-near infrared radiation. *Photochem. Photobiol. 80*, 366–372.

[22] Karu, T.I., Pyatibrat, L.V. and Afanasyeva, N.I. (2005). Cellular effects of low power laser therapy can be mediated by nitric oxide. *Lasers Surg. Med. 36*, 307-314.

[23] Karu, T.I. (2003). Cellular mechanism of low power laser therapy: new questions. In: *Lasers in Medicine and Dentistry, Vol. 3.*, Ed by Z. Simunovic, Vitrgraf: Rieka, , pp. 77-100.

[24] Abdvakhitova, A.K., Grigorieva, L.N. and Parkhomenko, I.M. (1982). Effect of laser radiation on chinese hamster cells cultured *in vitro*. *Radiobiology (Moscow) 22*, 40–43.

[25] Karu, T., Pyatibrat, L. and Kalendo, G. (1994). Irradiation with He-Ne laser can influence the cytotoxic response of HeLa cells to ionizing radiation. *Int. J. Radiat. Biol. 65*, 691–697.

[26] Joyce, K.M., Downes, C.S. and Hannigan, B.M. (1999). Radioadaptation in Indian muntjac fibroblast cells induced by low intensity laser irradiation. *Mutation Res. 435*, 35–42.

[27] Lage, C., Teixeira, P.C. and Leitao, A.C. (2000). Non-coherent visible and infrared radiation increase survival to UV (254 nm) in *Escherichia coli* K12. *J. Photochem. Photobiol. B: Biol. 54*, 155–161.

[28] Dube, A., Bock, C. Bauer, E., Kohli, R., Gupta, P.K. and Greulich, K.O. (2001). He-Ne laser irradiation protects B-lymphoblasts from UVA-induced DNA damage. *Radiat. Environ. Biophys. 40*, 77–82.

[29] Kohli, R. and Gupta, P.K. (2003). Irradiance dependence of the He-Ne laser-induced protection against UVC radiation in *E. coli* strains. *J. Photochem. Photobiol. B: Biol. 69*, 161–167.

[30] Wong-riley, M.T., Bai, X., Buchman, E. and Whelan, H.T. (2001). Light-emitting diode treatment reverses the effect of TTX on cytochrome c oxidase in neurons. *Neuroreport 12*, 3033–3037.

[31] Wong-Riley, M.T., Liang, H.L., Eells, J.T., Chance, B., Henry, M.M., Buchmann, E., Kane, M. and Whelan, H.T. (2005). Photobiomodulation directly benefits primary neurons functionally inactivated by toxins: role of cytochrome c oxidase. *J. Biol. Chem. 280*, 4761–4771.

[32] Eells, J.T., Henry, M.M., Summerfelt, P., Wong-Riley, M.T., Buchmann, E.V., Kane, M., Whelan, N.T. and Whelan, H.T. (2003). Therapeutic photobiomodulation for methanol-induced retinal toxicity. *Proc. Natl. Acad. Sci. USA 100*, 3439–3444.

[33] Eells, J., Wong-Riley, M.T., Verhoeve, J., henry, M., Buchman, E.V., Kane, M.P., Gould, L.J., das, R., Jett, M., Hodgson, B.D., Margolis, D. and Whelan, H.T. (2004). Mitochondrial signal introduction in accelerated wound and retinal healing by near-infrared light therapy. *Mitochondrion 4*, 559–567.

[34] Hutter, E., Unterluggauer, H., Garedew, A., Jansen-durr, P. and Gnaiger, E. (2006). High-resolution respirometry – a modern tool in agening research. *Exp. Gerontol. 41*, 103–109.

[35] Sibille, B., Ronot, X., Filippi, C., Nogueira, V., Keriel, C. and Leverve, X. (1998). 2, 4-dinitrophenol-uncoupling effect on ΔΨ in living hepatocytes depends on reducing-equivalent supply. *Cytometry 32*, 102–108.

[36] Liu, Y., Fiskum, G. and Schubert, D. (2002). Generation of reactive oxygen species by the mitochondrial electron transport chain. *J. Neurochem. 80*, 780–787.

[37] Moncada, S., Palmer, R.M.J. and Higgs, E.E. (1991). Nitric oxide: physiology, pathophysiology and pharmacology. *Pharmacol. Rev. 43*, 109–142.

[38] Brown, G.C. (2001). Regulation of mitochondrial respiration by nitric oxide inhibition of cytochrome c oxidase. *Biochim. Biophys. Acta 1504*, 46–57.

[39] Brown, G.C. (2003). NO says yes to mitochondria. *Science 299*, 838–839.

[40] Cooper, C. E. (2002). Nitric oxide and cytochrome oxidase: substrate, inhibitor or effector? *Trends Biochem. Sci. 27*, 33–39.

[41] Sarti, P., giuffre, A., Barone, M.C., Forte, E., Mastronicola, D. and Brunori, M. (2003). Nitric oxide and cytochrome c oxidase: reaction mechanisms from the enzyme to the cell. *Free Radic. Biol. Med. 34*, 509–520.

[42] Mastronicola, D., genova, M.L., arese, M., barone, M.C., giuffre, A., Bianchi, C., Brunori, M., Lenaz, G. and Sarti, P. (2003). Control of respiration by nitric oxide in keilin-Hartree particles, mitochondria and SH-SY5Y neuroblastoma cells. *Cell. Mol. Life Sci. 60*, 1752–1759.

[43] Brown, G.C. and Cooper, C.E. (1994). Nanomolar concentrations of nitric oxide reversibly inhibit synaptosomal cytochrome c oxidase respiration by competiting with oxygen at cytochrome c oxidase. *FEBS Lett. 356*, 295–298.

[44] Mason, M.G., Nicholls, P., Wilson, MT. and Cooper, C.E. (2006). Nitric oxide inhibition of respiration involved both competitive (heme) and noncompetitive (cooper) binding to cytochrome c oxidase. *Proc. Natl. Acad. Sci. USA 103*, 708–713.

[45] Wink, D.A., Cook, J.A., Krishna, M.C., Hanbauer, I., DeGraff, W., Gamson, J. and Mitchell, I.B. (1995). Nitric oxide protects against alkyl peroxide-mediated cytotoxicity: further insights into the role nitric oxide plays in oxidative stress. *Arch. Biochem. Biophys. 319*, 402–407.

[46] Karu, T.I. (1989). *Photobiology of Low-Power Laser Therapy*. Chur, London: Harwood Acad. Publ.

[47] Feelisch, M. (1998). The use of nitric oxide donors in pharmacological studies. *Naunyn Schmiedebergs Arch. Pharmacol. 358*, 113–122.

[48] Yamamato, T. and Bing, R. (2000). Nitric oxide donors. *Proc. Soc. Exp. Biol. Med. 225*, 200–206.

[49] Adami, A., Crivellente, F., De Prati, A.C., Cavalieri E., Cuzzolin, L., Tommasi, M., Suzuki, H. and Benoni, G. (1998). Biotransformation and cytotoxic properties of NO-donors on MCF7 and U251 cell lines. *Life Sci. 63*, 2097–2105.

[50] Ignarro, L.J., Napoli, C. and Loscalzo, J. (2002). Nitric oxide donors and cardiovascular agents modulating the bioactivity of nitric oxide: an overview. *Circ. Res. 90*, 21–28.

[51] Yoshikawa, S. Shinzawa-Itoh, K., Nakashima, R., Yaono, R., Yamashita, E., Inoue, N., Yao, M., Fei, M.J., Libeu, C.P., Tsunehiro, M., Yamaguchi, H., Tomizaki, T. and Tsukihara, T. (1998). Redox-coupled crystal structural changes in bovine heart cytochrome c oxidase. *Science 280*, 1723-1729.

[52] Xie, Z., Kometiani, P., Liu, J., Li, J., Shapiro, J. and Askari A. (1999). Intracellular reactive oxygen species mediate the linkage of Na^+/K^+-ATPase to hyperthrophy and its marker genes in cardiac myocytes. *J. Biol. Chem. 274*, 19323–19328.

[53] Carcia-Canero, R., Trilla, C., Perez de Diego, J., Diaz-Gil, J.J. and Cobo, J.M. (1999). Na^+/H^+ exchange inhibition induces intracellular acidosis and differentially impairs cell growth and viability of human and rat hepatocarcinoma cells. *Toxicol. Lett. 106*, 215–228.

[54] Xie, Z. and Askari, A. (2002). Na^+/K^+-ATPase as a signal transducer. *Eur. J. Biochem. 269*, 2434–2439.

[55] Haddad, J.J. (2005). Amiloride and regulation of NF-κB: an unsung crosstalk and missing link between fluid dynamics and oxidative stress-related information-controversy or pseudo-controversy? *Biochem. Biophys. Res. Comm. 372*, 373–381.

[56] Contreras, R. G., Shoshani, L., Flores-Maldonado, C., Lazaro, A. and Cereijido, M. (1999). Relationship between Na^+-K^+-ATPase and cell attachment. *J. Cell Sci. 112*, 4223–4232.

[57] Belusa, R., Aizman, O., Andersson, R. M. and Aperia, A. (2002). Changes in Na^+,K^+-ATPase activity influence cell attachment to fibronectin. *Am. J. Physiol. Cell Physiol. 282*, C302–C309.

[58] Margolis, L. B., Rozovskaya, I. A. and Cragoe, E. (1988). Intracellular pH and cell adhesion to solid substrata. *FEBS Lett. 234*, 449–450.

[59] Levinson, C. and Hempling, H. G. (1967). The role of ion transport in the regulation of respiration in the Ehrlich mouse ascites-tumor cell. *Biochem. Biophys. Acta 135*, 306–318.

[60] Karu, T. I., Afanasyeva, N.I., Kolyakov, S.F., Pyatibrat, L.V. and Welser, L. (2001). changes in absorbance of monolayer of living cells induced by laser radiation at 633, 670, and 820 nm. IEEE Select. *Topics Quantum Electr. 7*, 982–988.

[61] Karu, T.I., Kolyakov, S.F., Pyatibrat, L.V., Mikhailov, E.L. and Kompanets, O.N. (2001). Irradiation with a diode 820 nm induces changes in circular dichroism spectra (250-780 nm) of living cells. IEEE Select. *Topics Quantum Electr. 7*, 976–981.

[62] Brookes, P. and Darley-Usmar, V. M. (2002). Hypothesis: the mitochondrial NO signaling pathway, and the transduction of nitrosative to oxidative cell signals: an alternative function for cytochrome c oxidase. *Free Radic. Biol. Med. 32*, 370–374.

[63] Jorgensen, N.K., Petersen, S.F., Damgaard, I., Schousboe, A. and Hoffmann, E. K. (1999). Increase in $[Ca^{2+}]_i$ and changes in intracellular pH during chemical anoxia in mouse neocortical neurons in primary culture. *J. Neurosci. Res. 56*, 358–370.

[64] Shaw, R.W., Hansen, R.E. and Beinert, H. (1979). The oxygen reactions of reduced cytochrome c oxidase. Position of a form with an unusual epr signal in the sequence of early intermediates. *Biochim. Biophys. Acta 548*, 386–396.

[65] Beinert, H., Shaw, R.W., Hansen, R.E. and Hartzell, C.R. (1980). Studies on the origin of the near-infrared (800-900) absorption of cytochrome c oxidase. *Biochim. Biophys. Acta 591*, 458–470.

[66] Hill, B. C. and Greenwood, C. (1984). The reaction of fully reduced cytochrome c oxidase with oxygen studied by flow-flash spectrophotometry at room temperature. Evidence for new pathways of electron transfer. *Biochem. J. 218*, 913–921.

[67] Szundi, I., Liao, G.-L. and Einarsdóttir, O. (2001). Near-infrared time-resolved optical absorption studies of the reaction of fully reduced cytochrome c oxidase with dioxygen. *Biochemistry 40*, 2332–2339.

[68] Chance, B. and Leigh, J.S. (1977). Oxygen intermediates and mixed valence states of cytochrome oxidase: infrared absorption difference spectra of compounds A, B, and C of cytochrome oxidase and oxygen. *Proc. Natl. Acad. Sci. USA 74*, 4777–4780.

[69] Mitchell, R., Mitchell, P. and Rich, P. (1991). The assignment of the 655 nm spectral band of cytochrome c oxidase. *FEBS Lett. 280*, 321–324.

[70] Blackmore, R.S., Greenwood, C. and Gibson, H. (1991). Studies of the primary oxygen intermediates in the reaction of fully reduced cytochrome oxidase. *J. Biol. Chem. 266*, 19245–19249.

[71] Van Eps, N., Szundi, I. and Einarsdóttir, O. (2003). pH dependence of the reduction of dioxygen to water by cytochrome c oxidase. 1. The P(R) state is a pH-dependent mixture of three intermediates, A, P, and F. *Biochemistry 42*, 5065–5073.

[72] Einarsdóttir, O. and Szundi, I. (2004). Time-resolved optical absorption studies of cytochrome c oxidase dynamics. *Biochim. Biophys. Acta 1655*, 263–273.

[73] Brunori, M., Antonini, E., and Wilson, M.T. (1981). *Cytochrome c oxidase: an overview of recent work. in: Metal Ions in Biological Systems*. Ed. by H. Siegel, Basel: M. Dekker, N.Y. vol. 13, pp. 187–228.

[74] Morgan, J. E. and Wikström, M. (1991). Steady-state redox behaviour of cytochrome c, cytochrome a, and Cu_A of cytochrome c oxidase in intact rat liver mitochondria. *Biochem. 30*, 948–958.

[75] Clore, M. (1980). Characterization of the intermediates in the reaction of membrane-bound mixed-valence-state cytochrome oxidase with oxygen at low temperatures by optical spectroscopy in the visible region. *Biochem. J. 180*, 617–622.

[76] Butow, R.A. and Avadhani, N.G. (2004)Mitochondrial signaling: the retrograde response. *Mol. Cell. 14*, 1-15.

[77] Ryan, M.T. and Hoogenraad, N.J. (2007). Mitochondrial-nuclear communications. *Annu. Rev. Biochem. 76*, 701-722.

[78] Liu, Z. and Butow, R.A. (2007). Mitochondrial retrograde signaling. *Annu. Rev. genet. 40*, 159-185.

[79] Rhoads, D.M. and Subbaiah, C.C. (2007). Mitochondrial retrograde regulation in plants. *Mitochondrion 7*, 177-194.

[80] Biswas, G., Adebanjo, O.A., freedman, B.D., Anandatheerthavarada, H.K., Vijayasarathy, C., Zaidi, M., kotlikoff, M. and Avadhani, N.G. (1999). Retrograde Ca^{2+} signaling in C2C12 skeletal myocytes in response to mitochondrial genetic and metabolic stress: a novel mode of inter- organelle crosstalk. *The EMBO J. 18*, 522-533.

[81] Erol, A. (2005). retrograde regulation due to mitochondrial dysfunction may be an important mechanism for carcinogenesis. *Med. hypoth. 65*, 525-529.

[82] Schroeder, P., pohl, C., Calles, C., Marks, C., Wild, S. and Krutmann, J. (2007). Cellular response to infrared radiation involves retrograde mitochondrial signaling. *Free Rad. Biol. Med. 43*, 128-135.

[83] Karu, T. (2008). Mitochondrial signaling in mammalian cells activated by red and near IR radiation. *Photochem. Photobiol., 86*(5), (in the press).

[84] Karu, T. (1990). Effects of visible radiation on cultured cells. *Photochem. Photobiol. 52*, 1089-1099.

[85] Karu, T.I. (1989). Photobiology of low-power laser effects. *Health Phys. 56*, 691-704.

[86] Karu, T.I. (1987). Photobiological fundamentals of low-power laser therapy *IEEE J. Quant. Electr. QE-23*, 1703-1717.

[87] Karu, T. (2007). *Ten Lectures on Basic Science of Laser Phototherapy*. Grängesberg (Sweden), Prima Books AB.

[88] Zhang, Y., Song, S., Fong, C.-C., Tsang, C.-H., Yang, Z., and Yang, M. (2003). cDNA microarray analysis of gene expression proteins in human fibroblast cells irradiation with red light. *J. Invest. Dermatol. 120*, 849-857.

[89] Pastore, D., Greco, M. and Passarella, S. (2000). Specific He-Ne laser sensitivity of the purified cytochrome c oxidase. *Int. J. Rad. Biol. 76*, 863-870.

[90] Karu, T.I., Pyatibrat, L.V., Kolyakov, S.F. and Afanasyeva, N.I. (2005). Absorption measurements of a cell monolayer relevant to phototherapy: reduction of cytochrome c oxidase under near IR radiation. *J. Photochem. Photobiol.* B: Biol. *81*, 98-106.

[91] Karu, T. I., Pyatibrat, L. V., Kolyakov, S. F. and Afanasyeva, N. I. (2008). Absorption measurements of cell monolayers relevant to laser phototherapy: reduction or oxidation of cytochrome c oxidase under laser radiation at 632.8 nm. *J. Photochem. Photobiol.* B:Biol. (in the press)

[92] Karu, T. (1998). Introduction. In: *The Science of Low-Power Laser Therapy, Gordon and Breach Science Publishers*, Amsterdam, pp. XIII-XVIII.

[93] Kondrashova, M.N. (1970). Control of mitochondria respiration at increasing influences on the cell. *Biophysics (Moscow) 15*, 312-323.

[94] Selye, H. (1950). Stress and the general adaptation syndrome. *Br. Med. J. 1*, 1382-1392.

[95] Gordon, S.A. and Surrey, K. (1960). Red and far-red light action on oxidative phosphorylation. *Radiat. Res. 12*, 325–339.

[96] Kato, M., Sinizawa, K. and Yoshikawa, S. (1981). Cytochrome c oxidase is a possible photoacceptor in mitochondria. *Photochem. Photobiophys. 2*, 263-269.

[97] Passarella, S., Casamassima, E., Molinari, S., Pastore, D., Quagliariello, E., Catalano, I.M. and Cingolani, A. (1984). Increase of proton electrochemical potential and ATP synthesis in rat liver mitochondria irradiated *in vitro* by helium-neon laser. *FEBS Lett. 175*, 95-99.

[98] Vekshin, N.L. (1991). Light-dependent ATP synthesis in mitochondria. *Biochem. Int. 25*, 603-611.

[99] Vekshin, N.L. and Mironov, G.P. (1982). Flavin-dependent oxygen uptake in mitochondria under illumination. *Biophysics (Moscow) 27*, 537–539.

[100] 100. Passarella, S., Perlino, E., Quagliariello, E., Baldassarre, L., Catalano, I.M. and Cingolani, A. (1983). Evidence of changes induced by He-Ne laser irradiation in the biochemical properties of rat liver mitochondria. *Bioelectrochem. Bioenerg. 10*, 185–198.

[101] Hu, W.-P., J.J. Wang, C.-L. Yu, C.-C.E. Lan, G.S. Chen and H.-S. Yu (2007) He-Ne laser irradiation stimulates cell proliferation through photostimulatory effects in mitochondria. *J. Invest. Dermatol. 127*, 2048-2057.

[102] Passarella, S., Ostuni, A., Atlante, A. and Quagliariello, E. (1988). Increase in the ADP/ATP exchange in rat liver mitochondria irradiated *in vitro* by He-Ne laser. *Biochem. Biophys. Res. Comm. 156*, 978-986.

[103] Greco, M., Guida, G., Perlino, E., Marra, E. and Quagliariello, E. (1989). Increase in RNA and protein synthesis by mitochondria irradiated with He-Ne laser. *Biochem. Biophys. Res. Comm. 163*, 1428-1434.

[104] Vacca, R.A., Marra, E., Quagliariello, E. and Greco, M. (1993). Activation of mitochondrial DNA replication by He-Ne laser irradiation. *Biochem. Biophys. Res. Comm. 195*, 704-709.

[105] Vacca, R.A., Marra, E., Quagliariello, E. and Greco, M. (1993). Increase of both transcription and translation activites following separate irradiation of the *in vitro* system components with He-Ne laser. *Biochem. Biophys. Res. Comm. 203*, 991-997.

[106] Vacca, R.A., Marra, E., Passarella, S., Petragallo, V.A. and Greco, M. (1996). Increase in cytosolic and mitochondrial protein synthesis in rat hepatocytes irradiated *in vitro* by He-Ne laser. *J. Photochem. Photobiol. B: Biol. 34*, 197-202.

[107] Pastore, D., Greco, M., Petragallo, V.A. and Passarella, S. (1994). Increase in H^+/e^- ratio of the cytochrome c oxidase reaction in mitochondria irradiated with He-Ne laser. *Biochem. Mol. Biol. Int. 34*, 817-826.

[108] Morimoto, Y., Arai, T., Kikuchi, M., Nakajama, S. and Nakamura, H. (1994). Effect of low-intensity argon laser irradiation on mitochondrial respiration. *Lasers Surg. Med. 15*, 191–199.

[109] Yu, W., Naim, J. O., McGowan, H., Ippolito, K. and Lanzafame, R. J. (1997). Photomodulation of oxidative metabolism and electron chain enzymes in rat liver mitochondria. *Photochem. Photobiol. 66*, 866–871.

[110] Alexandratou, E., Yova, D., handris, P., Kletsas, D. and loukas, S. (2002). Human fibroblast alterations induced by low power laser irradiation at the single cell level using confocal microscopy. *Photochem. Photobiol. Sci. 1*, 547-552.

[111] Gavish, L., Asher, Y., Becker, Y. and Kleinman, Y. (2004). Low level laser irradiation stimulates mitochondrial membrane potential and disperses subnuclear promyelocytic leukemia protein. *Lasers Surg. Med. 35*, 369-376.

[112] Zhang, D.X. and gutterman, D.D. (2006). Mitochondrial reactive oxygen species-mediated signaling in endothelial cells. *Am. J. Physiol. Heart circ. Physiol. 292*, H2023-H2031.

[113] Camello-Almaraz, C., Gomez-pinilla, P.J., Pozo, M. and Camello, P.J. (2006). mitochondrial reactive oxygen species and Ca^{2+} signaling. *Am. J. Physiol. Cell Physiol. 291*, C1082-C1088.

[114] Schicke, S.M. and Finkel, T. (2006). Mitochondrial signaling, TOR, and life span. *Biol. Chem. 387*, 1357-1361.

[115] Kudin, A.P., Debska-Vilhaber, G. and Kunz, W.S. (2005). Characterization of superoxide production sites in isolated rat brain and skeletal muscle mitochondria. *Biomed. Pharmacother. 59*, 163-168.

[116] Brookes, P.S., Yoon, Y., Robotham, J.L., Anders, M.W. and Sheu, S.-S. (2004). Calcium, ATP, and ROS: a mitochondrial love-hate triangle. *Am. J. Physiol. Cell Physiol. 287*, C817-C833.

[117] Karu, T.I., Pyatibrat, L.V., Kalendo, G.S. and Esenaliev, R.O. (1996). Effects of monochromatic low-intensity light and laser irradiation on adhesion of HeLa cells in vitro. *Lasers Surg. Med., 18*, 171-177.

[118] Tan, D.X, Manchester, L.C., Reiter, R.J., Qi, W.B., Karbownik, H. and Calvo, J.R. (2000). Significance of melatonin in antioxidative defence system: reactions and products. *Biol. Signals Recept. 9*, 137–159.

[119] Martin, M., Macias, M., Escames, G., Leon, J. and Acuna-Castroviejo, D. (2000). Melatonin but not vitamins C and E maintains glutathione homeostasic in t-butyl hydroperoxide-induced mitochondrial oxidative stress. *FASEB J. 14*, 1677–1679.

[120] Acuna-Castroviejo, D., Martin, M., Macias, M., Excames, G., Leon, J., Khaldy, H. and Reiter, J. (2001). Melatonin, mitochondria, and cellular bioenergetics. *J. Pineal Res. 30*, 65–74.

[121] Karu, T., Kurchikov, A., Letokhov, V. and Mokh, V. (1996). He-Ne laser radiation influences single-channel ionic currents through cell membranes: a patch-clamp study. *Lasers Life Sci. 7*, 35–48.

[122] Cortassa, S., Aon, M.A., Winslow, R.L. and O'Rourke, B.(2004). A mitochondrial oscillator dependent on reactive oxygen species. *Biophys. J. 87*, 2060-2073.

[123] Aon, M.A., Cortassa, S., Marban, E. and O'Rourke, B. (2003). Synchronized whole cell oscillations in mitochondria metabolism triggered by a local release of reactive oxygen species in cardiac myocytes. *J. Biol. Chem. 278*, 44735-44744.

[124] Smolyaninova, N.K., Karu, T.I., Fedoseyeva, G.E. and Zelenin, A.V. (1991). Effect of He-Ne laser irradiation on chromatin properties and synthesis of nucleic acids synthesis of human blood lymphocytes. *Biomed. Sci. 2*, 121-126.

[125] Lubart, R., Friedmann, H., Levinshal, T., Lavie, R. and Breitbart, H. (1992). Effect of light on calcium transport in bull sperm cells. *J. Photochem. Photobiol. B: Biol. 15*, 337-341.

[126] Cohen, N., Lubart, R., Rubinstein, S. and Breitbart, H. (1998). Light irradiation of mouse spermatozoa: stimulation of in vitro fertilization and calcium signals. *Photochem. Photobiol. 68*, 407-413.

[127] Breitbart, H., Levinshal, T., Cohen, N., Friedmann, H. and Lubart, R. (1996). Changes in calcium transport in mammalian sperm mitochondria and plasma membrane irradiated at 633 nm (He-Ne laser). *J. Photochem. Photobiol. B: Biol. 34*, 117-121.

[128] Lubart, R., Friedmann, H., Sinyakov, H., Cohen, N. and Breitbart, H. (1997). Changes in calcium transport in mammalian sperm mitochondria and plasma membranes caused by 780 nm irradiation. *Lasers Surg. Med. 21*, 493-499.

[129] Vacca, R.A., Moro, L., Petragallo, V.A., Greco, M., Fontana, F. and Passarella, S. (1997). The irradiation of hepatocytes with He-Ne laser causes an increase of cytosolic free calcium concentration and an increase of cell membrane potential, correlated with it, both increases taking place in an oscillatory manner. *Biochem. Mol. Biol. Int. 43*, 1005-1014.

[130] Schwartz, F., Brodie, C., Appel, E., Kazimirsky, G. and Schainberg, A. (2002). Effect of helium/neon laser irradiation on nerve growth factor synthesis and secretion in skeletal muscle cultures. *J. Photochem. Photobiol. B: Biol. 66*, 195-200.

[131] Yang, W.-Z., Chen, J.-Y., Yu, J.T. and Zhou, L.-W. (2007). Effects of low power laser irradiation on intracellular calcium and histamine release in RBL-2H3 mast cells. *Photochem. Photobiol. 83*, 979-984.

[132] Cooper, C. E. (2002). Nitric oxide and cytochrome oxidase: substrate, inhibitor or effector? *Trends Biochem. Sci. 27*, 33–39.

[133] Brookes, P.S. and Darley-Usmar, V.M. (2002). Hypothesis: the mitochondrial $NO^{\bullet}$ signaling pathway, and the transduction of nitrosative to oxidative cell signaling: an alternative function for cytochrome c oxidase. *Free Radic. Biol. Med. 32*, 370-374.

[134] Giulivi, C., Pedroso, J.J. and Boveris, A. (1998). Production of nitric oxide by mitochondria. *J. Biol. Chem. 273*, 11038-11043.

[135] Brookes, P.S., Levonen, A.-L., Shiva, S., Sarti, P.S. and Darley-Usmar, V. M. (2002). Mitochondria: regulators of signal transduction by reactive oxygen and nitrogen species. *Free Radic. Biol. Med. 33*, 755–764.

[136] Karu T.I. (1988). Molecular mechanism of the therapeutic effect of low-intensity laser radiation. *Lasers Life Sci. 2*, 53-74.

[137] Tiphlova, O. and Karu, T. (1991). Action of low-intensity laser radiation on *Escherichia coli*. *Crit. Rev. Biomed. Eng. 18*, 387-412.

[138] Quickenden, T.I., Daniels, L.L. and Burne, L.J. (1995). Does low intensity He-Ne laser radiation affect the intracellular pH of intact *Escherichia coli* cells? *SPIE Proc. 2391*, 535-544.

[139] Trimm, J.L., salama, G. and Abramson, Y.J. (1986). Sylphydryl oxidation induces rapid calcium release from sarcoplasmic reticulum vesicles. *J. Biol. Chem. 261*, 16092-16098.

[140] Wang, T., Zhang, X. and Li, J.J. (2002). The role of NF-kB in the regulation of cell stress responses. *Int. Immunopharmacol. 2*, 1509-1520.

[141] Liu, X., Zhou, B., Mi, S., Xue, L., Shih, J., Lee, J., Chau, J., Un, F. and Yen, Y. (2007). An increase of cytochrome c oxidase mediated disruption of gemcitabine incorporation into DNA in a resistant κB clone. *Biochem. Pharmacol.73*, 1927-1938.

[142] Pueyo, M.E., Gonzales, W., Nicoletti, A., Savoie, F., Arnal, J.-F. and Michel, J.-B. (2000). Angiotensin II stimulates endothelial vascular cell adhesion molecule-1 via nuclear factor-κB activation induced by intracellular oxidative stress. *Arterioscler. Thromb. Vasc. Biol. 20*, 645-651.

[143] Nicoli, E. and carruba, M.o. (2006). nitric oxide and mitochondrial biogenesis. *J. Cell Sci. 119*, 2855-2862.

[144] Passarella, S., Roncali, L., Cicero, R. and Quagliariello, E. (1988). New ultrastructural conformations of mitochondria irradiated in vitro with a He-Ne laser. *Lasers Life Sci. 2*, 161-171.

[145] Greco, M., Perlino, E., Pastore, D., Grida, G., Marra, E. and Quagliariello, E. (1991). He-Ne laser irradiation of rat level mitochondria gives rise to a new subpopulation of mitochondria: isolation and first biochemical characterization. *J. Photochem. Photobiol. B:Biol. 10*, 71-78.

[146] Manteifel, V.M., Bakeeva, L.E. and Karu, T.I. (1999). He-Ne laser radiation causes changes in mitochondrial ultrastructure in successive generations of yeast cells. *Doklady Biophysics (Moscow) 364-366*, 58-60.

[147] Manteifel, V.M. and Karu, T.I. (2005). Structure of mitochondria and activity of their respiratory chain in successive generations of yeast cells exposed to He-Ne laser light. *Biol. Bull. 32*, 556-566.

[148] Jazwinski, S.M. (2005). The retrograde response links metabolism with stress responses, chromatin-dependent gene activation, and genome stability in yeast aging. *Gene 354, 22-27.*

[149] Liang, H.L., Whelan, H.T., Eells, J.T., Meng, H., Buchmann, E., Lerch-Gaggl, A. and Wong-Riley, M.T. (2006). Photobiomodulation partially rescues visual cortical neurons from cyanide-induced apoptosis. *Neuroscience 139,* 639–649.

[150] Karu, T.I. and Kolyakov, S.F. (2005). Exact action spectra for cellular responses relevant to phototherapy. *Photomed. and Laser Surg. 23*, 355-361.

In: New Cell Adhesion Research
Editors: Patrick Nott and Matthew Temple

ISBN 978-1-60692-378-8

Chapter II

FORCES AT ADHESIVE CONTACTS

Thuc-Nghi Nguyen and Soichiro Yamada***

Department of Biomedical Engineering, University of California, Davis, USA.

ABSTRACT

The regulation of cell-extracellular matrix and cell-cell adhesion is essential for tissue development and homeostasis. Stable cell junctions maintain normal tissue architecture, but, during embryogenesis, tissue regeneration and cancer cell migration, adhesive contacts loosen and become traction sites for cellular rearrangement. Previous studies have identified essential molecules for the assembly and maintenance of cell adhesion. Yet, we know very little about the mechanical role of cell adhesion as a site of force transmission. Cellular traction forces are generated by the actin cytoskeleton, and the remodeling of the actin network is thought to regulate the strength and dynamics of cell adhesion. The precise molecular linkage between the adhesion complex and the actin cytoskeleton, however, remains elusive. This review discusses novel experimental techniques that probe cellular forces exerted at the sites of adhesive contacts, and how these forces can explain the roles of cell adhesion in orchestrating multi-cellular behavior.

INTRODUCTION

Cell migration plays a central role during embryogenesis, tissue regeneration, cancer cell invasion, and wound healing. To navigate through a dense extracellular matrix and between neighboring cells, migrating cells generate and exert traction forces on the substrate. This

* Correspondence concerning this article should be addressed to: Soichiro Yamada, Department of Biomedical Engineering, University of California, Davis, Genome and Biomedical Sciences Facility, 451 Health Sciences Dr. Davis, CA 95616. Tel: 530-754-7251; Fax: 530-754-5739; Email: syamada@ucdavis.edu.

requires careful regulation of force generation and cell adhesions to the substrate for proper and efficient cellular rearrangement.

Cells are anchored to the extracellular matrix by protein complexes referred to as focal adhesions. The primary adhesion proteins at focal adhesions are integrins, a superfamily of heterodimers consisting of α and β subunits. Depending on the subunit pairs, integrin can bind to various components of the extracellular matrix (e.g., collagen, fibronectin, or laminin) or mediate leukocyte-endothelial cell interactions (Hynes, 2002). At its cytoplasmic domain, integrin indirectly interacts with the actin cytoskeleton to promote the formation and stabilization of focal adhesions (Burridge and Chrzanowska-Wodnicka, 1996; Chrzanowska-Wodnicka and Burridge, 1996; Delon and Brown, 2007; Geiger et al., 2001).

Similar to cell-extracellular matrix junctions, cell-cell junctions are also mediated by adhesive protein complexes. In many tissues, the predominant cell-cell adhesion proteins are calcium dependent. This includes classical cadherins such as E-, N-, or VE-cadherins. Cadherins form homophilic bonds at their extracellular domain and bind to β-catenin and p120 catenin at their cytoplasmic domain. As with focal adhesions, the strength of cell-cell adhesion also depends on the actin cytoskeleton (Angres et al., 1996; Chu et al., 2004; Imamura et al., 1999; Matsuzaki et al., 1990).

The interactions between cytoskeleton and adhesion molecules have been extensively studied through biochemical and mutational analysis. The results of these studies revealed the molecules and genetic interactions required for cell-extracellular and cell-cell adhesion. However, the mechanical role of cell adhesions is still not clear. Recent studies have shown that the actin cytoskeleton is not stably linked to sites of focal adhesion (Brown et al., 2006; Hu et al., 2007) or cell-cell adhesion (Drees et al., 2005; Yamada et al., 2005). These results suggest the dynamic connection between adhesion complex and the actin cytoskeleton is the prevailing interaction in many adhesive contacts. This type of interactions will likely to have a significant impact on how migrating cells regulate the traction forces at adhesive contacts.

Regulation of Organized Cell Movements

Organized cell movements, such as the sorting of germ layers during embryogenesis, require cell adhesion and force generation. In an *in vitro* model, a heterogeneous population of cells can preferentially rearrange into layers of a single cell type (Townes and Holtfreter, 1955). The cellular basis of sorting has been attributed to the cadherin type (Niessen and Gumbiner, 2002; Nose et al., 1988), cadherin expression level (Foty and Steinberg, 2005; Friedlander et al., 1989; Steinberg and Takeichi, 1994) and cortical tension of individual cells (Krieg et al., 2008). Since the cadherin expression level and cortical tension of individual cells are proportional to the overall surface tension of cellular aggregates, differential surface tension may be a guiding principle of cell sorting (Lecuit and Lenne, 2007).

What then is the underlying mechanism that determines the overall surface tension of a multi-cellular cluster? Surface tension is typically measured by compression of cell mass using a force-sensing probe. The mechanical response against the compressive force is provided by the actin cytoskeleton. Since the actin network drastically reorganizes upon the engagement of cell-cell adhesion proteins (Vasioukhin et al., 2000; Yamada and Nelson,

2007), it is likely that the number and type of cadherins and strength of cell-cell adhesion will have significant effects on the organization of actin and surface tension of cell aggregates. Cortical tension of single cells also depends on the actin cytoskeleton. However, since the actin network reorganizes upon the formation of cell-cell contacts, how the cortical tension of single cells determines the overall surface tension of a cell cluster is not clear. Nevertheless, these results imply that differences in surface tension, therefore cell sorting, could arise due to the regulation of the actin cytoskeleton.

Force transmission at the sites of cell adhesion is implicated in the organized migration of cell clusters. During embryonic development, cells migrate past each other to form an elongated and polarized body plan in a process known as convergent-extension. These cell movements are driven by internally generated forces that are independent of other tissues or external substrates (Keller, 2002). The forces are generated by the actin cytoskeleton and transmitted via cell adhesion complexes. Similarly, wound-healing cells migrate in a highly organized fashion. Cells surrounding a wound form a continuous ring of cortical actin at the exposed cell edge, and constriction of the actin ring is thought to pull the cells toward the center of the wound (Bement et al., 1999; Martin and Lewis, 1992; Redd et al., 2004; Tamada et al., 2007; Wood et al., 2002). This process requires coordinated contraction of the actin ring around multiple cells connected via cell-cell junctions, leading to global contraction and coordinated cell movement of the epithelial monolayer. Together, these results suggest that organized cell movements require cell adhesion junctions to act as sites of force transmission.

Traction Force Analysis

The basis of cell motility is the ability of cells to generate and exert traction force on the substrate. The analysis of cellular traction force is, however, an experimentally challenging problem. This is partly because the quantification of cellular traction requires a highly sensitive, biocompatible, and micron-sized force sensor. The first technique used to analyze cellular traction consisted of a flexible silicone membrane that deformed when fibroblasts attached and pulled on the membrane (Harris et al., 1981; Harris et al., 1980). The large deformation of the silicone substrate appeared as wrinkles and was interpreted as cellular traction force exerted by individual fibroblasts. The quantitative analysis of traction force, however, was limited due to non-linear deformation of the silicone substrates.

To prevent excessive deformation, the substrate was stiffened and fiducial surface markers were added to visualize the surface deformation (Balaban et al., 2001; Dembo et al., 1996). By quantitatively analyzing the displacements of the fiducial markers, a strain field of the elastic substrate can be reconstructed. The question remained as to how traction force is distributed at the cell- surface contact.

The distribution of traction force in the cell-surface contact area can be estimated using rigorous mathematical analysis. There are two complementary algorithms for solving the distribution of traction forces at adhesive contacts. The algorithms differ in the assignment of force vectors in the cell-surface contact area. One approach is to divide the cell adhesive area into smaller regions and iteratively solve for the magnitudes and directions of force vectors in each region. A solution is obtained when the reconstructed force field reproduces the observed displacement field (Dembo et al., 1996; Dembo and Wang, 1999). This algorithm generates a fitted force field in an unbiased manner without *a priori* assumptions about locations of force generation at the cell-surface contact. Alternatively, discrete focal adhesion spots were assumed as the point sources of force generation, and based on the strain field, traction forces were assigned to each focal adhesion (Balaban et al., 2001). Since this approach relies on immuno-labeling of focal adhesion proteins, the force distribution will be biased toward more prominent focal adhesions and may ignore the contributions from smaller but functionally important focal contacts (Geiger et al., 2001).

Another approach to measure cell traction force is to use micro-cantilevers that bend when cells crawl over them (Galbraith and Sheetz, 1997; Tan et al., 2003). Unlike the elastic substrate in which the deformation field is the result of multiple force vectors, the deflection of each cantilever is due to force generation by a single adhesive contact at the cantilever tip. With this method, calculation of traction force is direct. According to linear beam theory, the magnitude of force is proportional to the stiffness of the cantilever and the displacement of the cantilever tip. An early version of micro-cantilevers was produced by micromachining silicon and glass into cantilever beams (Galbraith and Sheetz, 1997). One drawback of the device was the horizontal orientation of the cantilevers, which can only measure force in one direction.

A simpler but more effective method to measure cell traction forces was developed using elastic micro-pillar arrays (Tan et al., 2003). These vertically oriented cantilevers are made by casting polymeric silicone onto a microfabricated mold. Cells are subsequently plated on top of the pillar array, and the active processes of cell attachment and migration cause the pillar tips to deflect in any possible direction. Additionally, the diameter, height, and spacing of the pillars can be easily manipulated to give the desired stiffness and spatial resolution. This approach has been applied to measure traction forces of muscle cell contraction (Zhao and Zhang, 2006) and multicellular traction force (du Roure et al., 2005).

CELL MIGRATION AND TRACTION FORCES

What have we learned from traction force analysis? Surprisingly, cell speed does not increase with cellular traction force. On soft rubber substrates, slow migratory fibroblasts cause more wrinkles than highly migratory cell types such as leukocytes and neural growth cones (Harris et al., 1981). Subsequently, more quantitative traction force analysis revealed migration speed is inversely related to traction force. For example, traction force is typically smaller for fast moving keratocytes (cell velocity of up to 30 μm/min, Anderson and Cross, 2000) than slow moving fibroblasts (mean velocities of less than 1 μm/min, Rotsch et al., 1999). Using a soft elastic substrate, the traction forces for keratocytes and fibroblasts are 25 nN and 100-200 nN, respectively. Likewise, using bead-embedded stiff elastic substrate, traction forces for keratocytes and fibroblasts are 140 nN and 2000 nN, respectively (Table 1). Despite the discrepancy between two experimental methods (see below for possible explanations), there is a consistent difference between slow and fast moving cells. Collectively, these results imply that cell migration require relatively little traction force.

Table 1. Cell traction forces determined using various methods

Cell type	Substrate	Surface coating	Max force (nN)	Mean force (nN)	Mean stress ($nN/\mu m^2$)	Reference
Chick heart fibroblast	Wrinkling elastic			100-200*	10 nN/μm	(Harris et al., 1980)
Keratocyte	Wrinkling elastic			25*		(Oliver et al., 1994)
Rat cardiac myocyte	Micropatterned elastic			70	2-5	(Balaban et al., 2001)
Human fibroblast	Micropatterned elastic			20	5.5	(Balaban et al., 2001)
Keratocyte	Bead-embedded elastic			140*	0.22	(Dembo et al., 1996)
3T3 fibroblast	Bead-embedded elastic			2000*	2	(Dembo and Wang, 1999)
Chick embryo fibroblast	Microcantilever	laminin	80 (tail)	2-80	< 4	(Galbraith and Sheetz, 1997)
Artery smooth muscle	Micropillar array	fibronectin	75	20-40	16	(Tan et al., 2003)
C2 myogenic cells	Micropillar array	fibronectin	100	30	9.5	(Ganz et al., 2006)
C2 myogenic cells	Micropillar array	N-cadherin	50	15	4.8	(Ganz et al., 2006)
GT1-7	Micropillar array	N-cadherin	60	14	4.5	(Ganz et al., 2006)
MDCK	Micropillar array	fibronectin	30	5	1.6	(du Roure et al., 2005)

Notes: * Total force for entire cell

How can cells migrate with such little force? One possibility is that highly motile cells may have weaker cell-extracellular matrix adhesion that avoids stable junctions and minimizes transmission of traction force to the substratum. The weakest molecular linkage of cell-extracellular matrix adhesion may reside at the extracellular or intracellular domain of adhesion receptors. For integrin-mediated cell adhesion, the adhesive strength of the extracellular domain can affect migration speed (Palecek et al., 1997), and the cytoplasmic proteins can regulate intrinsic bond strength of integrins (Luo et al., 2007; Springer, 1994). Similarly, the intracellular linkage has been postulated to act as a clutch to engage or disengage internal force generation machinery (Lin and Forscher, 1995; Mitchison and

Kirschner, 1988), thereby regulating the transmission of traction forces across cell junctions. Alternatively, highly motile cells may not have the ability to generate strong traction forces, and this is consistent with the lack of extensive stress fibers in highly migratory leukocytes or keratocytes.

Another possibility is that migratory cells have highly polarized distribution of traction forces from leading edge to trailing edge. This asymmetry in force generation allows for persistent cell movement in the forward direction. The reconstructed force field from elastic substrate studies indicates that the leading lamellipodium of fibroblasts has higher force contribution than the trailing edge (Dembo and Wang, 1999). In the cantilever study (Galbraith and Sheetz, 1997), however, the tail region of fibroblasts exerts greater traction force than the leading lamellipodium (Table 1). These contradictory data suggest that the magnitudes of traction forces may rapidly fluctuate in migrating cells. Since the distribution of traction force is not uniform, independent and localized measurement of forces for well-defined adhesion areas should be used to accurately analyze traction forces.

One caveat of traction force analysis is the effect of substrate stiffness on cellular processes. For example, comparing the results from the wrinkling substrate and the bead-embedded substrate, the traction forces differ by an order of magnitude (Table 1). These two force sensors have a considerable difference in substrate stiffness due to their methods of preparation. Both are made of a silicone polymer, but one is crosslinked only at the top layer, leaving a viscous constituent underneath (Harris et al., 1980) while the other is crosslinked uniformly throughout (Dembo et al., 1996). The mechanical properties of substrates have been shown to alter cell behavior including cell migration (Lo et al., 2000), acini formation in a three dimensional matrix (Paszek et al., 2005), and stem cell lineage (Engler et al., 2006). To work around these differences, relative comparison between experiments using the same method can still give meaningful conclusions.

Traction Forces at Cell-Cell Contacts

The sites of cell-cell contacts are constantly exposed to various forces. This is evident from the observation that spontaneous detachment of N-cadherin-mediated cell-cell junctions caused the cytoplasm to retract from the site of cell-cell contact (Ragsdale et al., 1997). This result suggests that cell-cell junctions are under constant tension. In some instances, these forces can enhance the local recruitment of cadherins (Riveline et al., 2001). Since neighboring cells exert these forces, traction forces at cell-cell junctions are more difficult to experimentally measure without disrupting the adhesion sites. A theoretical model has been developed to correlate the forces with morphology at cell-cell junctions (Brevier et al., 2007), but without direct experimental analysis of forces, the model remains unverified.

One way to directly measure force at cell-cell contacts is to use a micro-pillar array coated with cadherins, thus simulating the sites of cell-cell contacts. When coated with N-cadherin, the magnitudes of traction force at N-cadherin-mediated cell-cell contacts were similar to that of cell-extracellular matrix contacts (Table 1, Ganz et al., 2006). These results suggest that the same contractile machinery may act on molecularly distinct junctions.

Actin-myosin Contraction

The actin-myosin bundle is the molecular machinery that generates traction force. In non-muscle cells, stress fibers contain similar molecular components as sarcomeres in muscle cells, albeit lacking the higher order organization of sarcomeres. Early studies showed that stress fibers consist of actin filaments crosslinked by α-actinin (Lazarides and Burridge, 1975), with intercalating arrays of non-muscle myosin II lying periodically along the bundle length (Weber and Groeschel-Stewart, 1974). Since the bi-polar myosin complex translocates toward the plus ends of actin filaments, stress fiber contraction occurs when myosin moves along actin filaments of opposite polarities (Pellegrin and Mellor, 2007).

Evidence of stress fiber contraction has been demonstrated *in vitro* and in live cells. Isolated stress fibers from cultured fibroblasts and endothelial cells were contractile since shortening of the fibers occurred without actin filament depolymerization (Katoh et al., 1998). The contractile activity was regulated by calcium, MLCK and Rho kinase (Katoh et al., 2001). Using a laser nanoscissor, the contractile activity of stress fibers was also demonstrated in living cells. By laser ablating a small section of stress fiber, the severed ends of the stress fiber progressively moved further apart (Kumar et al., 2006). Since the fiducial marker on the stress fiber remained and did not disappear due to actin depolymerization, shortening of the fiber was attributed to being released from tension (Kumar et al., 2006). Together, these studies confirm the contractility of actin-myosin bundles and the ability of stress fibers to generate tension.

The organization of actin filament network is critical for the generation of traction forces. Depending on cell type and subcellular location, the actin organization and polarity of the bundles may vary. In fibroblasts, actin bundles can be organized in several ways (Pellegrin and Mellor, 2007; Small et al., 1998). Most stress fibers lie along the base of the cell with both ends terminated at focal adhesions, but some stress fibers are anchored to a focal adhesion at one end while the other end extends into the cell interior (Hotulainen and Lappalainen, 2006). The ends of these focal adhesion stress fibers can elongate by actin polymerization (Chan et al., 1998; Hotulainen and Lappalainen, 2006), which suggest that if the actin assembly occurs only at the plus end, these stress fibers will have uniform polarity. It is not clear how stress fibers with uniform polarity can actually contract to generate traction forces. Instead, they may serve as a precursor to the formation of other actin bundles (Pellegrin and Mellor, 2007). Similarly, some actin bundles are formed by the association of short actin bundles, giving them an alternating polarity similar to sarcomeres (Cramer et al., 1997; Hotulainen and Lappalainen, 2006).

Unique to motile fibroblasts, graded polarity actin bundles have uniform but opposite polarities at the ends of the bundle and mixed polarity in the central region (Cramer et al., 1997). In addition, the organization of myosin II is irregular in motile cells, whereas in non-motile cells, myosin II is distributed along actin bundles as punctas at regular intervals (Cramer et al., 1997). These differences in bundle size, polarity, and myosin II distribution suggest that the individual organization of actin-myosin bundles may have distinct mechanisms of force generation.

How does actin organization modulate traction force? Studies of motile keratocytes have indicated that contraction of actin bundles result in traction force on the substrate.

Specifically, for keratocytes, actin bundles at the lamellipodia-cell body transition zone is formed as a result of lateral contraction of a meshed network of actin filaments that propels the cell body forward (Svitkina et al., 1997). The lateral contraction results in wrinkles on elastic substrates perpendicular to the direction of contraction (Burton et al., 1999). Since traction force is a measure of actin-myosin contraction, analysis of the magnitude, direction, and spatial distribution of traction force may provide clues as to how the actin organization plays a role in regulating cell motility.

Linkage Between Adhesion Molecules and the Cytoskeleton

Adhesive contacts recruit proteins of all kind to mediate chemical and mechanical signal transduction. The actin cytoskeleton indirectly interacts with adhesion receptors, which transmits the tension generated by actin-myosin contraction to substratum. For cell-extracellular matrix adhesion, the observed traction forces linearly correlated with the intensity and size of focal adhesions (Balaban et al., 2001). This is consistent with the notion that a larger focal adhesion complex can interact with more actin filaments, thereby producing greater traction force. Intuitively, to maximize the force transmission, this linkage should be static. Recent experiments, however, dispute this notion and provide better descriptions of the molecular linkage between actin and adhesion proteins at adhesive contacts.

Due to a myriad of proteins residing at sites of cell adhesion, multiple protein interactions are expected. To analyze these interactions, high-resolution fluorescence live-cell imaging emerged as a useful tool for investigating protein-protein interactions. The diffusive or convective motion of fluorescently tagged proteins can be determined by sparse fluorescent labeling of proteins (Fluorescence Speckle Microscopy) or selective photobleaching at the site of interest (Fluorescence Recovery After Photobleaching). With both techniques, the protein-protein interactions can be deduced from quantitative analysis of relative protein movement.

In slow migrating cells with a rapid actin retrograde flow, the motion of actin binding proteins, α-actinin, vinculin, and talin, was highly correlated with actin retrograde flow (Brown et al., 2006; Hu et al., 2007). In contrast, the motion of intermediate proteins, which bind to the actin-binding proteins and integrin, was less correlated with actin movement (Brown et al., 2006; Hu et al., 2007). The correlation of speed and direction between actin and integrin movement was even less, which may be due to immobilization of integrins to the extracellular matrix. These results suggest that actin and actin-binding focal adhesion proteins are highly dynamic at focal adhesions and are not stably bound to immobilized integrins.

How then are traction forces transmitted? Interestingly, the flow of actin and vinculin became more synchronized as a focal adhesion complex detached from the matrix. Perhaps the dissociation of vinculin from less mobile focal adhesion components weakens the focal adhesion, allowing it to disengage from the matrix (Hu et al., 2007). In contrast, the movement of actin and vinculin became less synchronized when the focal adhesion stabilized and reattached to the matrix, suggesting that vinculin may be a molecular clutch between actin and focal adhesions (Hu et al., 2007). The consequence of this dynamic linkage is the fluctuations in traction force. In fact, the cantilever-based traction force analysis showed that traction force could oscillate periodically, indicating active force regulation during cell movement (Galbraith and Sheetz, 1997). These studies suggest that transient interactions between the adhesion complex and the actin cytoskeleton may be part of a regulatory mechanism to modulate traction force transmitted to the substrate.

Similarly, at sites of cadherin-mediated cell-cell adhesion, the link between the actin cytoskeleton and cadherin complexes may also be dynamic. The molecular linkage of the cadherin complex to the actin cytoskeleton has been thought to be α-catenin, a homolog of vinculin. The stability of adherens junctions of epithelial cells was assumed to be provided by the stable linkage of cadherin to actin via β-catenin and α-catenin (Burridge, 2006; Weis and Nelson, 2006). Contrary to this dogma, α-catenin cannot simultaneously bind to the cadherin complex and actin filaments, but can only bind to one or the other depending on its molecular state: monomer or homodimer (Drees et al., 2005; Yamada et al., 2005). Furthermore, the fluorescence recovery after photobleaching of E-cadherin along with β-catenin and α-catenin is slower than that of actin, suggesting that actin is not stably linked to the cadherin complex in polarized epithelial cells (Yamada et al., 2005).

To complicate the matter even further, actin bundles at the sites of cell-cell adhesion have been shown to be rapidly polymerizing (Vasioukhin et al., 2000; Yamada and Nelson, 2007). For cell-cell junctions between keratinocytes, this actin polymerization was mediated by an actin nucleator, formin (Kobielak et al., 2004). Similar preferential actin polymerization was also observed for actin bundles terminating at focal adhesions (Chan et al., 1998). Collectively, these experimental data suggest that highly dynamic protein interactions and rapid actin assembly may be tightly coupled to the regulation of traction force at adhesive contacts.

The question remains as to how the actin-myosin contractile force is transmitted via adhesion sites if the actin cytoskeleton is actively polymerizing and dynamically linked to the integrin or cadherin complexes. It is possible that the actin-adhesion protein linkage may be stabilized upon actin-myosin contraction or application of external force (Burridge, 2006). This force sensitive linkage has not been experimentally demonstrated, but remains an attractive model. Recently, a focal adhesion protein, p130Cas, has been shown to undergo a conformational change as a mechanical response to cell stretching (Sawada et al., 2006). The combined analysis of cellular force and protein interactions has a promise to unravel the mechanism and regulation of force generation at adhesive contacts.

CONCLUSION

Studies up until now have revealed differences in adhesion proteins, actin stress fiber structures, and molecular interactions that may explain how forces are generated. In light of recent studies that demonstrated the highly dynamic actin networks at sites of cell adhesion, a new paradigm must be developed to reconcile all experimental data. To better understand how cells control traction and their movement, force measurements should be an integral part of experimental analysis. By combining high-resolution imaging and micron-sized force sensors, the forces at adhesive contacts can be probed while simultaneously examining molecular interactions. Such devices will define how adhesion complexes regulate force transmission at cell adhesion sites, and ultimately enable us to predict how cells migrate in physiological and pathological processes.

REFERENCES

Anderson, K.I., and Cross, R. (2000). Contact dynamics during keratocyte motility. *Curr Biol 10*, 253-260.

Angres, B., Barth, A., and Nelson, W.J. (1996). Mechanism for transition from initial to stable cell-cell adhesion: kinetic analysis of E-cadherin-mediated adhesion using a quantitative adhesion assay. *J Cell Biol 134*, 549-557.

Balaban, N.Q., Schwarz, U.S., Riveline, D., Goichberg, P., Tzur, G., Sabanay, I., Mahalu, D., Safran, S., Bershadsky, A., Addadi, L., *et al.* (2001). Force and focal adhesion assembly: a close relationship studied using elastic micropatterned substrates. *Nat Cell Biol 3*, 466-472.

Bement, W.M., Mandato, C.A., and Kirsch, M.N. (1999). Wound-induced assembly and closure of an actomyosin purse string in Xenopus oocytes. *Curr Biol 9*, 579-587.

Brevier, J., Vallade, M., and Riveline, D. (2007). Force-extension relationship of cell-cell contacts. *Phys Rev Lett 98*, 268101.

Brown, C.M., Hebert, B., Kolin, D.L., Zareno, J., Whitmore, L., Horwitz, A.R., and Wiseman, P.W. (2006). Probing the integrin-actin linkage using high-resolution protein velocity mapping. *J Cell Sci 119*, 5204-5214.

Burridge, K. (2006). Cell biology: a break in the chain? *Nature 440*, 38-39.

Burridge, K., and Chrzanowska-Wodnicka, M. (1996). Focal adhesions, contractility, and signaling. *Annu Rev Cell Dev Biol 12*, 463-518.

Burton, K., Park, J.H., and Taylor, D.L. (1999). Keratocytes generate traction forces in two phases. *Mol Biol Cell 10*, 3745-3769.

Chan, A.Y., Raft, S., Bailly, M., Wyckoff, J.B., Segall, J.E., and Condeelis, J.S. (1998). EGF stimulates an increase in actin nucleation and filament number at the leading edge of the lamellipod in mammary adenocarcinoma cells. *J Cell Sci 111 (Pt 2)*, 199-211.

Chrzanowska-Wodnicka, M., and Burridge, K. (1996). Rho-stimulated contractility drives the formation of stress fibers and focal adhesions. *J Cell Biol 133*, 1403-1415.

Chu, Y.S., Thomas, W.A., Eder, O., Pincet, F., Perez, E., Thiery, J.P., and Dufour, S. (2004). Force measurements in E-cadherin-mediated cell doublets reveal rapid adhesion strengthened by actin cytoskeleton remodeling through Rac and Cdc42. *J Cell Biol 167*, 1183-1194.

Cramer, L.P., Siebert, M., and Mitchison, T.J. (1997). Identification of novel graded polarity actin filament bundles in locomoting heart fibroblasts: implications for the generation of motile force. *J Cell Biol 136*, 1287-1305.

Delon, I., and Brown, N.H. (2007). Integrins and the actin cytoskeleton. *Curr Opin Cell Biol 19*, 43-50.

Dembo, M., Oliver, T., Ishihara, A., and Jacobson, K. (1996). Imaging the traction stresses exerted by locomoting cells with the elastic substratum method. *Biophys J 70*, 2008-2022.

Dembo, M., and Wang, Y.L. (1999). Stresses at the cell-to-substrate interface during locomotion of fibroblasts. *Biophys J 76*, 2307-2316.

Drees, F., Pokutta, S., Yamada, S., Nelson, W.J., and Weis, W.I. (2005). Alpha-catenin is a molecular switch that binds E-cadherin-beta-catenin and regulates actin-filament assembly. *Cell 123*, 903-915.

du Roure, O., Saez, A., Buguin, A., Austin, R.H., Chavrier, P., Silberzan, P., and Ladoux, B. (2005). Force mapping in epithelial cell migration. *Proc Natl Acad Sci U S A 102*, 2390-2395.

Engler, A.J., Sen, S., Sweeney, H.L., and Discher, D.E. (2006). Matrix elasticity directs stem cell lineage specification. *Cell 126*, 677-689.

Foty, R.A., Pfleger, C.M., Forgacs, G., and Steinberg, M.S. (1996). Surface tensions of embryonic tissues predict their mutual envelopment behavior. *Development 122*, 1611-1620.

Foty, R.A., and Steinberg, M.S. (2005). The differential adhesion hypothesis: a direct evaluation. *Dev Biol 278*, 255-263.

Friedlander, D.R., Mege, R.M., Cunningham, B.A., and Edelman, G.M. (1989). Cell sorting-out is modulated by both the specificity and amount of different cell adhesion molecules (CAMs) expressed on cell surfaces. *Proc Natl Acad Sci U S A 86*, 7043-7047.

Galbraith, C.G., and Sheetz, M.P. (1997). A micromachined device provides a new bend on fibroblast traction forces. *Proc Natl Acad Sci U S A 94*, 9114-9118.

Ganz, A., Lambert, M., Saez, A., Silberzan, P., Buguin, A., Mege, R.M., and Ladoux, B. (2006). Traction forces exerted through N-cadherin contacts. *Biol Cell 98*, 721-730.

Geiger, B., Bershadsky, A., Pankov, R., and Yamada, K.M. (2001). Transmembrane crosstalk between the extracellular matrix--cytoskeleton crosstalk. *Nat Rev Mol Cell Biol 2*, 793-805.

Harris, A.K., Stopak, D., and Wild, P. (1981). Fibroblast traction as a mechanism for collagen morphogenesis. *Nature 290*, 249-251.

Harris, A.K., Wild, P., and Stopak, D. (1980). Silicone rubber substrata: a new wrinkle in the study of cell locomotion. *Science 208*, 177-179.

Hotulainen, P., and Lappalainen, P. (2006). Stress fibers are generated by two distinct actin assembly mechanisms in motile cells. *J Cell Biol 173*, 383-394.

Hu, K., Ji, L., Applegate, K.T., Danuser, G., and Waterman-Storer, C.M. (2007). Differential transmission of actin motion within focal adhesions. *Science 315*, 111-115.

Hynes, R.O. (2002). Integrins: bidirectional, allosteric signaling machines. *Cell 110*, 673-687.

Imamura, Y., Itoh, M., Maeno, Y., Tsukita, S., and Nagafuchi, A. (1999). Functional domains of alpha-catenin required for the strong state of cadherin-based cell adhesion. *J Cell Biol 144*, 1311-1322.

Katoh, K., Kano, Y., Amano, M., Onishi, H., Kaibuchi, K., and Fujiwara, K. (2001). Rho-kinase--mediated contraction of isolated stress fibers. *J Cell Biol 153*, 569-584.

Katoh, K., Kano, Y., Masuda, M., Onishi, H., and Fujiwara, K. (1998). Isolation and contraction of the stress fiber. *Mol Biol Cell 9*, 1919-1938.

Keller, R. (2002). Shaping the vertebrate body plan by polarized embryonic cell movements. *Science 298*, 1950-1954.

Kobielak, A., Pasolli, H.A., and Fuchs, E. (2004). Mammalian formin-1 participates in adherens junctions and polymerization of linear actin cables. *Nat Cell Biol 6*, 21-30.

Krieg, M., Arboleda-Estudillo, Y., Puech, P.H., Kafer, J., Graner, F., Muller, D.J., and Heisenberg, C.P. (2008). Tensile forces govern germ-layer organization in zebrafish. *Nat Cell Biol 10*, 429-436.

Kumar, S., Maxwell, I.Z., Heisterkamp, A., Polte, T.R., Lele, T.P., Salanga, M., Mazur, E., and Ingber, D.E. (2006). Viscoelastic retraction of single living stress fibers and its impact on cell shape, cytoskeletal organization, and extracellular matrix mechanics. *Biophys J 90*, 3762-3773.

Lazarides, E., and Burridge, K. (1975). Alpha-actinin: immunofluorescent localization of a muscle structural protein in nonmuscle cells. *Cell 6*, 289-298.

Lecuit, T., and Lenne, P.F. (2007). Cell surface mechanics and the control of cell shape, tissue patterns and morphogenesis. *Nat Rev Mol Cell Biol 8*, 633-644.

Lin, C.H., and Forscher, P. (1995). Growth cone advance is inversely proportional to retrograde F-actin flow. *Neuron 14*, 763-771.

Lo, C.M., Wang, H.B., Dembo, M., and Wang, Y.L. (2000). Cell movement is guided by the rigidity of the substrate. *Biophys J 79*, 144-152.

Luo, B.H., Carman, C.V., and Springer, T.A. (2007). Structural basis of integrin regulation and signaling. *Annu Rev Immunol 25*, 619-647.

Martin, P., and Lewis, J. (1992). Actin cables and epidermal movement in embryonic wound healing. *Nature 360*, 179-183.

Matsuzaki, F., Mege, R.M., Jaffe, S.H., Friedlander, D.R., Gallin, W.J., Goldberg, J.I., Cunningham, B.A., and Edelman, G.M. (1990). cDNAs of cell adhesion molecules of different specificity induce changes in cell shape and border formation in cultured S180 cells. *J Cell Biol 110*, 1239-1252.

Mitchison, T., and Kirschner, M. (1988). Cytoskeletal dynamics and nerve growth. *Neuron 1*, 761-772.

Niessen, C.M., and Gumbiner, B.M. (2002). Cadherin-mediated cell sorting not determined by binding or adhesion specificity. *J Cell Biol 156*, 389-399.

Nose, A., Nagafuchi, A., and Takeichi, M. (1988). Expressed recombinant cadherins mediate cell sorting in model systems. *Cell 54*, 993-1001.

Oliver, T., Lee, J., and Jacobson, K. (1994). Forces exerted by locomoting cells. *Semin Cell Biol 5*, 139-147.

Palecek, S.P., Loftus, J.C., Ginsberg, M.H., Lauffenburger, D.A., and Horwitz, A.F. (1997). Integrin-ligand binding properties govern cell migration speed through cell-substratum adhesiveness. *Nature 385*, 537-540.

Paszek, M.J., Zahir, N., Johnson, K.R., Lakins, J.N., Rozenberg, G.I., Gefen, A., Reinhart-King, C.A., Margulies, S.S., Dembo, M., Boettiger, D., *et al.* (2005). Tensional homeostasis and the malignant phenotype. *Cancer Cell 8*, 241-254.

Pellegrin, S., and Mellor, H. (2007). Actin stress fibres. *J Cell Sci 120*, 3491-3499.

Ragsdale, G.K., Phelps, J., and Luby-Phelps, K. (1997). Viscoelastic response of fibroblasts to tension transmitted through adherens junctions. *Biophys J 73*, 2798-2808.

Redd, M.J., Cooper, L., Wood, W., Stramer, B., and Martin, P. (2004). Wound healing and inflammation: embryos reveal the way to perfect repair. *Philos Trans R Soc Lond B Biol Sci 359*, 777-784.

Riveline, D., Zamir, E., Balaban, N.Q., Schwarz, U.S., Ishizaki, T., Narumiya, S., Kam, Z., Geiger, B., and Bershadsky, A.D. (2001). Focal contacts as mechanosensors: externally applied local mechanical force induces growth of focal contacts by an mDia1-dependent and ROCK-independent mechanism. *J Cell Biol 153*, 1175-1186.

Rotsch, C., Jacobson, K., and Radmacher, M. (1999). Dimensional and mechanical dynamics of active and stable edges in motile fibroblasts investigated by using atomic force microscopy. *Proc Natl Acad Sci U S A 96*, 921-926.

Sawada, Y., Tamada, M., Dubin-Thaler, B.J., Cherniavskaya, O., Sakai, R., Tanaka, S., and Sheetz, M.P. (2006). Force sensing by mechanical extension of the Src family kinase substrate p130Cas. *Cell 127*, 1015-1026.

Small, J.V., Rottner, K., Kaverina, I., and Anderson, K.I. (1998). Assembling an actin cytoskeleton for cell attachment and movement. *Biochim Biophys Acta 1404*, 271-281.

Springer, T.A. (1994). Traffic signals for lymphocyte recirculation and leukocyte emigration: the multistep paradigm. *Cell 76*, 301-314.

Steinberg, M.S. (2007). Differential adhesion in morphogenesis: a modern view. *Curr Opin Genet Dev 17*, 281-286.

Steinberg, M.S., and Takeichi, M. (1994). Experimental specification of cell sorting, tissue spreading, and specific spatial patterning by quantitative differences in cadherin expression. *Proc Natl Acad Sci U S A 91*, 206-209.

Svitkina, T.M., Verkhovsky, A.B., McQuade, K.M., and Borisy, G.G. (1997). Analysis of the actin-myosin II system in fish epidermal keratocytes: mechanism of cell body translocation. *J Cell Biol 139*, 397-415.

Tamada, M., Perez, T.D., Nelson, W.J., and Sheetz, M.P. (2007). Two distinct modes of myosin assembly and dynamics during epithelial wound closure. *J Cell Biol 176*, 27-33.

Tan, J.L., Tien, J., Pirone, D.M., Gray, D.S., Bhadriraju, K., and Chen, C.S. (2003). Cells lying on a bed of microneedles: an approach to isolate mechanical force. *Proc Natl Acad Sci U S A 100*, 1484-1489.

Townes, P.L., and Holtfreter, J. (1955). Directed movements and selective adhesion of embryonic amphibian cells. *J Exp Zoolog A: Comp Exp Biol 128*, 53-120.

Vasioukhin, V., Bauer, C., Yin, M., and Fuchs, E. (2000). Directed actin polymerization is the driving force for epithelial cell-cell adhesion. *Cell 100*, 209-219.

Weber, K., and Groeschel-Stewart, U. (1974). Antibody to myosin: the specific visualization of myosin-containing filaments in nonmuscle cells. *Proc Natl Acad Sci U S A 71*, 4561-4564.

Weis, W.I., and Nelson, W.J. (2006). Re-solving the cadherin-catenin-actin conundrum. *J Biol Chem 281*, 35593-35597.

Wood, W., Jacinto, A., Grose, R., Woolner, S., Gale, J., Wilson, C., and Martin, P. (2002). Wound healing recapitulates morphogenesis in Drosophila embryos. *Nat Cell Biol 4*, 907-912.

Yamada, S., and Nelson, W.J. (2007). Localized zones of Rho and Rac activities drive initiation and expansion of epithelial cell-cell adhesion. *J Cell Biol 178*, 517-527.

Yamada, S., Pokutta, S., Drees, F., Weis, W.I., and Nelson, W.J. (2005). Deconstructing the cadherin-catenin-actin complex. *Cell 123*, 889-901.

Zhao, Y., and Zhang, X. (2006). Cellular mechanics study in cardiac myocytes using PDMS pillars array. *Sensors and Actuators A: Physical 125*, 398-404.

In: New Cell Adhesion Research
Editors: Patrick Nott and Matthew Temple

ISBN 978-1-60692-378-8

Chapter III

Cell Adhesion Molecules for Uterine Receptivity to Human Embryo Implantation

Maryam Kabir-Salmani and Michiko N. Fukuda
National Institute of Genetic Engineering and Biotechnology, Iran (MKS)
Burnham Institute for Medical Research, USA (MNF).

Abstract

Morphological observations of human and monkey embryo implantation sites indicate that blastocyst trophectoderm cells that have adhered to the uterus proliferate and invade, whereas trophectoderm not in contact with the uterine epithelium remains a monolayer. This finding suggests that the initial adhesion triggers activation of cells in trophectoderm. Adhesion molecules which are specifically up-regulated in the plasma membrane during early pregnancy are the subject of much interest because they may function as receptors for the blastocyst. This review described the cell adhesion molecule trophinin uniquely involved in the process of human embryo implantation. This review also covers other cell adhesion molecules, integrins, cadherins, selectins, basigin, galectins and tissue transglutaminase.

Introduction

Morphological observations of human and monkey embryo implantation sites indicate that blastocyst trophectoderm cells that have adhered to the uterus proliferate and invade, whereas trophectoderm not in contact with the uterine epithelium remains a monolayer [1,2]. This finding suggests that the initial adhesion triggers activation of cells in trophectoderm. Adhesion molecules which are specifically up-regulated in the plasma membrane during early pregnancy are the subject of great interest because they may function as receptors for the blastocyst activation. The cell adhesion molecules (CAMs) are mainly composed of four members known as integrins, cadherins, selectins and immunoglobulins. In addition to these well-known CAMs, we found a unique cell adhesion molecule, trophinin [3-6]. Identification of trophinin was based solely on its unique cell adhesion activity: trophinin mediates apical adhesion between cell types representing trophectoderm cells and endometrial epithelial cells [3,7]. Although trophinin gene knockout mouse did not show a failure of embryo implantation [8], evidence suggests that trophinin is uniquely involved in human embryo implantation [9-12]. This chapter summarizes the role of CAMs in the process of human embryo implantation focusing to the maternal side or uterine receptivity to embryo.

Integrins:

Integrins are a family of heterodimeric transmembrane glycoproteins, formed by the association of two, non-covalently linked α and β subunits, and are expressed on the cell surface to adhere to extracellular matrix through tripeptide arginine-glycine-aspartic acid (RGD) sequence [13,14]. These subunits contain extracellular, transmembranal and intracellular domains. The extracellular domain enables integrins to act as a receptor to extracellular matrix components. The intracellular domain, on the other hand, is able to interact with the cytoskeleton and other cytoplasmic proteins.

A variety of integrins are expressed by the luminal and glandular endometrial epithelium [15-17]. While the majority of the integrins are constitutively expressed throughout the entire menstrual cycle, some integrins exhibit expression patterns dependent on hormonal cycle [16,18]. Integrins whose expression is increased in the mid-luteal phase were proposed as markers for the frame of the window of implantation [19,20]. Three cycle-specific integrins are co-expressed by the human endometrium defined histologically on days 20–24 of the human menstrual cycle: α1β1, α4β1 and αvβ3, but only the β3 mRNA subunit expression was shown to increase after day 19 and is not detected beforehand. αvβ3 integrin as well as its ligand osteopontin was detected by immunohistochemistry on the endometrial luminal epithelial surface, which may interact with the trophoblast [21]. The cycle-specific expression pattern of endometrial integrin is suggestive of hormonal regulation. Indeed, αvβ3 integrin expression is orchestrated in the human endometrium both by positive and negative factors [22,23]. During the proliferative phase, high estrogen levels act via the estrogen receptor-α to inhibit integrin expression. The elevation of luteal progesterone downregulates the number of progesterone receptors, thus indirectly suppressing the inhibitory effects of E2 on integrins. Probably, progesterone also acts positively by increasing paracrine stromal factors to induce epithelial β3 integrin expression that serves as the rate-limiting step in αvβ3 formation [24]. However, there are some controversies in the *in vitro* assays [25]. In regard to its expression pattern along with its epithelial localization, αvβ3 has been proposed as a potential receptor for embryonic attachment [24]. Moreover, signalling through αvβ3 has been reported to be important to maintain a balance between cell proliferation and apoptosis, along with the modulation of inflammatory responses of decidual cells [26].

Selectins:

In both human and mouse, there are three selectins: E-selectin, P-selectin, and L-selectin. E-selectin is expressed on the endothelial surface at inflammation site, P-selectinis is expressed on the activated platelet, and L-selectin is constitutively expressed on the lymphocytes. Although previously it was thought that selectins are expressed only in the hematopoietic cells, L-selectin was found in human blastocysts [27]. Furthermore, L-selectin ligand oligosaccharides were detected widely on luminal and glandular endometrial epithelia in the human uterus by immunohistochemistry using MECA-79 and HECA-452 antibodies, whose epitope structures are closely related to L-selectin ligand [28,29]. The expression of theses ligands are up-regulated in human endometrium during the period receptive to embryo implantation. MECA79 antigen is considered as a predictive marker of human endometrial receptivity [30]. Furthermore, it has been suggestedd that interactions between L-selectin on human blastocysts and oligosaccharide ligands on endometrial epithelia enable the initial adhesion of human embryo for implantation [27].

L-selectin expressed on leukocytes interacts with their carbohydrate-ligands on the blood vessel endothelial cells. This interaction allows the rolling of leukocytes on vascular endothelium prior to their firm adhesion and extravasation. A parallel was made between the leukocyte 'rolling' on vascular wall and the blastocyst apposition to the endometrial epithelium [27,31]. When the blastocyst enters uterus, while it may move freely in the uterine cavity it looks for the best spot for implantation. The 'rolling' may be strictly regulated to ensure that the blastocyst will settle in the proper spot in the uterus and in the correct orientation. To prevent the blastocyst from adhering to an area with poor chances of pregnancy, the major component of glycokalyx MUC-1 plays an important role. In rodents, MUC-1 is down-regulated during implantation in entire uterus. However, in humans MUC-1 increases before and during implantation, suggesting human endometrium actively prevents the embryo from adhering except to the very spot for implantation. MUC-1 expressed in human endometrium carries MECA79 antigen throughout the secretory phase of the menstrual cycle [32]. Interestingly, MUC-1 downregulated in the endometrial surface epithelia where the blastocyst adhered [33]. Such observation suggests that although MUC-1 represents a potential ligand for L-selectin expressed by human blastocysts, MUC-1 is excluded from the endometrial epithelial to allow strong adhesion mediated by protein-protein binding between two epithelial cells.

Trophinin

Trophinin was identified by expression cDNA cloning from cDNA library constructed from human teratocrcinoma cell line HT-H [3,34]. Teratocarcinomas are tumors composed of undifferentiated embryonal carcinoma (EC) cells and variously differentiated cell types [35,36]. Both human and mouse EC lines show characteristics of early embryonic cells [35,37,38]. Human EC lines have the tendency to differentiate into trophoblastic cells [38]. Trophoblastic cells derived from EC cells are assumed to mimic trophoblastic cells at very early embryonic stage, such as those in trophectoderm of the blastocyst. HT-H is a human EC line isolated from a testicular teratocarcinoma tumor [39]. HT-H cells spontaneously differentiate into syncytiotrophoblasts *in vitro* and secrete progesterone and hCG [34]. When trophoblastic HT-H cells were added to human endometrial adenocarcinoma SNG-M cells [40], HT-H cells instantly adhere to SNG-M cells [3]. HT-H cells also adhere to themselves but do not adhere to epithelial cells derived from other cell types, such as colon and lung. These observations suggest the existence of cell type specific apical cell adhesion molecule in HTH and SNG-M cells.

To identify this predicted apical adhesion molecule, we employed expression cDNA cloning. Based solely on the apical cell adhesion activity, we have identified trophinin [3,7]. Trophinin mediates homophilic cell adhesion through trophinin-trophinin binding. When trophoblastic cells expressing trophinin were overlayed on the apical surface of trophinin-expressing endometrial epithelia, both cell types adhered instantly with strong affinity. Trophinin protein was found in the trophectoderm cells in the monkey blastocyst, particularly in the cells at embryonic pole where the blastocyst implantation takes place [3]. A monoclonal antibody specific to human trophinin showed positive immunostaining in both trophoblast and maternal epithelia at embryo implantation sites in the human placenta [6].

In human, trophinin gene (*TRO*) is mapped to the short arm of X chromosome. This region of X chromosome is closely linked to the evolution of mammal: genes encoded in this region in one placental mammal are likely located on the X-chromosome in other mammals [41,42]. However, many of these genes are autosomal in marsupials and monotremes [43], which do not undertake proper implantation. The trophinin gene localization appears to correlate well with the evolution of eutherian mammals.

Substantial studies have shown that the processes of embryo implantation vary significantly among mammals [44]. Some mechanism of human embryo implantation is considered unique to humans. For example, in humans, more than 1% of all pregnancies are ectopic [45], but this condition has not been seen in animals [46]. While we were investigating trophinin expression in Fallopian tubal pregnancies, we found that trophinin was strongly expressed at the implantation site in both embryonic and maternal cells [9]; i. e., trophinin was expressed by the trophoblast in the chorionic villi and also by the maternal tubal epithelia adjacent to the chorionic villi. However, the tubal epithelia at a slight distance (5 mm) from the implantation site showed no trophinin. Therefore, it appears that expression of trophinin by maternal cells is implantation-dependent. We found that the β-chain of human chorionic gonadotropin (CGβ) could induce transcription of the *TRO* gene in the Fallopian tubal cells [9]. Furthermore, CGβ and IL-1β induce strong trophinin expression in human endometrial epithelial cells in primary culture [12]. Furthermore, CGβ and IL-1β treatment induced pinopodes, tall protrusions presented above the glycokalyx, found in the implantation sites [47,48]. Trophinin was found in the pinopodes [12].

Cell adhesion molecules expressed on the surface of the pinopode should allow direct interaction of trophectoderm cells of blastocysts. Unlike mouse blastocyst adhering in the narrowly enclosed uterine cavity, a human blastocyst adheres to pinopodes presented on the flat endometrial surface. Under such condition, the adhesion of human blastocyst to pinopodes can only be achieved by strong adhesion machinery. As trophinin binds to each other with high affinity, trophinin is a good candidate for this function. It is also possible that the strong cell adhesion requires more than one machinery. Gene knockout mouse experiments for trophinin [8] and L-selectin [49] and its ligand carbohydrates [50-55] indicate that neither trophinin nor L-selectin plays an essential role in embryo implantation in the mouse. By contrast, HB-EGF gene knockout mouse showed a failure in embryo implantation [56]. It is therefore possible that adhesion by ErbB4/HB-EGF is conserved throughout mammals and L-selectin and trophinin functions are acquired uniquely to human embryo implantation as additional mechanisms to ErbB4/HB-EGF. An integrated view of L-selectin and trophinin is presented in Figure 1.

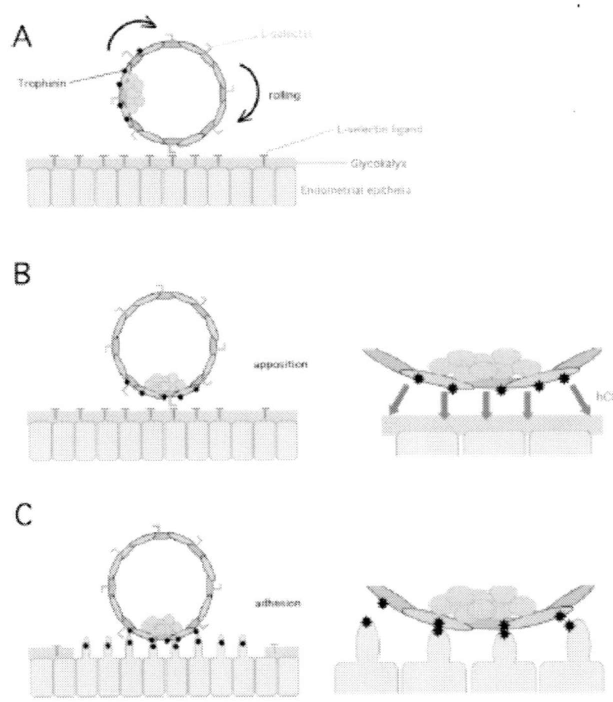

Figure 1. Proposed role of L-selectin and trophinin in human embryo implantation (57). (a) A human blastocyst entering the uterine cavity is prevented from adhering to endometrial surface epithelia by heavily glycosylated MUC1, except for epithelia that express L-selectin ligand (T). The human blastocyst expresses L-selectin (L), and 'rolls' on the endometrial surface covered by glycocalyx (cell surface layer shown by grey color). (b) The blastocyst weakly interacts on the glycocalyx. During this interaction, human chorionic gonadotropin (hCG) secreted from the blastocyst acts locally on endometrial surface epithelia to induce trophinin expression in the endometrial epithelial cells. (c) Trophinin expressed by endometrial epithelia is enriched in the pinopodes, the structure extended above the glycocalyx (12). MUC1, which is the major component of endometrial glycocalyx and carries L-selectin ligand, is down-regulated from the endometrial surface epithelia underneath the blastocyst (33), allowing direct contact of blastocyst trophectoderm cells and pinopodes. Trophectoderm cells and pinopodes then bind strongly together by trophinin-trophinin binding.

Potential involvement of trophinin in human embryo implantation has been strengthened by recent findings relevant to signaling triggered by trophinin-mediated cell adhesion [10,11]. When HB-EGF binds to ErbB4 on trophoblastic HT-H cells, ErbB4 is not activated because ErbB4 is arrested by trophinin and bystin. When trophinin-mediated cell adhesion occurs, bystin is released from trophinin, allowing activation of ErbB4 protein tyrosine kinase to promote proliferation and invasion of trophoblastic cells. Thus trophinin-mediated cell adhesion acts as a molecular switch for trophoblast activation. This mechanism explains how trophectoderm cells of human blastocyst adhering to the uterine epithelia grow and invade, while those at non-adhesive sites remain silent [58,59].

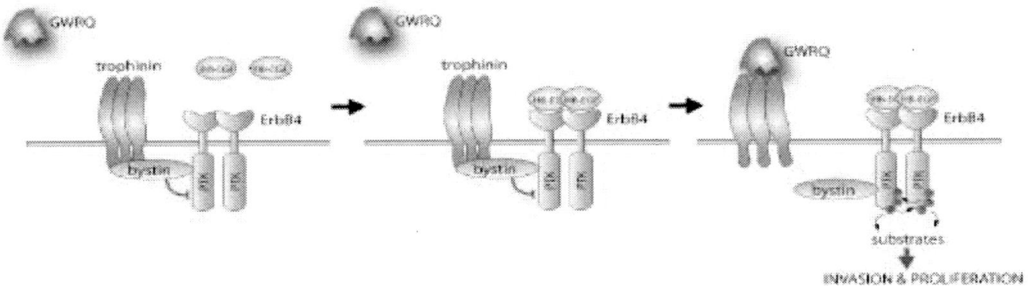

Figure 2. Interaction of trophinin, bystin, and ErbB4 in trophoblastic cells before and after trophinin-mediated cell adhesion (10). Trophinin-binding peptide GWRQ mimics trophinin-mediated cell adhesion in trophoblastic cells and promote cell proliferation and invasion. (Left) Trophinin and bystin bound together in a cytoplasmic complex, which further interacts with ErbB4 and suppresses its tyrosine kinase activity. (Center) HB-EGF binds to ErbB4 but protein kinase activity is suppressed by trophinin/bystin. (Right) Trophinin-mediated cell adhesion is mimicked by binding of GWRQ peptide to trophinin on the cell surface, resulting in dissociation of trophinin from bystin, leading to activation of ErbB4.

Basigin and Immunoglobulin Family Proteins:

Basigin, also known as CD147 or EMMPRIN, is a transmembrane glycoprotein belonging to the immunoglobulin super family [60-62]. Extradomain of basigin has binding sites for multiple binding partners, including extracellular matrix metalloproteinases (MMPs) inducer or activator of their activity. Basigin is expressed most prominently at the lateral epithelial surface of endometrial epithelial cells in human and rodent. It is also expressed on pre-implantation embryos. Basigin regulates the production of metaloprotease (MMP)-1, -2, and -3 by uterine fibroblast cells [63], which is thought to play important role in extensive remodeling of the uterine lining. Fertility of the mutant mice lacking basigin gene is severely compromised [64,65].

Intracellular adhesion molecule-I (ICAM-I or CD54), another immunoglobulin superfamily protein, is implicated in embryo implantation [66,67]. ICAM-1 is expressed on the cell surface of a variety of cell types, including fibroblasts, leukocytes, endothelial and epithelial cells. It is well established that the endometrium, under normal conditions, contains a wide population of leukocytes, which expresses ICAM-1 within the endometrium. Moreover, ICAM-1 was found, throughout the menstrual cycle, to the apical surface of the glandular and luminal endometrial epithelial cells as well as in the stroma. On the other hand, MUC1 that is expressed by early gestation trophoblasts is known to bind to ICAM-1 in other systems [68]. Since ICAM-1 is expressed in both stromal and epithelial endometrial cells, it is suggested that ICAM-1 may involve in the trophoblast adhesion to uterine endothelial cells and in trophoblast transendothelial migration [68].

Cadherins:

Cadherin-mediated cell-cell interaction is tightly regulated in response to cellular context and signaling [69]. Cadherin regulation is likely to reflect the interplay between a range of fundamental cellular processes, including surface organization of receptors, cytoskeletal organization and cell trafficking [69]. In the processes of embryo implantation, E-cadherin is located in the adherens junctions on the lateral side of the plasma membrane [70,71]. During the window of implantation, a major re-organization take place in the membrane proteins of endometrial epithelium, which has been suggested to represent a loss of polarity in these cells and this thought has been extended to suggest further that an epithelial-mesenchymal transition may occur in the cells during this time [72].

Intracellular calcium concentrations affect epithelial cell adhesiveness and polarity by triggering redistribution of cell adhesion molecules [73]. *In vitro* experiments on cultured endometrial adenocarcinoma Ishikawa cells demonstrated that a transient rise in intracellular calcium, triggered by calcitonin, suppresses E-cadherin expression at cellular contact sites [74]. Interestingly, calcitonin, a potential regulator and biomarker of endometrial receptivity [75,76], is induced by progesterone in the human endometrial epithelium specifically during the mid-secretory phase of the menstrual cycle [77]. Progesterone could regulate E-cadherin, probably via endometrial calcitonin induction leading to increased intracellular calcium. Thus, it is possible that cadherin expression at the lateral cell surface is required to maintain the polarity of endometrial epithelial cells, whereas E-cadherin may be down-regulated to enable epithelial cells dissociation to accept blastocyst invasion. The upregulation of cadherin and catenin in the epithelial cells of peri-implantation uteri and the down regulation of cadherin, catenin and Ca2+ in invasive trophoblast appear to be associated with embryo-uterine interactions during early pregnancy [78]. The suppression of E-cadherin at cellular contact sites is illustrated in Figure 3.

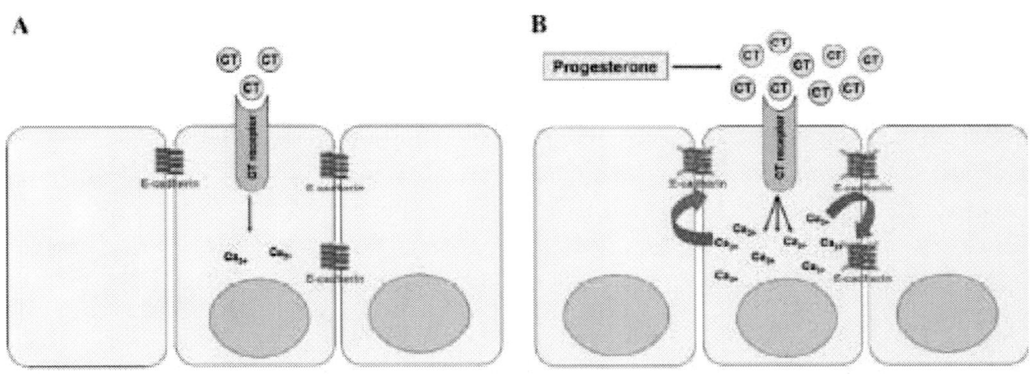

Figure 3. Role of E-cadherin in the endometrial epithelia before and during embryo implantation (79). (A) Epithelial cell adhesiveness by E-cadherin is controlled by intracellular calcium. (B) Elevation of progesterone levels induce calcitonin (CT) expression and thus increase the concentration of intracellular calcium, which then suppresses E-cadherin expression at cellular contact sites.

Galectins:

The galectins are a family of β-galactoside-binding proteins with a conserved carbohydrate recognition domain. Because carbohydrates surpass any other class of biomolecule in their capacity to store biological information [80], there are strong interest in the role of cell surface carbohydrates and carbohydrate-binding proteins in embryo implantation, including L-selectin and its carbohydrate ligand described above. Several studies focused on a family of galectins for their roles in cell growth and adhesion [81,82], which parallels in the function of galectins and the regulation of human endometrium. First, galectin-1 and -3 mediate inflammation in hematopoietic cells [83]. As endometrial function and implantation involve many inflammatory mediators, galectins might contribute to the modulation of the endometrial immune system. Second, galectins have been shown to play an important role in cell adhesion, migration and chemotaxis [83]. As adequate endometrial function is dependent on leukocyte migration and immune cells accumulate around the implantation site, galectins is thought to play a role in the regulation of endometrial leukocytes. Finally, galectins have been characterized as major players in the defense against invading micro-organisms [83]. As the endometrial immune system is constantly challenged by bacteria ascending through the cervix, galectins might contribute to the protection of the endometrium against bacterial infection. Allison Gray *et al.*, [84] suggested galectins as the marker for the ovine endometrial receptivity. Galectin-1 was present in the tissue compartments of the uterus except for the luminal and glandular epithelium. After implantation of the embryo, the endometrium including the decidualized stromal cells and the glandular epithelium showed immunohistochemical staining for galectin-1. Galectin-3 expression was low in non-pregnant mouse uteri and increased in the primary decidual zone and in the uterine epithelial cells adjacent to the implanting blastocyst immediately after implantation. Recent study on the localization of galectin in human endometrium revealed a similar pattern: galectin-1 is strongly expressed in decidualized stromal cells and galectin-3 is strongly expressed in secretory-phase glandular epithelial cells [85]. Galectin-9 was suggested as a new endometrial epithelial marker for the mid- and late-secretory and decidual phases, which exclusively expressed in the endometrial epithelial cells [81]. Galectin-9 associates with glycodelin (PP14) which is strictly regulated during the menstrual cycle, with a significantly increased expression during the mid-secretory phase [86]. Because of pro-adhesive as well as angti-adhesive activity of extracellular galectins, extracellular galectins may be a novel class of adhesion-modulating proteins collectively known as matricellular proteins [87]. Thus, a question remains whether galectins might be involved in cellular adhesion during implantation, as previously suggested for integrin and its ligand osteopontin.

Tissue Transglutaminase (tTG):

tTG is a unique member of the transglutaminase family, as it is both a transamidating enzyme and a GTPase [88]. Although tTG is localized mostly in the cytosol, it is also found in the nucleus, the extracellular matrix (ECM), and in the plasma membrane [89,90]. A complex of tTG and integrins is formed in the cytoplasm and accumulates on the cell surface and focal adhesions (FAs) [91]. Human endometrial tTG is developmentally regulated. The levels of human tTG are 10- fold higher in the luteal phase than in the follicular phase [92], which implies that progesterone regulates the expression of tTG in vivo [93]. Furthermore, the importance of tTG expression during decidualization in human endometrial stromal cells [94] has been suggested: Kabir-Salmani et al. [95] demonstrated the expression of tTG in pinopodes in human endometrium and suggested that tTG actively participates in adhesion events at the embryo-maternal interface through its interaction with FN by activating integrin signaling pathways. As tTG lacks a transmembrane domain, activation of intracellular signaling pathways may be achieved by an indirect mechanism through a membrane protein associated with tTG [96]. Given that integrins span the cell membrane and are physically and functionally associated with tTG [91], it is tempting to hypothesize that an integrin-tTG-FN axis exists in these cells. Accordingly, a model is proposed for the ternary adhesion complexes concerning these interactions (Figure 4). In this model, tTG as a bridge between integrin and fibronectincould strengthen these adhesion complexes through its higher affinity for fibronectin and by simultaneously providing the second integrin molecule with the binding site for the fibronectin chain.

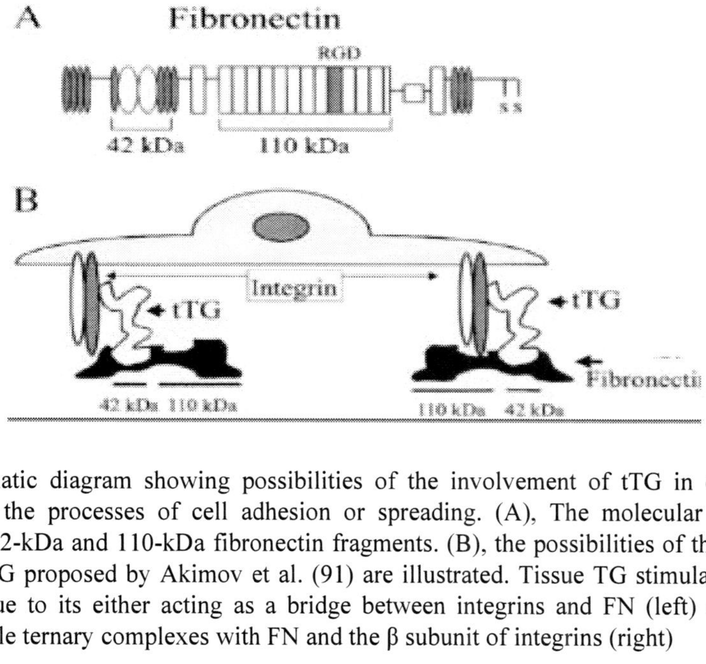

Figure 4. Schematic diagram showing possibilities of the involvement of tTG in conjunction with integrins during the processes of cell adhesion or spreading. (A), The molecular structure of the fibronectin and 42-kDa and 110-kDa fibronectin fragments. (B), the possibilities of the associations of integrins with tTG proposed by Akimov et al. (91) are illustrated. Tissue TG stimulates cell adhesion and spreading due to its either acting as a bridge between integrins and FN (left) or mediating the formation of stable ternary complexes with FN and the β subunit of integrins (right)

REFERENCES

[1] Hertig, A. T. and Rock, J. Searching for early fertilized human ova. *Gynecol Invest, 4:* 121-139, 1973.

[2] Enders, A. C. Cytology of human early implantation. *Res Reprod, 8:* 1-2, 1976.

[3] Fukuda, M. N., Sato, T., Nakayama, J., Klier, G., Mikami, M., Aoki, D., and Nozawa, S. Trophinin and tastin, a novel cell adhesion molecule complex with potential involvement in embryo implantation. *Genes Dev, 9:* 1199-1210, 1995.

[4] Suzuki, N., Zara, J., Sato, T., Ong, E., Bakhiet, N., Oshima, R. G., Watson, K. L., and Fukuda, M. N. A novel cytoplasmic protein, bystin, interacts with trophinin, tastin and cytokeratin, and may be involved in trophinin mediated cell adhesion between trophoblast and endometrial epithelial cells. *Proceedings of the National Academy of Sciences,USA, 95:* 5027-5032, 1998.

[5] Fukuda, M. N., Nadano, D., Suzuki, N., and Nakayama, J. Potential Involvement of Trophinin, Bystin and Tastin in embryo implantation. *In:* D. D. Carson (ed.), *Embryo Implantation: Molecular, Cellular and Clinical Aspects*., pp. 132-140. Norwell, MA: Serono Symposia USA, 1999.

[6] Suzuki, N., Nakayama, J., Shih, I. M., Aoki, D., Nozawa, S., and Fukuda, M. N. Expression of trophinin, tastin, and bystin by trophoblast and endometrial cells in human placenta. *Biol Reprod, 60:* 621-627, 1999.

[7] Fukuda, M. N. and Nozawa, S. Trophinin, tastin, and bystin: a complex mediating unique attachment between trophoblastic and endometrial epithelial cells at their respective apical cell membranes. *Semin Reprod Endocrinol, 17:* 229-234, 1999.

[8] Nadano, D., Sugihara, K., Paria, B. C., Saburi, S., Copeland, N. G., Gilbert, D. J., Jenkins, N. A., Nakayama, J., and Fukuda, M. N. Significant differences between mouse and human trophinins are revealed by their expression patterns and targeted disruption of mouse trophinin gene. *Biol Reprod, 66:* 313-321, 2002.

[9] Nakayama, J., Aoki, D., Suga, T., Akama, T. O., Ishizone, S., Yamaguchi, H., Imakawa, K., Nadano, D., Fazleabas, A. T., Katsuyama, T., Nozawa, S., and Fukuda, M. N. Implantation-Dependent Expression of Trophinin by Maternal Fallopian Tube Epithelia during Tubal Pregnancies: Possible Role of Human Chorionic Gonadotrophin on Ectopic Pregnancy. *Am J Pathol, 163:* 2211-2219, 2003.

[10] Sugihara, K., Sugiyama, D., Byrne, J., Wolf, D. P., Lowitz, K. P., Kobayashi, Y., Kabir-Salmani, M., Nadano, D., Aoki, D., Nozawa, S., Nakayama, J., Mustelin, T., Ruoslahti, E., Yamaguchi, N., and Fukuda, M. N. Trophoblast cell activation by trophinin ligation is implicated in human embryo implantation. *Proc Natl Acad Sci U S A, 104:* 3799-3804, 2007.

[11] Fukuda, M. N. and Sugihara, K. Signal transduction in human embryo implantation. *Cell Cycle, 6:* 1153-1156, 2007.

[12] Sugihara, K., Kabir-Salmani, M., Byrne, J., Wolf, D. P., Lessey, B., Iwashita, M., Aoki, D., Nakayama, J., and Fukuda, M. N. Induction of trophinin in human endometrial surface epithelia by CGbeta and IL-1beta. *FEBS Lett, 582:* 197-202, 2008.

[13] Ruoslahti, E. Integrins. *J Clin Invest, 87:* 1-5, 1991.

[14] Hynes, R. O. Integrins: bidirectional, allosteric signaling machines. *Cell, 110:* 673-687, 2002.

[15] Mangale, S. S. and Reddy, K. V. Expression pattern of integrins and their ligands in mouse feto-maternal tissues during pregnancy. *Reprod Fertil Dev, 19:* 452-460, 2007.

[16] Lessey, B. A., Damjanovich, L., Coutifaris, C., Castelbaum, A., Albelda, S. M., and Buck, C. A. Integrin adhesion molecules in the human endometrium. Correlation with the normal and abnormal menstrual cycle. *J.Clin.Invest., 90:* 188-195, 1992.

[17] Klentzeris, L. D., Bulmer, J. N., Trejdosiewicz, L. K., Morrison, L., and Cooke, I. D. Beta-1 integrin cell adhesion molecules in the endometrium of fertile and infertile women. *Hum Reprod, 8:* 1223-1230, 1993.

[18] Coutifaris, C., Omigbodun, A., and Coukos, G. Integrins, endometrial maturation, & human embryo implantation. *Semin Reprod Endocrinol, 16:* 219-229, 1998.

[19] Lessey, B. A., Castelbaum, A. J., Wolf, L., Greene, W., Paulson, M., Meyer, W. R., and Fritz, M. A. Use of integrins to date the endometrium. *Fertil Steril, 73:* 779-787, 2000.

[20] Reddy, K. V. and Mangale, S. S. Integrin receptors: the dynamic modulators of endometrial function. *Tissue Cell, 35:* 260-273, 2003.

[21] Apparao, K. B., Murray, M. J., Fritz, M. A., Meyer, W. R., Chambers, A. F., Truong, P. R., and Lessey, B. A. Osteopontin and its receptor alphavbeta (3) integrin are coexpressed in the human endometrium during the menstrual cycle but regulated differentially. *J Clin Endocrinol Metab, 86:* 4991-5000, 2001.

[22] Yoshimura, Y. Integrins: expression, modulation, and signaling in fertilization, embryogenesis and implantation. *Keio J Med, 46:* 16-24, 1997.

[23] Somkuti, S. G., Yuan, L., Fritz, M. A., and Lessey, B. A. Epidermal growth factor and sex steroids dynamically regulate a marker of endometrial receptivity in Ishikawa cells. *J Clin Endocrinol Metab, 82:* 2192-2197, 1997.

[24] Lessey, B. A. Two pathways of progesterone action in the human endometrium: implications for implantation and contraception. *Steroids, 68:* 809-815, 2003.

[25] Sillem, M., Prifti, S., Schmidt, M., Rabe, T., and Runnebaum, B. Endometrial integrin expression is independent of estrogen or progestin treatment in vitro. *Fertil Steril, 67:* 877-882, 1997.

[26] Mangale, S. S., Modi, D. N., and Reddy, K. V. Identification of genes regulated by an interaction between alphavbeta3 integrin and vitronectin in murine decidua. *Reprod Fertil Dev, 20:* 311-319, 2008.

[27] Genbacev, O. D., Prakobphol, A., Foulk, R. A., Krtolica, A. R., Ilic, D., Singer, M. S., Yang, Z. Q., Kiessling, L. L., Rosen, S. D., and Fisher, S. J. Trophoblast Lselectin-mediated adhesion at the maternal-fetal interface. *Science, 299:* 405-408, 2003.

[28] Lai, T. H., Shih Ie, M., Vlahos, N., Ho, C. L., Wallach, E., and Zhao, Y. Differential expression of L-selectin ligand in the endometrium during the menstrual cycle. *Fertil Steril, 83 Suppl 1:* 1297-1302, 2005.

[29] Shamonki, M. I., Kligman, I., Shamonki, J. M., Schattman, G. L., Hyjek, E., Spandorfer, S. D., Zaninovic, N., and Rosenwaks, Z. Immunohistochemical expression of endometrial L-selectin ligand is higher in donor egg recipients with embryonic implantation. *Fertil Steril, 86:* 1365-1375, 2006.

[30] Foulk, R. A., Zdravkovic, T., Genbacev, O., and Prakobphol, A. Expression of Lselectin ligand MECA-79 as a predictive marker of human uterine receptivity. *J Assist Reprod Genet, 24:* 316-321, 2007.

[31] Dominguez, F., Pellicer, A., and Simon, C. The chemokine connection: hormonal and embryonic regulation at the human maternal-embryonic interface--a review. *Placenta, 24 Suppl B:* S48-55, 2003.

[32] Carson, D. D., Julian, J., Lessey, B. A., Prakobphol, A., and Fisher, S. J. MUC1 is a scaffold for selectin ligands in the human uterus. *Front Biosci, 11:* 2903-2908, 2006.

[33] Meseguer, M., Aplin, J. D., Caballero-Campo, P., O'Connor, J. E., Martin, J. C., Remohi, J., Pellicer, A., and Simon, C. Human endometrial mucin MUC1 is upregulated by progesterone and down-regulated in vitro by the human blastocyst. *Biol Reprod, 64:* 590-601, 2001.

[34] Izhar, M., Siebert, P., Oshima, R. G., DeWolf, W. C., and Fukuda, M. N. Trophoblastic differentiation of human teratocarcinoma cell line HT-H. *Developmental Biology, 116:* 510-518, 1986.

[35] Martin, G. R. Teratocarcinomas as a model system for the study of embryogenesis and neoplasia. *Cell, 5:* 229-243, 1975.

[36] Martin, G. R. Teratocarcinomas and mammalian embryogenesis. *Science, 209:* 768-776, 1980.

[37] Andrews, P. W., Goodfellow, P. N., and Damjanov, I. Human teratocarcinoma cells in culture. *Cancer Surveys, 2:* 41-73, 1983.

[38] Andrews, P. W., Bronson, D. L., Benham, F., Strickland, S., and Knowles, B. B. A comparative study of eight cell lines derived from human testicular teratocarcinoma. *Int J Cancer, 26:* 269-280, 1980.

[39] Wang, N., Trend, B., Bronson, D. L., and Fraley, E. E. Nonrandom abnormalities in chromosome 1 in human testicular cancers. *Cancer Res, 40:* 796-802, 1980.

[40] Ishiwata, I., Nozawa, S., Inoue, T., and Okumura, H. Development and characterization of established cell lines from primary and metastatic regions of human endometrial adenocarcinoma. *Cancer Res, 37:* 1777-1785, 1977.

[41] Ohno, S. *Sex chromosomes and sex linked genes*. Berlin: Springer ,, 1967.

[42] Ohno, S. Patterns in genome evolution. *Curr.Opin.Genet.Dev., 3:* 911-914, 1993.

[43] Mandel, J.-L., Monaco, A. P., Nelson, D. L., Schlessinger, D., and Willard, H. Genome analysis and the human X chromosome. *Science, 258:* 103-109, 1992.

[44] Carson, D. D., Bagchi, I., Dey, S. K., Enders, A. C., Fazleabas, A. T., Lessey, B. A., and Yoshinaga, K. Embryo implantation. *Developmental Biology, 223:* 217- 237, 2000.

[45] Graczykowski, J. W. and Mishell, D. R. J. Ectopic pregnancy. *In:* R. A. Lobo, D. R. J. Mishell, R. J. Paulson, and D. Shoupe (eds.), *Mishell's Textbook of Infertility, Contraception, and reproductive Endocrinology*, pp. 623-637. Masachusetts, USA: Blackwell Science, 1997.

[46] Corpa, J. M. Ectopic pregnancy in animals and humans. *Reproduction, 131:* 631- 640, 2006.

[47] Nikas, G., Drakakis, P., and Loutradis, D. Uterine pinopodes as markers of the "nidation window" in cycling women receiving exogenous estradiol and progesterone. *Human Reproduction, 10:* 1208-1213, 1995.

[48] Bentin-Ley, U., Sjogren, A., Nilsson, L., Hamberger, L., Larsen, J. F., and Horn, T. Presence of uterine pinopodes at the embryo-endometrial interface during human implantation in vitro. *Hum Reprod, 14:* 515-520, 1999.

[49] Frenette, P. S. and Wagner, D. D. Insights into selectin function from knockout mice. *Thromb Haemost, 78:* 60-64, 1997.

[50] Mal, P., Thall, A. D., Petryniak, B., Rogers, C. E., Smith, P. L., Marks, R. M., Kelley, R. J., Gersten, K. M., Cheng, G., Saunders, T. L., Camper, S. A., Camphausen, R. T., Sullivan, F. X., Isogai, Y., Hindsgaul, O., Von Andrian, V. H., and Lowe, J. B. The alpha(1, 3) fucosyltransferase FucTVII controls leukocyte trafficking through an essential role in L-, E-, and P-selectin ligand biosynthesis. *Cell, 86:* 643-653, 1996.

[51] Ellies, L. G., Tsuboi, S., Petryniak, B., Lowe, J. B., Fukuda, M., and Marth, J. D. Core 2 oligosaccharide biosynthesis distinguishes between selectin ligands essential for leukocyte homing and inflammation. *Immunity, 9:* 881-890, 1998.

[52] Hemmerich, S., Bistrup, A., Singer, M. S., van Zante, A., Lee, J. K., Tsay, D., Peters, M., Carminati, J. L., Brennan, T. J., Carver-Moore, K., Leviten, M., Fuentes, M. E., Ruddle, N. H., and Rosen, S. D. Sulfation of L-selectin ligands by an HEV-restricted sulfotransferase regulates lymphocyte homing to lymph nodes. *Immunity, 15:* 237-247, 2001.

[53] Kawashima, H., Petryniak, B., Hiraoka, N., Mitoma, J., Huckaby, V., Nakayama, J., Uchimura, K., Kadomatsu, K., Muramatsu, T., Lowe, J. B., and Fukuda, M. Nacetylglucosamine-6-O-sulfotransferases 1 and 2 cooperatively control lymphocyte homing through L-selectin ligand biosynthesis in high endothelial venules. *Nat Immunol, 6:* 1096-1104, 2005.

[54] Hiraoka, N., Kawashima, H., Petryniak, B., Nakayama, J., Mitoma, J., Marth, J. D., Lowe, J. B., and Fukuda, M. Core 2 branching beta1,6-Nacetylglucosaminyltransferase and high endothelial venule-restricted sulfotransferase collaboratively control lymphocyte homing. *J Biol Chem, 279:* 3058-3067, 2004.

[55] Mitoma, J., Bao, X., Petryanik, B., Schaerli, P., Gauguet, J. M., Yu, S. Y., Kawashima, H., Saito, H., Ohtsubo, K., Marth, J. D., Khoo, K. H., von Andrian, U. H., Lowe, J. B., and Fukuda, M. Critical functions of N-glycans in L-selectinmediated lymphocyte homing and recruitment. *Nat Immunol, 8:* 409-418, 2007.

[56] Xie, H., Wang, H., Tranguch, S., Iwamoto, R., Mekada, E., Demayo, F. J., Lydon, J. P., Das, S. K., and Dey, S. K. Maternal heparin-binding-EGF deficiency limits pregnancy success in mice. *Proc Natl Acad Sci U S A, 104:* 18315-18320, 2007.

[57] Fukuda, M. N. and Sugihara, K. An integrated view of L-selectin and trophinin function in human embryo implantation. *J Obstet Gynaecol Res, 34:* 129-136, 2008.

[58] Lindenberg, S. Ultrastructure in human implantation: transmission and scanning electron microscopy. *Baillieres Clin Obstet Gynaecol, 5:* 1-14, 1991.

[59] Lindenberg, S., Hyttel, P., Lenz, S., and Holmes, P. V. Ultrastructure of the early human implantation in vitro. *Hum Reprod, 1:* 533-538, 1986.

[60] Kanekura, T., Miyauchi, T., Tashiro, M., and Muramatsu, T. Basigin, a new member of the immunoglobulin superfamily: genes in different mammalian species, glycosylation changes in the molecule from adult organs and possible variation in the N-terminal sequences. *Cell Struct Funct, 16:* 23-30, 1991.

[61] Fan, Q. W., Kadomatsu, K., Uchimura, K., and Muramatsu, T. Embigin/basigin subgroup of the immunoglobulin superfamily: different modes of expression during

mouse embryogenesis and correlated expression with carbohydrate antigenic markers. *Dev Growth Differ*, *40:* 277-286, 1998.

[62] Muramatsu, T. and Miyauchi, T. Basigin (CD147): a multifunctional transmembrane protein involved in reproduction, neural function, inflammation and tumor invasion. *Histol Histopathol*, *18:* 981-987, 2003.

[63] Braundmeier, A. G., Fazleabas, A. T., Lessey, B. A., Guo, H., Toole, B. P., and Nowak, R. A. Extracellular matrix metalloproteinase inducer regulates metalloproteinases in human uterine endometrium. *J Clin Endocrinol Metab*, *91:* 2358-2365, 2006.

[64] Kuno, N., Kadomatsu, K., Fan, Q. W., Hagihara, M., Senda, T., Mizutani, S., and Muramatsu, T. Female sterility in mice lacking the basigin gene, which encodes a transmembrane glycoprotein belonging to the immunoglobulin superfamily. *FEBS Lett*, *425:* 191-194, 1998.

[65] Igakura, T., Kadomatsu, K., Kaname, T., Muramatsu, H., Fan, Q. W., Miyauchi, T., Toyama, Y., Kuno, N., Yuasa, S., Takahashi, M., Senda, T., Taguchi, O., Yamamura, K., Arimura, K., and Muramatsu, T. A null mutation in basigin, an immunoglobulin superfamily member, indicates its important roles in periimplantation development and spermatogenesis. *Dev Biol*, *194:* 152-165, 1998.

[66] Matsumoto, H., Daikoku, T., Wang, H., Sato, E., and Dey, S. K. Differential expression of ezrin/radixin/moesin (ERM) and ERM-associated adhesion molecules in the blastocyst and uterus suggests their functions during implantation. *Biol Reprod*, *70:* 729-736, 2004.

[67] Blois, S., Tometten, M., Kandil, J., Hagen, E., Klapp, B. F., Margni, R. A., and Arck, P. C. Intercellular adhesion molecule-1/LFA-1 cross talk is a proximate mediator capable of disrupting immune integration and tolerance mechanism at the feto-maternal interface in murine pregnancies. *J Immunol*, *174:* 1820-1829, 2005.

[68] Thirkill, T. L., Cao, T., Stout, M., Blankenship, T. N., Barakat, A., and Douglas, G. C. MUC1 is involved in trophoblast transendothelial migration. *Biochim Biophys Acta*, *1773:* 1007-1014, 2007.

[69] Yap, A. S., Crampton, M. S., and Hardin, J. Making and breaking contacts: the cellular biology of cadherin regulation. *Curr Opin Cell Biol*, *19:* 508-514, 2007.

[70] Gumbiner, B. M. Cell adhesion: the molecular basis of tissue architecture and morphogenesis. *Cell*, *84:* 345-357, 1996.

[71] Huber, O., Bierkamp, C., and Kemler, R. Cadherins and catenins in development. *Curr Opin Cell Biol*, *8:* 685-691, 1996.

[72] Ghersi, G. Roles of molecules involved in epithelial/mesenchymal transition during angiogenesis. *Front Biosci*, *13:* 2335-2355, 2008.

[73] Gumbiner, B. M. Epithelial morphogenesis. *Cell*, *69:* 385-387, 1992.

[74] Li, Q., Wang, J., Armant, D. R., Bagchi, M. K., and Bagchi, I. C. Calcitonin down-regulates E-cadherin expression in rodent uterine epithelium during implantation. *J Biol Chem*, *277:* 46447-46455, 2002.

[75] Ding, Y. Q., Zhu, L. J., Bagchi, M. K., and Bagchi, I. C. Progesterone stimulates calcitonin gene expression in the uterus during implantation. *Endocrinology*, *135:* 2265-2274, 1994.

[76] Zhu, L. J., Cullinan-Bove, K., Polihronis, M., Bagchi, M. K., and Bagchi, I. C. Calcitonin is a progesterone-regulated marker that forecasts the receptive state of endometrium during implantation. *Endocrinology, 139:* 3923-3934, 1998.

[77] Kumar, S., Zhu, L. J., Polihronis, M., Cameron, S. T., Baird, D. T., Schatz, F., Dua, A., Ying, Y. K., Bagchi, M. K., and Bagchi, I. C. Progesterone induces calcitonin gene expression in human endometrium within the putative window of implantation. *J Clin Endocrinol Metab, 83:* 4443-4450, 1998.

[78] Jha, R. K., Titus, S., Saxena, D., Kumar, P. G., and Laloraya, M. Profiling of Ecadherin, beta-catenin and Ca(2+) in embryo-uterine interactions at implantation. *FEBS Lett, 580:* 5653-5660, 2006.

[79] Achache, H. and Revel, A. Endometrial receptivity markers, the journey to successful embryo implantation. *Hum Reprod Update, 12:* 731-746, 2006.

[80] Varki, A. Biological roles of oligosaccharides: all of the theories are correct. *Glycobiology, 3:* 97-130, 1993.

[81] Popovici, R. M., Krause, M. S., Germeyer, A., Strowitzki, T., and von Wolff, M. Galectin-9: a new endometrial epithelial marker for the mid- and late-secretory and decidual phases in humans. *J Clin Endocrinol Metab, 90:* 6170-6176, 2005.

[82] von Wolff, M., Classen-Linke, I., Heid, D., Krusche, C. A., Beier-Hellwig, K., Karl, C., and Beier, H. M. Tumour necrosis factor-alpha (TNF-alpha) in human endometrium and uterine secretion: an evaluation by immunohistochemistry, ELISA and semiquantitative RT-PCR. *Mol Hum Reprod, 5:* 146-152, 1999.

[83] Almkvist, J. and Karlsson, A. Galectins as inflammatory mediators. *Glycoconj J, 19:* 575-581, 2004.

[84] Allison Gray, C., Bartol, F. F., Taylor, K. M., Wiley, A. A., Ramsey, W. S., Ott, T. L., Bazer, F. W., and Spencer, T. E. Ovine uterine gland knock-out model: effects of gland ablation on the estrous cycle. *Biol Reprod, 62:* 448-456, 2000.

[85] von Wolff, M., Wang, X., Gabius, H. J., and Strowitzki, T. Galectin fingerprinting in human endometrium and decidua during the menstrual cycle and in early gestation. *Mol Hum Reprod, 11:* 189-194, 2005.

[86] Ish-Shalom, E., Gargir, A., Andre, S., Borovsky, Z., Ochanuna, Z., Gabius, H. J., Tykocinski, M. L., and Rachmilewitz, J. alpha2,6-Sialylation promotes binding of placental protein 14 via its Ca2+-dependent lectin activity: insights into differential effects on CD45RO and CD45RA T cells. *Glycobiology, 16:* 173-183, 2006.

[87] Elola, M. T., Wolfenstein-Todel, C., Troncoso, M. F., Vasta, G. R., and Rabinovich, G. A. Galectins: matricellular glycan-binding proteins linking cell adhesion, migration, and survival. *Cell Mol Life* Sci, *64:* 1679-1700, 2007.

[88] Lorand, L. and Graham, R. M. Transglutaminases: crosslinking enzymes with pleiotropic functions. *Nat Rev Mol Cell Biol, 4:* 140-156, 2003.

[89] Gaudry, C. A., Verderio, E., Jones, R. A., Smith, C., and Griffin, M. Tissue transglutaminase is an important player at the surface of human endothelial cells: evidence for its externalization and its colocalization with the beta(1) integrin. *Exp Cell Res, 252:* 104-113, 1999.

[90] Beninati, S. and Piacentini, M. The transglutaminase family: an overview: minireview article. *Amino Acids, 26:* 367-372, 2004.

[91] Akimov, S. S., Krylov, D., Fleischman, L. F., and Belkin, A. M. Tissue transglutaminase is an integrin-binding adhesion coreceptor for fibronectin. *J Cell Biol, 148:* 825-838, 2000.

[92] Hager, H., Gliemann, J., Hamilton-Dutoit, S., Ebbesen, P., Koppelhus, U., and Jensen, P. H. Developmental regulation of tissue transglutaminase during human placentation and expression in neoplastic trophoblast. *J Pathol, 181:* 106-110, 1997.

[93] Signorini, M., Pansini, F., Bonaccorsi, G., Mollica, G., Ferrari, C., and Bergamini, C. M. Regulation of endometrial transglutaminase activity during the menstrual cycle. *Biochem Int, 16:* 77-82, 1988.

[94] Fujimoto, M., Kanzaki, H., Nakayama, H., Higuchi, T., Hatayama, H., Iwai, M., Kaneko, Y., Mori, T., and Fujita, J. Requirement for transglutaminase in progesterone-induced decidualization of human endometrial stromal cells. *Endocrinology, 137:* 1096-1101, 1996.

[95] Kabir-Salmani, M., Shiokawa, S., Akimoto, Y., Sakai, K., Sakai, K., and Iwashita, M. Tissue transglutaminase at embryo-maternal interface. *J Clin Endocrinol Metab, 90:* 4694-4702, 2005.

[96] Gentile, V., Saydak, M., Chiocca, E. A., Akande, O., Birckbichler, P. J., Lee, K. N., Stein, J. P., and Davies, P. J. Isolation and characterization of cDNA clones to mouse macrophage and human endothelial cell tissue transglutaminases. *J Biol Chem, 266:* 478-483, 1991.

In: New Cell Adhesion Research
Editors: Patrick Nott and Matthew Temple

ISBN 978-1-60692-378-8

Chapter IV

Control of Cell-Cell Adhesion by Src Kinase: Implications for Cancer Progression

***Eleri Lloyd Davies*[1] *and Stephen Hiscox*[2,*]**
[1]Department of Surgery, C-Block Room 710, Royal Gwent Hospital, Newport, Wales, NP20 2UB, UK;
[2]Welsh School of Pharmacy, Cardiff University, Cardiff, Wales, CF10 3NB, UK.

Abstract

The ability of cancer cells to interact with their surrounding environment directly affects many aspects of their behaviour. In addition to interactions with extracellular matrix constituents that can promote cell survival, changes in intercellular adhesive contacts can facilitate cell dissociation and migration. The activity of Src, a nonreceptor protein tyrosine kinase, and associated Src family members, can regulate the signalling and function of cell-cell adhesion proteins; these kinases are frequently elevated in tumour tissue and cells where they are associated with a gain in aggressive cell features including increased migration and invasion. Subsequently, the ability of inhibitors of Src family kinases to positively regulate cell-cell adhesion and suppress cell migration and invasion has been recently demonstrated. Here, we review the current understanding of the interplay between Src and receptors that mediate cell-cell adhesion and the implications for tumour development and spread.

* Correspondence concerning this article should be addressed to: Dr. S. Hiscox, Welsh School of Pharmacy, Redwood Building, King Edward VII Avenue, Cardiff CF10 3NB. Tel: 029 20 870306; Fax: 029 20 875152; Email: HiscoxSE1@cf.ac.uk.

INTRODUCTION

In their natural environment, cells are in constant contact with protein components of the extracellular matrix as well as other cells of the same or different type. These cellular interactions are maintained through a number of different classes of cell surface adhesion receptors and enable the cell to sense its microenvironment and respond accordingly. Such interactions contribute to cell integrity, provide platforms to orchestrate cell shape changes and allow for migration upon appropriate stimulus. Coordinated regulation of intercellular adhesion and the cytoskeleton system underlying it plays a key role in epithelial–mesenchymal transitions (EMT), both in embryogenesis and during cancer metastasis, where it plays a central role as a mechanism to facilitate cellular migration.

Cell-cell adhesive interactions can be regulated through a broad range of cell surface receptors, many of which can be regulated by members of the Src family of intracellular kinases, with Src itself being implicated as a key component. This article focuses on the role that Src plays as a modulator of intercellular interactions through regulation of cadherin- and desmosome-mediated adhesion complexes where the ability to regulate such adhesion complexes can have significant consequences for the development of an aggressive cell phenotype and tumour progression.

SRC KINASE

As the first tyrosine-specific kinase identified, Src is the most well studied kinase (see [1] for a comprehensive review of the history of Src). Src itself is the prototypic member of a family of non-receptor tyrosine kinases, eight of which (Src, Fyn, Yes, Lyn, Lck, Hck, Blk and Fgr) are expressed in humans. Whereas Src, Fyn and Yes are ubiquitously expressed, with particularly high levels in brain tissue, osteoclasts and platelets, the remaining Src family kinases (SFKs) exhibit a more tissue-restricted distribution, being found primarily in cells of a hematopoietic nature. SFKs share a common structure comprising six distinct functional domains. These include an N-terminal SH4 domain which contains a lipid-modification sequence to promote membrane association of the protein, and a unique region which is poorly-conserved between the different family members. The SH3 and SH2 domains adjacent to the N-terminus facilitate protein-protein interactions between Src family kinases and their substrates, whilst the SH1 domain possesses the tyrosine kinase activity of these enzymes; indeed, the auto-phosphorylation of Y419 within this domain is required for maximum kinase activity of Src. The C-terminal negative-regulatory tail contains a conserved tyrosine (at position 530 in human Src), the phosphorylation of which renders the kinase inactive by promoting a change in protein conformation (see [2] for a comprehensive review of Src-family kinase structure and regulation).

Historically, the Src-family member most often associated with cancer is Src itself [3]. Elevated Src activity has been reported in many different cancer types [4-7], where it has been shown to correlate with disease recurrence [8] and a poor clinical prognosis [9]. Increased Src kinase activity in cancer is likely to arise from the deregulation of Src expression and/or activation mechanisms [10] rather than the presence of activating mutations, since genetic mutations of this kind are rarely reported in [11-14]. Interestingly, whereas constitutively activated forms of Src are transforming, wild type Src has a relatively low transformation potential. This, alongside current understanding of the cellular functions of Src, has prompted investigators to hypothesize that the role of Src in cancer is to facilitate intracellular signalling by regulating, either directly or indirectly, the activity of other signalling proteins, rather than it being a dominant transforming agent in its own right.

Src is able to interact with a large number of cellular proteins including receptor tyrosine kinases (e.g. the EGF receptor family), integrins, cell-cell adhesion molecules [4,10,15,16], in addition to STATs [17], FAK [18], the adaptor protein p130Cas [19] and GPCRs [20] (reviewed in [21]). The ability of Src to function as both an effector and regulator of receptor-induced signalling allows it to mediate cross-talk between normally distinct signalling pathways and thus regulate a wide variety of both normal and oncogenic processes, including proliferation, differentiation, survival, adhesion, motility, invasion and angiogenesis.

Although a single role cannot be assigned to Src, a common feature of many of the major Src substrates (including p120-ctn, p130CAS, FAK, Tensin, p110AFAP) is their ability to regulate the actin cytoskeleton. Thus, despite the number and complexity of its substrates, an overarching role of c-Src may be to coordinate actin-dependent cellular activities, such as adhesion and motility. Clearly, the ability of Src to promote a migratory phenotype could have major consequences in the context of tumour progression.

Src family kinases interact with both cadherins and catenins, regulating their phosphorylation and ultimately the integrity of the cadherin adhesion complex. It is not surprising; therefore, that a hallmark of aberrant Src activity in a variety of cell types is a profound disorganization of cadherin-dependent cell–cell contacts [22-26]. Indeed, given that increases in Src activity are associated with the loss of epithelial morphology and the acquisition of mesenchymal properties that confer an enhanced metastatic potential [27], and that these processes rely on the integrity of cell-cell and cell-matrix adhesions, Src plays a key role in modulation of cell adhesive interactions. Here we review some of the key interactions between Src and cell components of the cell's adhesive machinery that may promote an adverse tumour cell phenotype.

MODULATION OF CADHERIN-MEDIATED ADHESION BY SRC KINASE

The cadherins are a superfamily of calcium-dependent membrane glycoproteins that play prominent roles as mediators of homotypic cell adhesion, tissue organisation and cellular morphogenesis in both physiological and pathological situations. E-cadherin, the primary cadherin of epithelial cells, is a trans-membrane glycoprotein that forms Ca^{2+}-dependent homophilic adhesion with cadherin molecules on adjacent cells through their extracellular domains [28]. Although individual interactions between E-cadherin molecules are relatively weak, the local clustering of many cadherins act to increase the strength and stability of these cadherin-mediated intercellular adhesions [29]. The intracellular domains of classical cadherins interact both directly and indirectly with proteins of the catenin family (α, β, and p120-catenin) [30]; formation of these adhesion complexes allows tethering of surface cadherins to the actin cytoskeleton and are critical in regulating cell–cell adhesion [31]. Changes in cadherin function are implicated in a number of processes such as resistance to cell detachment and the control of tissue patterning and adhesion. Given this broad function it is not surprising that cadherin function is regulated by a wide variety of cellular signalling pathways. Many reports demonstrate that loss of E-cadherin in tumours correlates with infiltrative and metastatic ability [32-35] and it is proposed that impaired intercellular adhesion arising through E-cadherin loss is a prerequisite for tumor cell invasion and metastasis formation [36,37]. Since experimental restoration of functional cadherin complexes results in a reversion from an invasive, mesenchymal-like phenotype to a benign, epithelial-like phenotype of cultured tumor cells [38], the E-cadherin gene is widely considered to function as a tumour/invasion suppressor gene. Other mechanisms that result in loss of E-cadherin-mediated cell-cell adhesion arise independently form changes in E-cadherin expression. For example, loss of association between α- or β-catenin molecules and the cytoplasmic region of E-cadherin results in impaired E-cadherin function [39,40]; the relevance of this in tumour progression is illustrated in clinical studies which have demonstrated an association between reduced catenin expression, or catenin mislocalization within the cell, with poor prognostic features [41-44].

There are a wealth of data that demonstrate the impact of dysfunction/loss of cell-cell adhesion within the tumour on the outlook for the cancer patient. A significant difference in survival between normal and aberrant expression of components of the cadherin adhesion complex is seen clinically, with a reduction in expression of one of more components associating with poorer prognosis/reduced survival [41,45-50]. Dysfunction of epithelial cell-cell junctions can arise through three major mechanisms in which Src and/or SFKs are involved: alteration in the turnover/trafficking of cadherin molecules within the cell; shedding of cadherins from the cell surface and modulation of the expression of cadherins at the gene level.

E-Cadherin Turnover/Degradation

The downregulation of cadherin expression can be observed in cells expressing constitutively active Src. This is thought to occur through the Src-dependent modification of E-cadherin endocytosis, a pathway that can lead to its degradation within the lysosome [51]. Src may also modulate cadherin turnover through its interactions with ASAP1, an activator of small GTP-binding proteins of the ADP-ribosylation factor (Arf) family. Src physically interacts with ASAP1 [52], in turn modulating the activity of Arf6 which plays a central role in the clathrin dependent endocytosis of E-cadherin in MDCK cells [53] inducing cell-cell dissociation [54]. Further evidence of a role for Src in E-cadherin turnover comes from studies where the increased cell scattering observed in MDCK cells expressing v-Src can be inhibited by proteosome inhibitors that reduce the Src-induced turnover of E-cadherin in these cells [55]. The activation of cdc42 in response to Ca^{2+} depletion results in up-regulation of epidermal growth factor receptor (EGFR) signalling. The subsequent activation of Src, a key component of EGFR signalling pathway, leads to E-cadherin ubiquitination and lysosomal degradation [56]. A further mechanism through which E-cadherin may be degraded involves the E3-ubiquitin ligase, Hakai. Expression of Hakai increases both the ubiquitination and rate of E-cadherin endocytosis in a Src phosphorylation-dependent manner [57]. Such observations as these suggest that elevated Src activity within the cancer cell may bias the trafficking of cadherins towards a degradative fate thus promoting cell dissociation and establishing permissive conditions for subsequent cellular migration and/or invasion.

E-Cadherin Shedding

The loss of cadherin-mediated adhesion may also arise through the shedding of cadherin from the cell surface; this has been reported to occur through a mechanism involving both MMP9 and Src kinase [58]. Interestingly, soluble cadherins may represent novel diagnostic/prognostic biomarkers since they correlate with other serum indicators of a poor prognosis, e.g. PSA and CA 19-9, in sera of prostate and pancreatic cancer patients respectively [59].

E-Cadherin Expression

There is evidence that Src can act to modulate the expression of cadherin at the cell membrane and thus affect the integrity of cell-cell junctions. Src is able to promote the downregulation of E-cadherin mRNA [60] whilst inhibition of Src family kinases using the pharmacological agent, PP2, has been shown to induce E-cadherin expression in human prostate cells [61]. In addition to indirect regulation of E-cadherin expression at the genomic level, Src may also promote mislocalization of E-cadherin at the cell membrane. This is evidences by studies which demonstrate that, following its activation via TGFβ pathway, Src kinase promotes the phosphorylation of the focal adhesion protein, FAK and leads to a delocalization of membrane-bound E-cadherin and increased cell invasion [62].

SRC-INDUCED MODULATION OF CADHERIN ADHESION THROUGH MODULATION OF CATENIN FUNCTION

Essential to the correct functioning of cadherin-based adhesion complexes is the presence of catenin molecules (α, β and p120-catenin) within the cell cytoplasm and their interactions with the cadherin proteins themselves. As well as a loss of catenin expression, cell-cell adhesion can be modulated through the phosphorylation, specifically on tyrosine, of catenin proteins; indeed, it has been recently demonstrated that the increase in tyrosine phosphorylation of β- and p120-catenin seen in human colon adenocarcenoma cells expressing homeodomain transcription factors Cdx1 and Cdx2 is accompanied by a significant decline in cell-cell adhesion [63]. Phosphorylation of catanin proteins may arise though multiple stimuli including growth factors, transformation by ras and, importantly, Src family members as detailed below.

p120 Catenin

p120 catenin (p120ctn) is a scaffold protein that interacts with the cytoplasmic domain of cadherins and acts as a crucial component of the signalling pathways that regulate the cycle of adherens junction formation and disassembly. p120ctn was first reported as a Src substrate, the phosphorylation of which correlated with transformation (reviewed in [64]). Considerable evidence now shows that the core function of p120ctn within the cadherin complex is to stabilize the adhesion complexes [65], primarily through its ability to inhibit clathrin-mediated cadherin endocytosis [65,66-69]. Interestingly, p120ctn also has additional roles within the cell cytoplasm and nucleus and research to date suggests that loss, downregulation or mislocalization of p120ctn correlates with progression of different types of human tumours (reviewed in [70]). Although clinical studies have suggested an association between the loss and/or mislocalization of p120ctn and tumour stage, grade and local spread [71], the role of p120ctn in cancer progression is further complicated by the fact that there are several isoforms expressed, not all of which appear to correlate with poor prognosis [72]. However, the fact that p120ctn is physically or functionally linked to a broad range of oncogenes and tumour suppressors that include E-cadherin, β-catenin, receptor tyrosine kinases and Src kinase itself suggest that p120ctn may play key roles in cancer where it may act to regulate cell morphology, adhesion and migration. Since p120ctn binds to the cytoplasmic domain of E-cadherin, regulating its trafficking [73], modulation of p120ctn phosphorylation by Src family members can result in loss of E-cadherin function at cell-cell contacts [74] which may arise through degradation of E-cadherin. Indeed, several members of Src family kinases have been shown to physically associate with p120ctn at the cell membrane leading to the proposal that p120ctn may play a scaffolding role to regulate tyrosine kinase activity at cadherin complexes [75]. A number of tyrosine residues have been identified on p120ctn as preferred sites for v-Src phosphorylation [76]. Interestingly, the majority of these occur within the cadherin-binding region of p120ctn [77] which might suggest that Src-induced phosphorylation of p120ctn might affect its binding to cadherin. Such a hypothesis is supported by the observations that aggregation of L cells expressing E-cadherin constructs that can interact with p120ctn is reduced by v-Src expression which is in contrast to cells which express E-cadherin mutants lacking the ability to bind p120ctn [74].

β-Catenin

The transmembrane receptor tyrosine kinases, EGFR and c-Met, along with intracellular Src and Fer kinases, are known to target specific residues in β-catenin critical for its interaction with cadherins or α-catenin. It thus follows that activation of EGFR and c-Met, or overexpression of Src or Fer, correlate with enhanced tyrosine phosphorylation of β-catenin and are also associated with the loss of E-cadherin-mediated adhesion [78]. Three conserved tyrosine residues (Y142, Y489 and Y654) in β-catenin are proximate to the interface between β-catenin and the cadherin molecule and act to regulate the integrity of the core cadherin adhesion complex. Y142 is in the α-catenin binding region of β-catenin [79] and is targeted by the Src-related kinases, Fyn and Fer in addition to the c-Met receptor [80,81]. Subsequently, phosphorylation of β-catenin at Y142 by activated SFKs or Met results in loss of its association with α-catenin [80, 81] and loss of cadherin adhesion [82]. In contrast, phosphorylation of β-catenin at Y489 as a result of Abl kinase activity results in release of β-catenin from cadherin [84] whilst phosphorylation of β-catenin at Y654, primarily via EGFR or Src, produces a phosphorylated residue that is too large to accommodate the interaction between β-catenin and E-cadherin, resulting in a reduction in the affinity of β-catenin for cadherin [83].

α-Catenin

The importance of α-catenin in cell-cell adhesion has been demonstrated through experimental mutation in the α-catenin binding site of β-catenin in the fly which results in disruption of cell-cell adhesion, suggesting that the interaction of α-catenin with core cadherin adhesion complex components is essential in the physical process of intercellular adhesion. Although it is unclear as to whether Src can directly modulate α-catenin activity/phosphorylation status, the SFK-mediated phosphorylation of Y142 in β-catenin impairs the interaction between these two proteins leading to loss of cell-cell adhesion [75]. Interestingly, however, recent studies have demonstrated that tyrosine phosphorylation of α-catenin promotes its translocation to the plasma membrane and its subsequent interaction with β-catenin, leading to enhanced actin polymerization and stabilization of adherens junction-mediated intercellular adhesion [85]; conversely, SHP2-mediated de-phosphorylation of α-catenin occurs in response to growth factor activation and may thus favour loss of cell-cell adhesion [85]. In the latter case, the effect of growth factor receptors on α-catenin function may occur via SFK members which are central components of many growth factor initiated signalling pathways.

DESMOSOMES

Desmosomes are small (< 0.5μm in diameter) intercellular junctions found mainly in epithelial cells the primary function of which is to maintain tissue integrity through strong cell-cell adhesion, although they also have a role in cell signalling and wound repair [86,87]. Importantly, desmosomes attach to intermediate filaments and thus provide a means of linking the cell cytoskeleton between cells [86]. Desmosomes are comprised of desmosomal cadherins (desmocollin (Dsc) and desmoglein (Dsg) family proteins), linked to the armadillo family members, plakoglobin and plakophillin [88] and, like cadherin-mediated adherens junctions, desmosome assembly requires Ca^{2+}. However, whereas cadherin-mediated adhesion occurs through homophilic binding, desmosomes are formed through the heterotypic interactions of Dsc and Dsg members [87,89]. Although desmosomal cadherins can initiate and maintain cell–cell adhesion in the absence of any contribution from classical cadherins [90], adherens junction (E-cadherin-mediated) adhesion usually occurs first, with formation of desmosomes then acting to stabilise intercellular adhesion. Moreover, there is a reciprocal interdependence between desmosomes and adherens junction components as evidenced by the fact that suppression of desmosome expression subsequently disrupts adherens junction-mediated adhesion [91].

Given that the presence of desmosomes suppresses cell migration and invasion [90,92] it is not suprising to find that the loss of desmosomal adhesion appears to be a prerequisite for the transition from an epithelial to a mesenchymal phenotype, with such events being implicated in the conversion of early stage tumours to invasive cancers. Indeed, clinical studies have revealed that increased loss of expression of desmoplakin and/or desmoglein is observed in a number of tumour types compared with normal tissue [93,94] with reduced desmosome formation baing associated with invasive tumour behaviour and local progression and spread [95-97].

Src has been implicated as playing a role in the disassembly of desmosomes [98] whilst a link between Src and plakoglobin expression is suggested by the fact that the pharmacological Src inhibitor, PP2, can modulate plakoglobin expression in cancer cell lines [99]. Moreover, immunofluorescence studies on xenografts from dominant negative-Src expressing tumours show that these cells are characterized by large numbers of desmosomes at the cell surface in contrast to tumours from cells expressing high levels of activated Src where desmoplakin immunostaining is located in a punctuate cytoplasmic pattern suggestive of internalization [27]. Interestingly, these latter tumours also appear to have a high metastatic potential in vivo. Although a definitive and direct link between Src, of SFKs, and desmosome regulation remains to be established, is has been observed that Src can activate the desmosomal component, plakoglobin. Plakoglobin can substitute for β-catenin in adherens junctions, providing a connection between E-cadherin and alpha-catenin in its absence. Src-mediated phosphorylation of plakoglobin at Y643 both decreases its interaction with E-cadherin and alpha-catenin whilst increasing its interaction with desmoplakin [100]. Thus, whilst the function of one adhesion system may decrease, there may be a degree of compensation arising from increases in other mechanisms that regulate cell-cell adhesion.

Conclusion

Expression of activated mutants of Src in epithelial cells can induce tumorigenicity with such cells undergoing dramatic morphological changes and displaying disruption of cell-cell contacts and increased migration/invasion Furthermore, elevated Src activity is frequently reported in tumour cells where it can correlate with increased migratory/invasive capacity in vitro and tumour spread in vivo. In the light of the wealth of data implicating Src as a key regulator of cell adhesive interactions, this may present a key mechanism through which Src can modulate the aggressive features of cancer cells. With emerging studies demonstrating the effectiveness of inhibitors of Src kinase as suppressers of cell invasion, such evidence points to the potential use of Src inhibitors clinically, where, through the ability to restore cell-cell adhesion, they may present effective agents to suppress tumour spread.

REFERENCES

[1] Martin, G. S. (2004) The road to Src. *Oncogene 23*, 7910-7

[2] Thomas, S. M. and Brugge, J. S. (1997) Cellular functions regulated by Src family kinases. *Annu Rev Cell Dev Biol 13*, 513-609

[3] Yeatman, T. J. (2004) A renaissance for SRC. *Nat Rev Cancer 4*, 470-80

[4] Irby, R. B. and Yeatman, T. J. (2000) Role of Src expression and activation in human cancer. *Oncogene 19*, 5636-42

[5] Summy, J. M. and Gallick, G. E. (2003) Src family kinases in tumor progression and metastasis. *Cancer Metastasis Rev 22*, 337-58

[6] Rosen, N., Bolen, J. B., Schwartz, A. M., Cohen, P., DeSeau, V. and Israel, M. A. (1986) Analysis of pp60c-src protein kinase activity in human tumor cell lines and tissues. *J Biol Chem 261*, 13754-9

[7] Verbeek, B. S., Vroom, T. M., Adriaansen-Slot, S. S., Ottenhoff-Kalff, A. E., Geertzema, J. G., Hennipman, A. and Rijksen, G. (1996) c-Src protein expression is increased in human breast cancer. An immunohistochemical and biochemical analysis. *J Pathol 180*, 383-8

[8] Wilson, G. R., Cramer, A., Welman, A., Knox, F., Swindell, R., Kawakatsu, H., Clarke, R. B., Dive, C. and Bundred, N. J. (2006) Activated c-SRC in ductal carcinoma in situ correlates with high tumour grade, high proliferation and HER2 positivity. *Br J Cancer 95*, 1410-4

[9] Aligayer, H., Boyd, D. D., Heiss, M. M., Abdalla, E. K., Curley, S. A. and Gallick, G. E. (2002) Activation of Src kinase in primary colorectal carcinoma: an indicator of poor clinical prognosis. *Cancer 94*, 344-51

[10] Biscardi, J. S., Ishizawar, R. C., Silva, C. M. and Parsons, S. J. (2000) Tyrosine kinase signalling in breast cancer: epidermal growth factor receptor and c-Src interactions in breast cancer. *Breast Cancer Res 2*, 203-10

[11] Irby, R. B., Mao, W., Coppola, D., Kang, J., Loubeau, J. M., Trudeau, W., Karl, R., Fujita, D. J., Jove, R. and Yeatman, T. J. (1999) Activating SRC mutation in a subset of advanced human colon cancers. *Nat Genet 21*, 187-90

[12] Sugimura, M., Kobayashi, K., Sagae, S., Nishioka, Y., Ishioka, S., Terasawa, K., Tokino, T. and Kudo, R. (2000) Mutation of the SRC gene in endometrial carcinoma. Jpn *J Cancer Res 91*, 395-8

[13] Wang, N. M., Yeh, K. T., Tsai, C. H., Chen, S. J. and Chang, J. G. (2000) No evidence of correlation between mutation at codon 531 of src and the risk of colon cancer in Chinese. *Cancer Lett 150*, 201-4

[14] Laghi, L., Bianchi, P., Orbetegli, O., Gennari, L., Roncalli, M. and Malesci, A. (2001) Lack of mutation at codon 531 of SRC in advanced colorectal cancers from Italian patients. *Br J Cancer 84*, 196-8

[15] Huveneers, S., van den Bout, I., Sonneveld, P., Sancho, A., Sonnenberg, A. and Danen, E. H. (2007) Integrin alpha v beta 3 controls activity and oncogenic potential of primed c-Src. *Cancer Res 67*, 2693-700

[16] Galliher, A. J. and Schiemann, W. P. (2006) Beta3 integrin and Src facilitate transforming growth factor-beta mediated induction of epithelial-mesenchymal transition in mammary epithelial cells. *Breast Cancer Res 8*, R42

[17] Silva, C. M. (2004) Role of STATs as downstream signal transducers in Src family kinase-mediated tumorigenesis. *Oncogene 23*, 8017-23

[18] Brunton, V. G., Avizienyte, E., Fincham, V. J., Serrels, B., Metcalf, C. A., 3rd, Sawyer, T. K. and Frame, M. C. (2005) Identification of Src-specific phosphorylation site on focal adhesion kinase: dissection of the role of Src SH2 and catalytic functions and their consequences for tumor cell behavior. *Cancer Res 65*, 1335-42

[19] Cowell, L. N., Graham, J. D., Bouton, A. H., Clarke, C. L. and O'Neill, G. M. (2006) Tamoxifen treatment promotes phosphorylation of the adhesion molecules, p130Cas/BCAR1, FAK and Src, via an adhesion-dependent pathway. *Oncogene 25*, 7597-607

[20] McGarrigle, D. and Huang, X. Y. (2007) GPCRs signaling directly through Src-family kinases. *Sci STKE* 2007, pe35

[21] Frame, M. C. (2002) Src in cancer: deregulation and consequences for cell behaviour. *Biochim Biophys Acta 1602*, 114-30

[22] Daniel, J. M., Reynolds, A. B., Anastasiadis, P. Z., Reynolds, A. B., Thoreson, M. A., Reynolds, A. B., Kowalczyk, A. P., Reynolds, A. B., Reynolds, A. B., Roczniak-Ferguson, A. and Reynolds, A. B. (1997) Tyrosine phosphorylation and cadherin/catenin function Regulation of Rho GTPases by p120-catenin Altered expression of the catenin p120 in human cancer: implications for tumor progression Protecting your tail: regulation of cadherin degradation by p120-catenin Emerging roles for p120-catenin in cell adhesion and cancer p120-catenin: Past and present. *Bioessays 19*, 883-91

[23] Takeda, H., Nagafuchi, A., Yonemura, S., Tsukita, S., Behrens, J., Birchmeier, W. and Tsukita, S. (1995) V-src kinase shifts the cadherin-based cell adhesion from the strong to the weak state and beta catenin is not required for the shift. *J Cell Biol* 131, 1839-1847

[24] Warren, S. L. and Nelson, W. J. (1987) Nonmitogenic Biological Activity of Pp60s-R-C on Multicellular Epithelial Structures. *Mol Cell Biol* 105, 306A

[25] Avizienyte, E., Wyke, A. W., Jones, R. J., McLean, G. W., Westhoff, M. A., Brunton, V. G. and Frame, M. C. (2002) Src-induced de-regulation of E-cadherin in colon cancer cells requires integrin signalling. *Nat Cell Biol* 4, 632-638

[26] Irby, R. B. and Yeatman, T. J. (2002) Increased Src activity disrupts cadherin/catenin-mediated homotypic adhesion in human colon cancer and transformed rodent cells. *Cancer Res 62*, 2669-74

[27] Boyer, B., Bourgeois, Y. and Poupon, M. F. (2002) Src kinase contributes to the metastatic spread of carcinoma cells. *Oncogene 21*, 2347-56

[28] Boggon, T. J., Murray, J., Chappuis-Flament, S., Wong, E., Gumbiner, B. M. and Shapiro, L. (2002) C-cadherin ectodomain structure and implications for cell adhesion mechanisms. *Science 296*, 1308-13

[29] Haussinger, D., Ahrens, T., Aberle, T., Engel, J., Stetefeld, J. and Grzesiek, S. (2004) Proteolytic E-cadherin activation followed by solution NMR and X-ray crystallography. *Embo J 23*, 1699-708

[30] Leckband, D. and Sivasankar, S. (2000) Mechanism of homophilic cadherin adhesion. *Curr Opin Cell Biol 12*, 587-92

[31] Takeichi, M., Nakagawa, S., Aono, S., Usui, T. and Uemura, T. (2000) Patterning of cell assemblies regulated by adhesion receptors of the cadherin superfamily. *Phil Trans R Soc Lond B Biol Sci* 355, 885-890

[32] Takeichi, M. (1993) Cadherins in cancer: implications for invasion and metastasis. *Curr Opin Cell Biol 5*, 806-11

[33] Onder, T. T., Gupta, P. B., Mani, S. A., Yang, J., Lander, E. S. and Weinberg, R. A. (2008) Loss of E-cadherin promotes metastasis via multiple downstream transcriptional pathways. *Cancer Res 68*, 3645-54

[34] Howard, E. W., Camm, K. D., Wong, Y. C. and Wang, X. H. (2008) E-cadherin upregulation as a therapeutic goal in cancer treatment. *Mini Rev Med Chem 8*, 496-518

[35] Hudson, L. G., Zeineldin, R. and Stack, M. S. (2008) Phenotypic plasticity of neoplastic ovarian epithelium: unique cadherin profiles in tumor progression. *Clin Exp Metastasis*

[36] Ghadimi, B. M., Behrens, J., Hoffmann, I., Haensch, W., Birchmeier, W. and Schlag, P. M. (1999) Immunohistological analysis of E-cadherin, alpha-, beta- and gamma-catenin expression in colorectal cancer: implications for cell adhesion and signaling. *Eur J Cancer 35*, 60-5

[37] Behrens, J., Behrens, J., Frixen, U., Schipper, J., Weidner, M. and Birchmeier, W. (1993) The role of cell adhesion molecules in cancer invasion and metastasis Cell adhesion in invasion and metastasis. *Breast Cancer Res Treat 24*, 175-84

[38] Vleminckx, K., Vakaet, L., Jr., Mareel, M., Fiers, W. and van Roy, F. (1991) Genetic manipulation of E-cadherin expression by epithelial tumor cells reveals an invasion suppressor role. *Cell 66*, 107-19

[39] Benjamin, J. M. and Nelson, W. J. (2008) Bench to bedside and back again: molecular mechanisms of alpha-catenin function and roles in tumorigenesis. *Semin Cancer Biol 18*, 53-64

[40] Brembeck, F. H., Rosario, M. and Birchmeier, W. (2006) Balancing cell adhesion and Wnt signaling, the key role of beta-catenin. *Curr Opin Genet Dev 16*, 51-9

[41] Hsu, P. K., Li, A. F., Wang, Y. C., Hsieh, C. C., Huang, M. H., Hsu, W. H. and Hsu, H. S. (2008) Reduced membranous beta-catenin protein expression is associated with metastasis and poor prognosis in squamous cell carcinoma of the esophagus. *J Thorac Cardiovasc Surg 135*, 1029-35

[42] Chen, S., Liu, J., Li, G., Mo, F., Xu, X., Zhang, T., Zhang, X., Li, J., Han, X. and Sun, Y. (2008) Altered distribution of beta-catenin and prognostic roles in colorectal carcinogenesis. *Scand J Gastroenterol 43*, 456-64

[43] Kim, M. J., Jang, S. J. and Yu, E. (2008) Loss of E-cadherin and cytoplasmic-nuclear expression of beta-catenin are the most useful immunoprofiles in the diagnosis of solid-pseudopapillary neoplasm of the pancreas. *Hum Pathol 39*, 251-8

[44] Chung, Y., Lam, A. K., Luk, J. M., Law, S., Chan, K. W., Lee, P. Y. and Wong, J. (2007) Altered E-cadherin expression and p120 catenin localization in esophageal squamous cell carcinoma. *Ann Surg Oncol 14*, 3260-7

[45] van Oort, I. M., Tomita, K., van Bokhoven, A., Bussemakers, M. J., Kiemeney, L. A., Karthaus, H. F., Witjes, J. A. and Schalken, J. A. (2007) The prognostic value of E-cadherin and the cadherin-associated molecules alpha-, beta-, gamma-catenin and p120ctn in prostate cancer specific survival: a long-term follow-up study. *Prostate 67*, 1432-8

[46] Kashibuchi, K., Tomita, K., Schalken, J. A., Kume, H., Takeuchi, T. and Kitamura, T. (2007) The prognostic value of E-cadherin, alpha-, beta- and gamma-catenin in bladder cancer patients who underwent radical cystectomy. *Int J Urol 14*, 789-94

[47] Shimazui, T., Schalken, J. A., Giroldi, L. A., Jansen, C. F. J., Akaza, H., Koiso, K., Debruyne, F. M. J. and Bringuier, P. P. (1996) Prognostic value of cadherin-associated molecules (alpha-, beta-, and gamma-catenins and p120(cas)) in bladder tumors. *Cancer Res* 56, 4154-4158

[48] Shimazui, T., Bringuier, P. P., van Berkel, H., Ruijter, E., Akaza, H., Debruyne, F. M., Oosterwijk, E. and Schalken, J. A. (1997) Decreased expression of alpha-catenin is associated with poor prognosis of patients with localized renal cell carcinoma. *Int J Cancer 74*, 523-8

[49] Kreizenbeck, G. M., Berger, A. J., Subtil, A., Rimm, D. L. and Gould Rothberg, B. E. (2008) Prognostic significance of cadherin-based adhesion molecules in cutaneous malignant melanoma. *Cancer Epidemiol Biomarkers Prev 17*, 949-58

[50] Salahshor, S., Naidoo, R., Serra, S., Shih, W., Tsao, M. S., Chetty, R. and Woodgett, J. R. (2008) Frequent accumulation of nuclear E-cadherin and alterations in the Wnt signaling pathway in esophageal squamous cell carcinomas. *Mod Pathol 21*, 271-81

[51] Palacios, F., Tushir, J. S., Fujita, Y. and D'Souza-Schorey, C. (2005) Lysosomal targeting of E-cadherin: a unique mechanism for the down-regulation of cell-cell adhesion during epithelial to mesenchymal transitions. *Mol Cell Biol* 25, 389-402

[52] Brown, M. T., Andrade, J., Radhakrishna, H., Donaldson, J. G., Cooper, J. A. and Randazzo, P. A. (1998) ASAP1, a phospholipid-dependent arf GTPase-activating protein that associates with and is phosphorylated by Src. *Mol Cell Biol 18*, 7038-51

[53] Kon, S., Tanabe, K., Watanabe, T., Sabe, H. and Satake, M. (2008) Clathrin dependent endocytosis of E-cadherin is regulated by the Arf6GAP isoform SMAP1. *Exp Cell Res 314*, 1415-28

[54] Palacios, F. and D'Souza-Schorey, C. (2003) Modulation of Rac1 and ARF6 activation during epithelial cell scattering. *J Biol Chem* 278, 17395-17400

[55] Tsukamoto, T. and Nigam, S. K. (1999) Cell-cell dissociation upon epithelial cell scattering requires a step mediated by the proteasome. *J Biol Chem 274*, 24579-84

[56] Shen, Y., Hirsch, D. S., Sasiela, C. A. and Wu, W. J. (2008) Cdc42 regulates E-cadherin ubiquitination and degradation through an epidermal growth factor receptor to Src-mediated pathway. *J Biol Chem* 283, 5127-5137

[57] Fujita, Y., Krause, G., Scheffner, M., Zechner, D., Leddy, H. E. M., Behrens, J., Sommer, T. and Birchmeier, W. (2002) Hakai, a c-Cbl-like protein, ubiquitinates and induces endocytosis of the E-cadherin complex. *Nat Cell Biol* 4, 222-231

[58] Symowicz, J., Adley, B. P., Gleason, K. J., Johnson, J. J., Ghosh, S., Fishman, D. A., Hudson, L. G. and Stack, M. S. (2007) Engagement of collagen-binding integrins promotes matrix metalloproteinase-9-dependent E-cadherin ectodomain shedding in ovarian carcinoma cells. *Cancer Res* 67, 2030-2039

[59] De Wever, O., Derycke, L., Hendrix, A., De Meerleer, G., Godeau, F., Depypere, H. and Bracke, M. (2007) Soluble cadherins as cancer biomarkers. *Clin Exp Metastasis 24*, 685-97

[60] Ulianich, L., Garbi, C., Treglia, A. S., Punzi, D., Miele, C., Raciti, G. A., Beguinot, F., Consiglio, E. and Di Jeso, B. (2008) ER stress is associated with dedifferentiation and an epithelial-to-mesenchymal transition-like phenotype in PC Cl3 thyroid cells. *J Cell Sci 121*, 477-86

[61] Wei, J. C., Xu, G., Wu, M. F., Zhang, Y. T., Li, Q., Liu, P., Zhu, T., Song, A. P., Zhao, L. P., Han, Z. Q., Chen, G., Wang, S. X., Meng, L., Zhou, J. F., Lu, Y. P., Wang, S. X. and Ma, D. (2008) Overexpression of vimentin contributes to prostate cancer invasion and metastasis via src regulation. *Anticancer Res* 28, 327-334

[62] Cicchini, C., Laudadio, I., Citarella, F., Corazzari, M., Steindler, C., Conigliaro, A., Fantoni, A., Amicone, L. and Tripodi, M. (2008) TGFbeta-induced EMT requires focal adhesion kinase (FAK) signaling. *Exp Cell Res 314*, 143-52

[63] Ezaki, T., Guo, R. J., Li, H., Reynolds, A. B. and Lynch, J. P. (2007) The homeodomain transcription factors Cdx1 and Cdx2 induce E-cadherin adhesion activity by reducing beta- and p120-catenin tyrosine phosphorylation. *Am J Physiol Gastrointest Liver Physiol 293*, G54-65

[64] Reynolds, A. B. (2007) p120-catenin: Past and present. *Biochem Biophys Acta* 1773, 2-7

[65] Ireton, R. C., Davis, M. A., van Hengel, J., Mariner, D. J., Barnes, K., Thoreson, M. A., Anastasiadis, P. Z., Matrisian, L., Bundy, L. M., Sealy, L., Gilbert, B., van Roy, F. and Reynolds, A. B. (2002) A novel role for p120 catenin in E-cadherin function. *J Cell Biol* 159, 465-476

[66] Xiao, K. Y., Garner, J., Buckley, K. M., Vincent, P. A., Chiasson, C. M., Dejana, E., Faundez, V. and Kowalczyk, A. P. (2005) p120-Catenin regulates clathrin-dependent endocytosis of VE-cadherin. *Mol Cell Biol* 16, 5141-5151

[67] Miyashita, Y. and Ozawa, M. (2007) Increased internalization of p120-uncoupled E-cadherin and a requirement for a dileucine motif in the cytoplasmic domain for endocytosis of the protein. *J Biol Chem* 282, 11540-11548

[68] Davis, M. A., Ireton, R. C. and Reynolds, A. B. (2003) A core function for p120-catenin in cadherin turnover. *J Cell Biol* 163, 525-534

[69] Xiao, K. Y., Allison, D. F., Buckley, K. M., Kottke, M. D., Vincent, P. A., Faundez, V. and Kowalczyk, A. P. (2003) Cellular levels of p120 catenin function as a set point for cadherin expression levels in microvascular endothelial cells. *J Cell Biol* 163, 535-545

[70] van Hengel, J. and van Roy, F. (2007) Diverse functions of p120ctn in tumors. *Biochim Biophys Acta 1773*, 78-88

[71] Silva Neto, B., Smith, G. L., Mandeville, J. A., Vanni, A. J., Wotkowicz, C., Rieger-Christ, K. M., Baumgart, E., Jacobs, M. A., Cohen, M. S., Zeheb, R., Loda, M.,

Libertino, J. A. and Summerhayes, I. C. (2008) Prognostic significance of altered p120 ctn expression in bladder cancer. *BJU Int 101*, 746-52

[72] Liu, Y., Xu, H. T., Dai, S. D., Wei, Q., Yuan, X. M. and Wang, E. H. (2007) Reduction of p120(ctn) isoforms 1 and 3 is significantly associated with metastatic progression of human lung cancer. *APMIS 115*, 848-56

[73] Peifer, M and Yap, A. S. (2003) Traffic control: p120 catenin acts as a gatekeeper to control the fate of classical cadherins in mammalian cells. *J Cell Biol 163,* 437-40.

[74] Ozawa, M. and Ohkubo, T. (2001) Tyrosine phosphorylation of p120(ctn) in v-Src transfected L cells depends on its association with E-cadherin and reduces adhesion activity. *J Cell Sci* 114, 503-512

[75] Piedra, J., Miravet, S., Castano, J., Palmer, H. G., Heisterkamp, N., Garcia de Herreros, A. and Dunach, M. (2003) p120 Catenin-associated Fer and Fyn tyrosine kinases regulate beta-catenin Tyr-142 phosphorylation and beta-catenin-alpha-catenin Interaction. *Mol Cell Biol 23*, 2287-97

[76] Mariner, D. J., Anastasiadis, P., Keilhack, H., Bohmer, F. D., Wang, J. and Reynolds, A. B. (2001) Identification of Src phosphorylation sites in the catenin p120(ctn). *J Biol Chem* 276, 28006-28013

[77] Alema, S. and Salvatore, A. M. (2007) p120 catenin and phosphorylation: Mechanisms and traits of an unresolved issue. *Biochim Biophys Acta 1773*, 47-58

[78] Lilien, J. and Balsamo, J. (2005) The regulation of cadherin-mediated adhesion by tyrosine phosphorylation/dephosphorylation of beta-catenin. *Curr Opin Cell Biol* 17, 459-465

[79] Aberle, H., Schwartz, H., Hoschuetzky, H. and Kemler, R. (1996) Single amino acid substitutions in proteins of the armadillo gene family abolish their binding to alpha-catenin *J Biol Chem 271*, 1520-6

[80] Piedra, J., Miravet, S., Castano, J., Palmer, H. G., Heisterkamp, N., de Herreros, A. G. and Dunach, M. (2003) p120 catenin-associated Fer and Fyn tyrosine kinases regulate beta-catenin Tyr-142 phosphorylation and beta-catenin-alpha-catenin interaction. *Mol Cell Biol* 23, 2287-2297

[81] Brembeck, F. H., Schwarz-Romond, T., Bakkers, J., Wilhelm, S., Hammerschmidt, M., Birchmeier, W., Harris, T. J. and Peifer, M. (2004) Essential role of BCL9-2 in the switch between beta-catenin's adhesive and transcriptional functions *Genes Dev 18*, 2225-30

[82] Ozawa, M. and Kemler, R. (1998) Altered cell adhesion activity by pervanadate due to the dissociation of alpha-catenin from the E-cadherin.catenin complex. *J Biol Chem 273*, 6166-70

[83] Roura, S., Miravet, S., Piedra, J., de Herreros, A. G. and Dunach, M. (1999) Regulation of E-cadherin/catenin association by tyrosine phosphorylation. *J Biol Chem* 274, 36734-36740

[84] Rhee, J., Mahfooz, N. S., Arregui, C., Lilien, J., Balsamo, J. and VanBerkum, M. F. (2002) Activation of the repulsive receptor Roundabout inhibits N-cadherin-mediated cell adhesion. *Nat Cell Biol 4*, 798-805

[85] Burks, J. and Agazie, Y. M. (2006) Modulation of alpha-catenin Tyr phosphorylation by SHP2 positively effects cell transformation induced by the constitutively active FGFR3. *Oncogene* 25, 7166-7179

[86] Garrod, D. and Chidgey, M. (2008) Desmosome structure, composition and function. *Biochim Biophys Acta 1778*, 572-87

[87] Garrod, D. R., Merritt, A. J. and Nie, Z. (2002) Desmosomal cadherins. *Curr Opin Cell Biol 14*, 537-45

[88] Kottke, M. D., Delva, E. and Kowalczyk, A. P. (2006) The desmosome: cell science lessons from human diseases. *J Cell Sci 119*, 797-806

[89] Marcozzi, C., Burdett, I. D., Buxton, R. S. and Magee, A. I. (1998) Coexpression of both types of desmosomal cadherin and plakoglobin confers strong intercellular adhesion. *J Cell Sci 111 (Pt 4)*, 495-509

[90] Tselepis, C., Chidgey, M., North, A. and Garrod, D. (1998) Desmosomal adhesion inhibits invasive behavior. *Proc Natl Acad Sci U S A 95*, 8064-9

[91] Vasioukhin, V., Bowers, E., Bauer, C., Degenstein, L. and Fuchs, E. (2001) Desmoplakin is essential in epidermal sheet formation. *Nat Cell Biol 3*, 1076-85

[92] Yin, T., Getsios, S., Caldelari, R., Kowalczyk, A. P., Muller, E. J., Jones, J. C. and Green, K. J. (2005) Plakoglobin suppresses keratinocyte motility through both cell-cell adhesion-dependent and -independent mechanisms. *Proc Natl Acad Sci U S A 102*, 5420-5

[93] Conn, I. G., Vilela, M. J., Garrod, D. R., Crocker, J. and Wallace, D. M. (1990) Immunohistochemical staining with monoclonal antibody 32-2B to desmosomal glycoprotein 1. Its role in the histological assessment of urothelial carcinomas. *Br J Urol 65*, 176-80

[94] Shinohara, M., Hiraki, A., Ikebe, T., Nakamura, S., Kurahara, S., Shirasuna, K. and Garrod, D. R. (1998) Immunohistochemical study of desmosomes in oral squamous cell carcinoma: correlation with cytokeratin and E-cadherin staining, and with tumour behaviour. *J Pathol 184*, 369-81

[95] Davies, E. L., Gee, J. M., Cochrane, R. A., Jiang, W. G., Sharma, A. K., Nicholson, R. I. and Mansel, R. E. (1999) The immunohistochemical expression of desmoplakin and its role in vivo in the progression and metastasis of breast cancer. *Eur J Cancer 35*, 902-7

[96] Sobolik-Delmaire, T., Katafiasz, D., Keim, S. A., Mahoney, M. G. and Wahl, J. K., 3rd. (2007) Decreased plakophilin-1 expression promotes increased motility in head and neck squamous cell carcinoma cells. *Cell Commun Adhes 14*, 99-109

[97] Davies, E. L., Cochrane, R. A., Hiscox, S., Jiang, W. G., Sweetland, H. M. and Mansel, R. M. (1997) The role of Desmoglein 2 (Dsg2) and E-cadherin in the invasion and metastasis of breast cancer cells. *International Journal of Oncology 11*, 415-419

[98] Chernyavsky, A. I., Arredondo, J., Kitajima, Y., Sato-Nagai, M. and Grando, S. A. (2007) Desmoglein versus non-desmoglein signaling in pemphigus acantholysis: characterization of novel signaling pathways downstream of pemphigus vulgaris antigens. *J Biol Chem 282*, 13804-12

[99] Chiang, G. J., Billmeyer, B. R., Canes, D., Stoffel, J., Moinzadeh, A., Austin, C. A., Kosakowski, M., Rieger-Christ, K. M., Libertino, J. A. and Summerhayes, I. C. (2005)

The src-family kinase inhibitor PP2 suppresses the in vitro invasive phenotype of bladder carcinoma cells via modulation of Akt. *BJU Int 96*, 416-22

[100] Miravet, S., Piedra, J., Castano, J., Raurell, I., Franci, C., Dunach, M. and Garcia de Herreros, A. (2003) Tyrosine phosphorylation of plakoglobin causes contrary effects on its association with desmosomes and adherens junction components and modulates beta-catenin-mediated transcription. *Mol Cell Biol 23*, 7391-402

In: New Cell Adhesion Research
Editors: Patrick Nott and Matthew Temple
ISBN 978-1-60692-378-8

Chapter V

NOVEL DATA ON CELL ADHESION

Adam Curtis
CCE University of Glasgow, Glasgow, Scotland G12 8QQ.

ABSTRACT

Recent publications and the theories arising from their results are surveyed for three areas of work on cell-substratum and cell-cell adhesion. These areas are 1. The roles of nano and microtopography in the substratum upon cell adhesion. 2. The effects of substratum and near-substratum mechanical properties on cell adhesion. 3. Mechanotransduction effects and cell adhesion. It is concluded that all three are important and that the patterns of force interaction with the cell may pay important roles not only in adhesion but also in cell spreading and gene expression activation.

INTRODUCTION

Explanations of the mechanisms of cell adhesion started with purely physicochemical explanations for example in terms of surface energies ([1] van Oss 1975) moved to the rather more specific Deryagin-Landau-Verwey-Overbeek (DLVO) explanation ([2] Curtis,1964) and then onto the concepts that the cells are bound to each other by cell adhesion molecules (CAMS) and cadherins, and for cell to substratum by fibronectin, laminin, vitronectin etc being bound by integrins ([3-5]). There is much evidence in favor of these theories though they may well refer to differing phases of attachment. We might nowadays refer to them as classical explanations. But in recent years two other main sets of data have begun to suggest that the situation is not quite as simple as we thought.

The discovery that microtopography and perhaps even more nanotopography affect cell attachment ([6-9]) leads to questions about mechanism. More recently a further almost unexplored territory has begun to open up which seems to lie in the mechanical properties of the substratum near to the cell or to tissue properties near to the cells whose adhesion is being examined. This article aims to summarize the evidence about the roles of topography and mechanical properties of the substratum discuss the mechanisms involved and outline the implications of all these recent findings.

SURFACE TOPOGRAPHY

In the 1970s and thereafter reports began to emerge that various micro-structures affected cell behaviors believed to be associated with cell adhesion such as cell orientation to substratum features, rates of cell migration ([10] Curtis) and rates of cell attachment. These topographies were initially made in hard materials such as silica, silicon, titanium, epoxy-resins or the same materials coated with metals including Au, Pt [11],[12]. Though there were some differences in the degrees of reaction to a particular scale of topography, basically all eukaryote animal cells (except the erythrocyte) and some fungal cells [13] gave similar responses to the same topographies. Latterly softer materials were included in the test systems and pDMS polymers which have elastic constants in the range 1-10 Mpa gave similar results.

Table 1 summarizes a number of results on effects of topography, see also Figure 1. Many of the substrata tested had 90° edges in the topography [14] suggesting that the cells responded to sharp edges in the test substrata rather than to exact dimensions of the topography. That conclusion appears to be at variance with several published claims (eg [15] Loesberg et al 2007) which stated that particular dimensions or dimensional limits applied to particular cell types or even generally. We would rather make the statement that an essential requirement is a well-defined sharp edge with two planes intersecting at 90° or at angles >60° which is at least 1 micron long [16]. These authors looked at the interesting question of how long a sharp edge was required and concluded that 0.4 micron of edge was the minimum necessary for an individual epithelial cell to react to an edge.

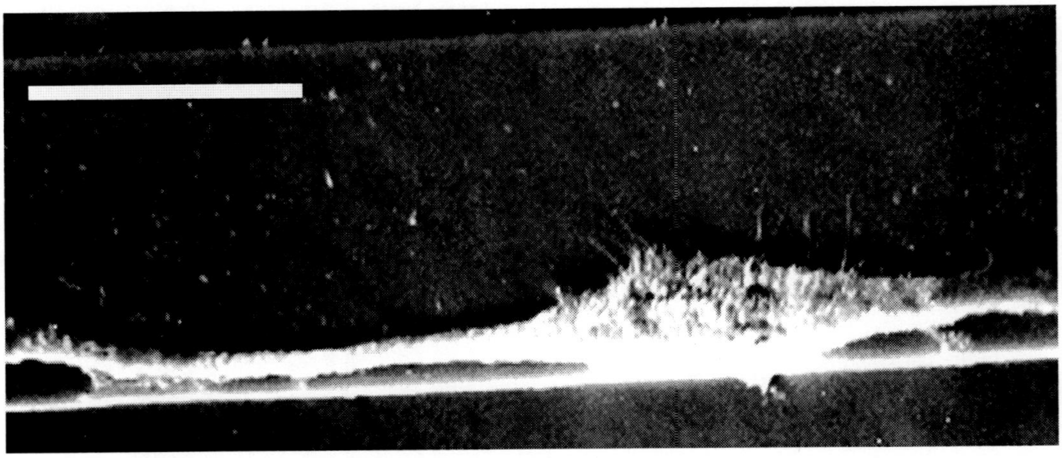

Figure 1. Perhaps the simplest type of topography to which a cell will react. A low cliff. Note alignment to cliff 1 micron high. A few microvilli can be seen on the side of the cell (BHK). Scale bar 36 micron. Courtesy of Dr P. Clark.

Towards the end of the 1990s electron beam lithography was well enough developed to make the relatively large areas of precise nanofeatured substrata which are necessary to test whether biological cells would react to topography on the nanoscale. This area should be large enough to allow several cells to spread out on the topography without having to occupy the same space or lie along or across the edges of the area. However, note that cells lying across edges may in themselves be an interesting type of reaction to topography.

Table 1 lists many of the effects of nanotopography that have been described in the literature. Results are not as clear cut as those found with microtopography, but in part this may be because Microfabrication tends to produce long continuous structures with ease while nanofabrication by electron beam methods or by colloidal lithographic methods often writes in ‘dots’ of exposure or uses colloidal spheres both of which are hard to make into continuous structures. The dots may be pits or domes or pillars of varying depth. Typically a dot might be 50-300nm in diameter and 5 to 50nm high or deep. In addition there are probably important questions about the minimum size of a feature which is likely to produce an obvious effect in a cell. Limitations of fabrication methods mean that at present really small features in the 10nm size range are still hard to make and even smaller ones exceedingly difficult.

To the surprise of many people, myself included initially, nano features appear to be more effective in altering features of cells than do micro-features. Reasons for this are discussed below. Examples are shown in Figure 2 and 3.

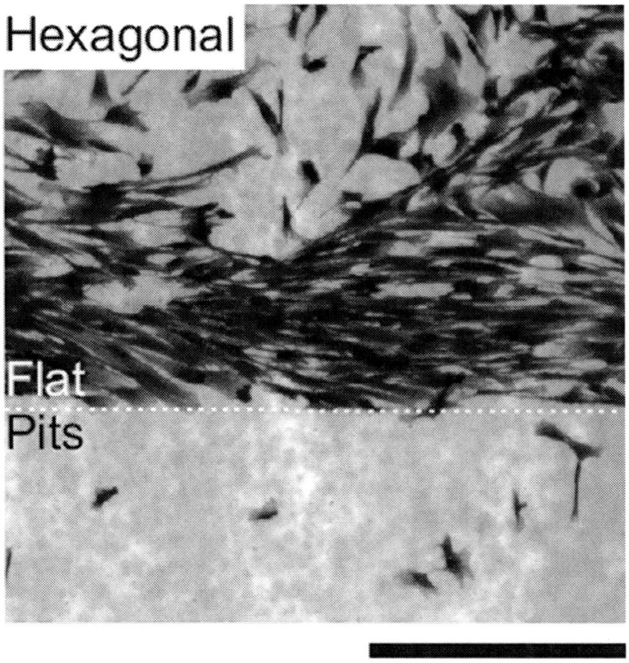

Figure 2. Cells reacting to a nanotopographic surface of hexagonally distributed pits 300 nm center to center, 120 nm diameter. Note the very sharp boundary and the alignment of cells on the flat (unfeatured) surface to that boundary. A few cells are found on the nanofeatured part of the surface. Scale bar 250 micron.

ADSORPTION OR DIRECT EFFECT OF TOPOGRAPHY

However, though it is obvious that micro- and nano-topographical features can have large effects on a range of cell activities it is perhaps questionable whether the effect is direct or otherwise. One suggestion is that the cells are responding to differences in protein etc adsorption to the nanosurface either in amount or kind and that this in turn affects cell adhesion etc. It would be expected that the various surfaces used silica, polymer, metals and the various nanoparticles used in colloidal lithography would bind macromolecules from the medium but if adsorption is important it is surprising that the effects seem to be specific to substrate shapes and dimension rather than to substrate bulk chemistry or to the medium which is being adsorbed. To be specific in the latter case different culture media containing rather different amounts and types of proteins were used in the many experiments listed in Table 1. This suggests that protein adsorption may not be an over-riding factor in determining cell reactions. Certainly more work should be done on nanosurfaces to test their adsorption properties and perhaps the conformation of any adsorbed proteins etc as well. Recently [17] Raffa et al (2008) have found that focused gallium ion beam fabrication of glass could give enhanced adsorption of fluorescently labeled albumin but whether this was related to pattern or not is not clear. [18] Bhushan et al (2005) wrote lines at spacings of 400 nm and depths of 15-20nm and widths close to 45nm and found increased binding of biotin revealed by streptavidin secondary binding but this is not good evidence for specific adsorption to topography since the binding was irregular. Furthermore if you consider experiments with nano- and microgrooves it would be expected that adsorption on the flat parts of these structures, ie. groove bottoms and tops would be identical for all dimensions for a given material and that any special adsorption on the edges would be similarly identical for all surfaces on a given material. Therefore it would be expected that a single cell type contacting a set of grooves of differing dimensions would show exactly the same effects on different dimensions. But this is not so.

[7] Clark et al (1990) found that there is an optimal groove depth for reaction to grooves. These groove depths were 2 micron deep for BHK cells. [9] Curtis showed evidence that the cells were only attached to the substratum at the top edges of the grooves and did not extend down the side walls. Thus the cells may be responding to a wall property which is not cellular. The work of [62] suggests that on flat surfaces this may be the mechanical properties of the substratum because they found that cells tended to move from substrates of low stiffness up the gradient to stiffer substrata. Perhaps this explains how deep grooves may affect cell reactions even though cells do not reach to the deeper parts of the walls.

These observations immediately suggest that mechanical sensing of the substratum might be an essential cell interaction with the surface and near surface. The cell exerts mechanical forces on the substratum [63-66] which deforms to an extent controlled by its physical properties as well as the force applied.

When we turn to consideration of the mechanisms that may be involved one set of mechanisms seems especially relevant. This is that mechanotransduction is involved.
The reasons for choosing this system are that:

1. Externally applied mechanical forces produce many of the same results as growing cells on nanostructures, for instance cell alignment and shape change, changes in gene expression.
2. Cells growing on planar and on microstructured surfaces exert mechanical forces, [67,68]
3. Both in vivo and in vitro many types of cells can strain their environment mechanically.
4. Because action and reaction are equal and opposite the stressed substratum acts back upon the cell.

This view could be followed as far as to suggest that micro- and nanotopography are only concomitants of the basic mechanotransduction effects. If so, the most likely explanation of their effects is that topography concentrates (or diffuses) mechanical stresses.

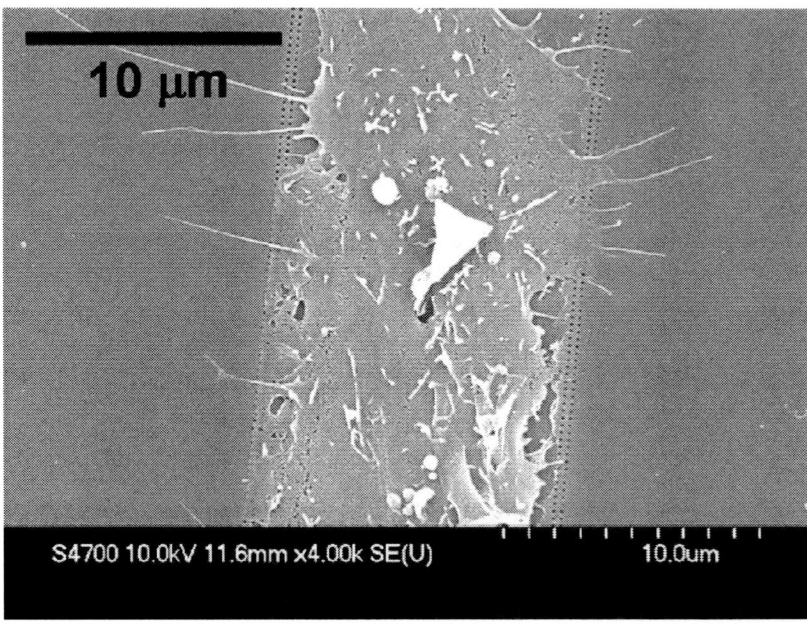

Figure 3. The confining barrier provided by double rows of nanopits on an otherwise chemically uniform flat surface. Note that microvill are not confined. SEM courtesy of Dr. K.Seunarine.

DIRECT EFFECTS

Many surfaces bearing nanotopography show the phenomena of superhydrophobicity or alternatively superhydrophilicity for water as a contact fluid. Since contact angles to surfaces of cells in an aqueous environment are good predictors of adhesive behavior it seems likely that nanotopography might determine adhesion by this direct route. [1,67-71] have examined this to check whether the superhydrophobicity of pillar and pits surfaces could be due to air trapping. They found that contact angles of water on 216nm wide 100 nm deep nanopits in polycaprolactone showed Cassie-Baxter behavior which is consistent with air trapping. These surfaces which do not allow appreciable cell attachment are superhydrophobic but the extent and persistence of air entrapment depends very considerably on how the water was 'added' to the nanofeatured surface, Martines (personal communication 2008) [20,21].

However, there are the interesting observations that regular arrays of nanopits and nanopillars tend to have low adhesion for cells. Random ones have adhesion not different from that of flat surfaces and orthogonally or hexagonally ordered arrays have adhesion notably lower while the 'near-orthogonal' arrays have higher adhesion, see [20] Curtis et al 2001, [21] Curtis et al 2004,. In the particular near-orthogonal arrays studied by [30] the pit centers in the program to produce orthogonal arrays were displaced randomly by 10 or 20 nm. Obviously many other array degradation procedures could be tried. It is interesting to consider the events that may be happening when 'degraded arrays' are used. They have a resemblance to the finding that edges, both of nanoarrays and of micrometric structures, appear to be preferred site for adhesion. All these phenomena suggest that a comparator system is used. Namely, the cell can compare reactions on two (perhaps more) differing surfaces when it happens to lie across a boundary, this is discussed later.

Though other mechanisms for direct effects can be considered there does not seem to be good experimental evidence for effects in which micro- and nanotopography are likely to modify adhesion and almost no literature on such matters.

Effects of Substratum Dimension and Pattern. Regularity and Irregularity

The remarkable sensitivity of cells to the dimensions and regularity of the features of the substrate on which they are growing or moving is such that feature spacings (x,y dimensions) or z-axis heights down to a few nm in some cases are detected. Cavalcanti-Adam et al [29,72-76] found sharp cut-offs in cell adhesion once adhesion site (RGD) patches rose above about 60 nm in patch to patch distance. Rather similar findings came from our own work with nanometric pillars or pits. In the first case the easy interpretation is that it is a chemical spacing though the RGD patches will have physical height. In our own work there are strong arguments that it is simply physical dimension. The reason for that conclusion is that much the same effect is found with the same topography on a variety of chemically different surfaces for a given cell type.

Cavalcanti–Adam et al [73,75,76] had prepared surfaces with gold nanodots bearing cyclic RGD peptides at various average spacings and found that stable attachment of rat fibroblast cells was high when the average spacing was 58nm but poor with a spacing of 168 nm. She and her colleagues interpreted this as being due to a need to match the RGD dot spacings to the integrin spacings on the cells, But it could also be that the dot arrays which were ordered but not of exact geometry closely resembled the less chemically specific arrays used by Biggs et al [30] (2007) which had high adhesion for cells. Unfortunately the ingenious gold nanodot technology with its RGD signal regions is yet to be matched with the exact e-beam techniques.

Relevant to these considerations are the fairly numerous suggestions in the literature that there are patterns either in the plasmalemmal lipids (Halegan et al 2004 [77]) or in the plasmalemmal proteins (especially focal contacts, see [76] Cavalcanti et al 2007) which control the reactions of cells to patterned substrata. Plausible though such concepts may seem they run into the objection that the fluidity of the membrane would tend to destroy this structure and that the elasticity of the membrane would allow appreciable distortions to take place.

A precise orthogonal array of cell surface features could mean that all positions of stress concentration in the substratum are close, symmetrical and tend to cancel each other out so that there is little mechanical reason to concentrate actin polymerization. This correlates with observation. With somewhat degenerate arrays stress concentration is more likely to take place [19] at discontinuities and thus stimulate focal contacts. It also accords well with the observations that stress concentrations lead to focal adhesion development [78] Pathak et al 2007, [78] Deshpande et al and [79] Nicolas and Safran 2006). Despite such concepts the fact that some nanotopographical patterns seem to inhibit or deter adhesion should not be overlooked.

Time Scales of Adhesive Events

One topic that has been given very little attention is the time scale for the formation of an adhesion. Probably four stages should be envisaged. 1. The initial capture event after which the surfaces have approached at least to at a point contact as close as they can and the contact has a lifetime sufficient to have a good probability of surviving detachment processes such as Brownian motion for a reasonable time period. So that the next process of cell spreading can start which enhances the stability of the adhesion. This initial process can be estimated from the work of [80] Pierres et al 2008 to have a lifetime in the 10s to 100s of milliseconds, far shorter than can be detected in many systems for measuring cell adhesion. 2. The cell-cell or cell-substratum adhesion spreads in extent and other limitations may or may not come into play. The first important event is the removal of medium from between the approaching surfaces and this drainage can slow down the rate of approach of one surface to another and thus the rate of spreading of the surfaces.

The density of adhesive bond components such as fibronectin and integrins and possibly their lateral mobility is the next important factor and if these are insufficient spreading will be inadequate. At some time early in this stage mechanical interactions will begin as force is exerted on the cell or substratum with which adhesion is being made. The process has no end-point that can be determined with precision because changes in cell spreading, ie shape change and movement are likely to continue for very long periods.

Dynamic Comparisons

Nearly all measurements of cell adhesion are run as end-point measurements it being assumed that the process has completed or is close to completion. Events that may be missed if dynamic measurements are not made are changes in the rate or extent of adhesion. Over very short time periods at the start of the process the cells are approaching the surface. The studies of cell adhesion under Couette (laminar) flow collisions (Curtis [81,82]) suggests that the interaction succeeds or fails in at most 100 milliseconds and may well show changes within this period. Later on as the contact zone spreads changes may be occurring over a time scale of minutes. The fact that the collision efficiency of the cell-cell interactions drops late in aggregation suggests that adhesive interactions are reducing in intensity.

When a cell adheres near a transition between the nanotopography of the surface the cell is likely to try to spread dynamically from one surface towards the other. Often the transition zone seems to be a barrier. An example of this is shown in Figure 2 where double lines of nanopits seem to prevent the cell spreading across them even though the flat substratum on either side of the double tows is identical. Note that the microvilli are not prevented from extending across the boundary even though the main body of the cell is.

Rigidity of the Substratum and Mechanotransduction

If the cell interacts mechanically with the substratum as it must do in order to spread or to move, the cell will be applying stresses to the substratum. Lo et al [62] found that cells tended to move from regions of lower substratum stiffness to those which were stiffer, A simple interpretation of this is that the cell can gain more traction on the stiffer material. Obviously on a very soft or compliant material the cell cannot achieve enough traction to move or perhaps even to spread. Thus adhesions may fail to spread if the mechanical properties of the substratum are unsuitable.

Subsequent to the pioneering work of Lo on substratum rigidity work has been done which micro- or nano topography has been changed as well as rigidity. For example [83] showed that substrate stiffness and micrometric features (lines or pillars) affected cell morphology and motility for 3T3 cells but a tumorous fibroblast was only affected by the topography.

Dr .M. Riehle suggested to me that these effects of rigidity might be such that when a cell was adhering to a low rigidity material the presence of a more rigid material just beneath parts of the surface of the weaker material would lead ro a pattern of rigidity which the cell could sense. An example of this is shown in Figure 4. The sensing mechanism would be likely to be due to mechanical forces exerted on the substratum by the cell, see Figure 5. In turn such effects could reflect back on the organization of the cell's internal components, see Figure 6.

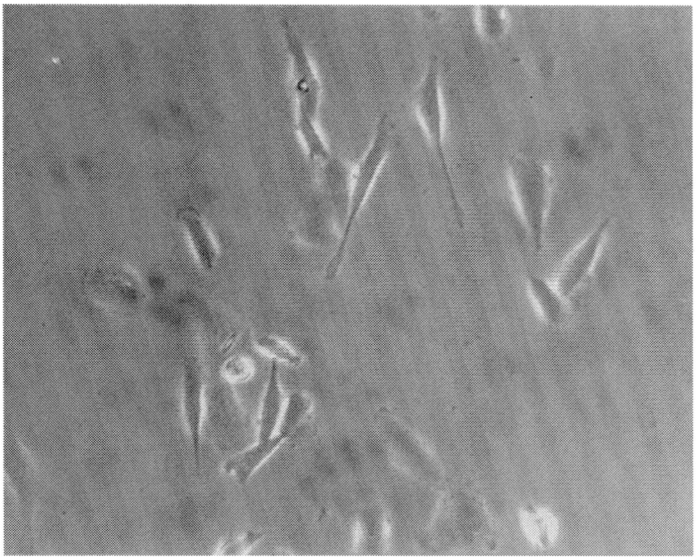

Figure 4. Cells can react to nanotopography buried under a thin (6 micron) layer of gelatin. Grooves/ ridges 10 micron wide, 150 nm deep/high. Phase contrast picture. Note fairly good alignment of cells to the buried silica topography.

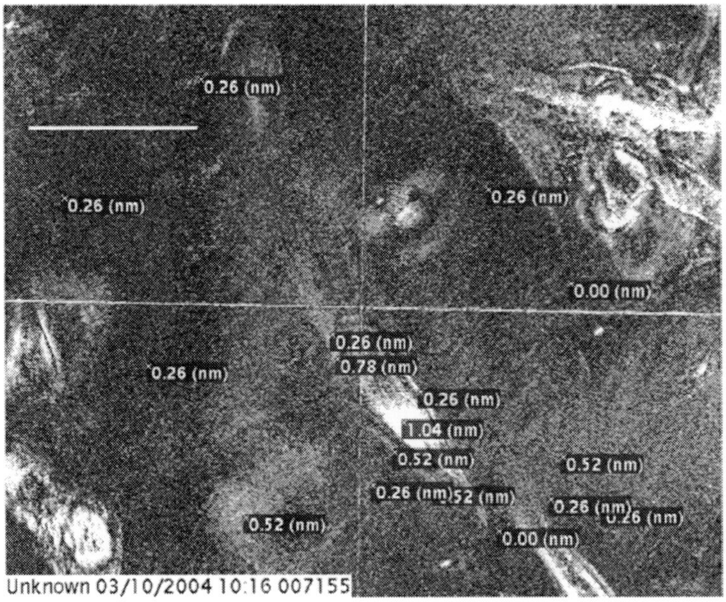

Figure 5. Cell mechanical forces visualised by birefringence of polydimethylsiloxane (pDMS) on which the cells are growing. Retardance image with retardances stamped at various positions. Note varying shades of grey between the cells indicating strained substratum. Cells with their many birefringent components are bright in this type of image. Scale bar 40 micron.

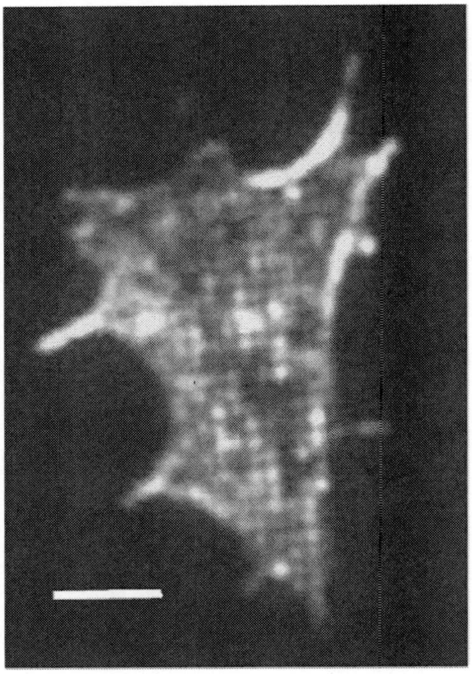

Figure 6. A platelet reacting to contact with a polycarbonate surface by arranging its actin (bright), as seen by fluorescence after staining with Texas Red phalloidin, to reproduce the underlying pattern of nanopits. Scale bar 1 micron. Courtesy of Dr N.Gadegaard.

Genomic and Phenotypic Effects

One of the most impressive results from the exposure of cells to nanotopographically patterned surfaces has been the finding that gene expression can be extensively changed in cells grown on such surfaces. Early experiments were confined to one or two proteins, eg. fibronectin [84] detected with Western blotting but even these first attempts suggested that topography might affect gene expression. But with the advent of large genomic arrays (eg. [85]) it became clear that many up and down changes in expression took place. These seem to particularly involve cell signalling pathways and especially the ERK pathway and perhaps the Wnt pathway seem to be important.. At the moment we cannot go much beyond speculation as to the mechanisms of this. Since rather similar studies on the effects of Mechanical stimulation have revealed changes in gene expression it is perhaps natural to start with a mechanical approach.

One line of thought has considered how mechanical stimulation might open up the genome to increased transcriptional events. Early work by [86] suggested and provided some experimental evidence that stretching the nucleus could alter the positions of the chromosomes. Since the cells are stretched on some topographies probably due to their own contractile activities or stretched by mechanical interactions with other cells (Curtis et al 2007 [68]) there could be changes in chromosomal packing and territories. This might produce an opening up the genome to transcription. This appears to have been found by [87]

Viewed very simply telomeric ends of chromosomes might be expected to be opened up for transcription while centromeric ends might be compacted so that expressions might be shifted towards the telomeric ends of the chromosomes [88]. This type of approach needs much further validation.

RHYTHMIC EVENTS

Rhythmic mechanical stimulations at about 1 Hz appears to stimulate general processes such as the cell cycle and a variety of protein syntheses [86,89,90,91]. Possibly the need for rhythmicity is due to 'pumping' characteristics needed to move nucleotides into and oligonucleotides out of the chromosomal complexes. Lamins might be expected to form part of the mechanical chain because they connect the chromosomes to the nuclear envelope which in turn is linked to cytoskeletal and even external events. The advantages of rhythmic stimulation may represent nothing more than the relaxation and creep events that may occur under continuous stimulation leading to decreases in the forces being applied and the resulting structural effects. It should also be noted that often rhythymic events imply a succession of transient simulations of the cell.

CONCLUSION

In this review paper I have attempted to examine three factors that are probably of great importance in cell adhesion but which have only emerged into the awareness of those working in the field recently. These are 1. The roles of surface topography at both micro- and nano-scale, 2 the importance of mechanical forces, both those derived from the cells themselves and from external forces and 3 the importance of the near-surface physical properties of the substratum. Any one of these factors can over-ride all other effects on cell adhesion such as surface chemistry.

ACKNOWLEDGMENTS

I should like to thank my colleagues Dr Matt Dalby, Dr Nikolaj Gadegaard and Dr Kris Seunarine and my former colleague Dr Peter Clark for the substantial assistance they gave me in the preparation of this article.

REFERENCES

[1] van Oss CJ, Gillman CF, Neumann AW. *Phagocytic engulfment and cell adhesiveness*. New York: Dekker 1975.

[2] Curtis ASG. The mechanism of adhesion of cells to glass. *JCell Biol*. 1964;20:199-215.

[3] Bearer CF. L1 cell adhesion molecule signal cascades: Targets for ethanol developmental neurotoxicity. *Neurotoxicology*. 2001;22(5):625-33.

[4] Issekutz AC, Burns JA, Yagita H, Issekutz TB. The VLA-4 and VLA-5 integrins mediate beta2 (CD11/CD18) integrin independent neutrophil recruitment to endotoxin lung inflammation. *Faseb Journal*. 2001;15(4):A57-A.

[5] van Wetering S, van den Berk N, van Buul JD, Mul FPJ, Lommerse I, Mous R, et al. VCAM-1-mediated Rac signaling controls endothelial cell-cell contacts and leukocyte transmigration. *American J. Physiology-Cell Physiology*. 2003;285(2):C343-C52.

[6] Clark P, Connolly P, Curtis ASG, Dow JAT, Wilkinson CDW. Topographical control of cell behaviour. I. Simple step cues. *Development*. 1987;99:439-48.

[7] Clark P, Connolly P, Curtis ASG, Dow JAT, Wilkinson CDW. Topographical control of cell behaviour II. Multiple grooved substrata. *Development*. 1990;108:635-44.

[8] den-Braber ET, Deruijter JE, Smits H, Ginsel LA, Vonrecum AF, Jansen JA. Effect of parallel surface microgrooves and surface-energy on cell- growth. *J.Biomedical Materials Research*. 1995;29:511-8.

[9] Curtis ASG, Wilkinson CDW. New depths in cell behaviour : the reactions of cells to nanotopography. In: Lacke JM, G.A.Dunn & G.E.Jones, ed. *Cell behaviour: control and Mechanism of motility*. London: Biochemical Society 1999:15-26.

[10] Curtis A, Wilkinson C, Wojciak-Stothard B. Accelerating cell movement. *JCellular Eng*. 1995;1:35-8.

[11] Greve F, Frerker S, Bittermann AG, Burkhardt C, Hierlemann A, Hall H. Molecular design and characterization of the neuron-microelectrode array interface. *Biomaterials*. 2007;28:5246-58.

[12] Bjornholm T, Greve DR, Reitzel N, Hassenkam T, Kjaer K, Howes PB, et al. Self-assembly of regular, amphiphilic polythiophenes into highly ordered pi-stacked conjugated polymer thin films and nanocircuits. *J. American Chemical Society*. 1998;120(30):7643-4.

[13] Perera T, Gregory DW, Marshall D, Gow N. Contact-sensing by hyphae of dermatophytic and saprophytic fungi. *J. Medical & Veterinary Mycology*. 1997;35:289-93.

[14] Curtis ASG, Clark P. The effects of topographic and mechanical properties of materials on cell behaviour. *Critical Reviews in Biocompatibility*. 1990;5:343-62.

[15] Loesberg WA, te Riet J, van Delft F, Schon P, Figdor CG, Speller S, et al. The threshold at which substrate nanogroove dimensions may influence fibroblast alignment and adhesion. *Biomaterials*. 2007;28(27):3944-51.

[16] Andersson AS, Olsson P, Lidberg U, Sutherland D. The effects of continuous and discontinuous groove edges on cell shape and alignment. *Experimental Cell Research*. 2003 Aug 1;288(1):177-88.

[17] Raffa V, Vittorio O, Pensabene V, Menciassi A, Dario P. FIB-nanostructured surfaces and investigation of bio/nonbio interactions at the nanoscale. *Ieee Transactions on Nanobioscience*. 2008 Mar;7(1):1-10.

[18] Bhushan B, Tokachichu DR, Keener MT, Lee SC. Morphology and adhesion of biomolecules on silicon based surfaces. *Acta Biomaterialia*. 2005 May;1(3):327-41.

[19] Casey BG, Monaghan W, Wilkinson CDW. Embossing of nanoscale features and environments. *Microelectronic Engineering*. 1997;35(1-4):393-6.

[20] Curtis ASG, Casey B, Gallagher JD, Pasqui D, Wood.M.A.W., Wilkinson CDW. Substratum nanotopography and the adhesion of biological cells. Are symmetry or reg.ularity of nanotopography important? *Biophysical Chemistry*. 2001;94:275-83.

[21] Curtis ASG, Gadeg.aard N, Dalby MJ, Riehle MO, Wilkinson CDW, Aitchison G. Cells react to nanoscale order and symmetry in their surroundings. *Ieee Transactions on Nanobioscience*. 2004 Mar;3(1):61-5.

[22] Dalby MJ, Childs S, Riehle MO, Johnstone HJH, Affrossman S, Curtis ASG. Fibroblast reaction to island topography: changes in cytoskeleton and morphology with time. *Biomaterials*. 2003;24(6):927-35.

[23] Dalby MJ, Giannaras D, Riehle MO, Gadegaard N, Affrossman S, Curtis ASG. Rapid fibroblast adhesion to 27 nm high polymer demixed nano- topography. *Biomaterials*. 2004 Jan;25(1):77-83.

[24] Dalby MJ, Riehle MO, Sutherland DS, Agheli H, Curtis ASG. Fibroblast response to a controlled nanoenvironment produced by colloidal lithography. *Journal of Biomedical Materials Research Part A*. 2004 May 1;69A(2):314-22.

[25] Cyster LA, Parker KG, Parker TL, Grant DM. The effect of surface chemistry and nanotopography of titanium nitride (TiN) films on primary hippocampal neurones. *Biomaterials*. 2004 Jan;25(1):97-107.

[26] Lenardi C, Perego C, Cassina V, Podesta A, D'Amico A, Gualandris D, et al. Adhesion and proliferation of fibroblasts on cluster-assembled nanostructured carbon films: The role of surface morphology. *Journal of Nanoscience and Nanotechnology*. 2006;6(12):3718-30.

[27] Meredith DO, Eschbach L, Riehle MO, Curtis ASG, Richards RG. Microtopography of metal surfaces influence fibroblast growth by modifying cell shape, cytoskeleton, and adhesion. *Journal of Orthopaedic Research*. 2007 Nov;25(11):1523-33.

[28] Meredith DO, Eschbach L, Wood MA, Riehle MO, Curtis ASG, Richards RG. Human fibroblast reactions to standard and electropolished titanium and Ti-6Al-7Nb, and electropolished stainless steel. *Journal of Biomedical Materials Research Part A*. 2005 Dec;75A(3):541-55.

[29] Cavalcanti-Adam EA, Bezler M, Tomakidi P, Spatz JP. Integrin ligands arranged in nanopatterns modulate focal adhesion assembly and signaling in osteoblasts. *Journal of Bone and Mineral Research*. 2004;19:S64-S.

[30] Biggs MJP, Richards RG, Gadegaard N, Wilkinson CDW, Dalby MJ. The effects of nanoscale pits on primary human osteoblast adhesion formation and cellular spreading. *Journal of Materials Science-Materials in Medicine*. 2007;18(2):399-404.

[31] Pattison MA, Webster TJ, Haberstroh KM. Select bladder smooth muscle cell functions were enhanced on three-dimensional, nano-structured poly(ether urethane) scaffolds. *Journal of Biomaterials Science-Polymer Edition*. 2006;17(11):1317-32.

[32] Wojciak-Stothard B, Curtis A, Monaghan W, Macdonald K, Wilkinson C. Guidance and activation of murine macrophages by nanometric scale topography. *Experimental Cell Research*. 1996;223:426-35.

[33] Wojciak-Stothard B, Curtis ASG, Monaghan W, McGrath M, Sommer I, Wilkinson CDW. Role of the cytoskeleton in the reaction of fibroblasts to multiple grooved substrata. *Cell Motility and the Cytoskeleton*. 1995;31:147-58.

[34] Hsu SH, Tang CM, Lin CC. Biocompatibility of poly(epsilon-caprolactone)/poly(ethylene glycol) diblock copolymers with nanophase separation. *Biomaterials*. 2004 Nov;25(25):5593-601.

[35] Yim EKF, Reano RM, Pang SW, Yee AF, Chen CS, Leong KW. Nanopattern-induced changes in morphology and motility of smooth muscle cells. *Biomaterials*. 2005 Sep;26(26):5405-13.

[36] Andersson AS, Backhed F, von Euler A, Richter-Dahlfors A, Sutherland D, Kasemo B. Nanoscale features influence epithelial cell morphology and cytokine production. *Biomaterials*. 2003 Sep;24(20):3427-36.

[37] Hu W, Yim EKF, Reano RM, Leong KW, Pang SW. Effects of nanoimprinted patterns in tissue-culture polystyrene on cell behavior. *Journal of Vacuum Science & Technology B*. 2005 Nov-Dec;23(6):2984-9.

[38] Johansson F, Carlberg P, Danielsen N, Montelius L, Kanje M. Axonal outgrowth on nano-imprinted patterns. *Biomaterials*. 2006 Mar;27(8):1251-8.

[39] Bettinger CJ, Orrick B, Misra A, Langer R, Borenstein JT. Micro fabrication of poly (glycerol-sebacate) for contact guidance applications. *Biomaterials*. 2006 Apr;27(12):2558-65.

[40] Zhu BS, Zhang QQ, Lu QH, Xu YH, Yin J, Hu J, et al. Nanotopographical guidance of C6 glioma cell alignment and oriented growth. *Biomaterials*. 2004 Aug;25(18):4215-23.

[41] Dalby MJ, Riehle MO, Johnstone H, Affrossman S, Curtis ASG. In vitro reaction of endothelial cells to polymer demixed nanotopography. *Biomaterials*. 2002;23:2945-54.

[42] Berry CC, Charles S, Wells S, Dalby MJ, Curtis ASG. The influence of transferrin stabilised magnetic nanoparticles on human dermal fibroblasts in culture. *International Journal of Pharmaceutics*. 2004 Jan 9;269(1):211-25.

[43] Dalby MJ, Giannaras D, Riehle MO, Gadegaard Nea. Rapid fibroblast adhesion to 27 nm high polymer demixed nano-topography. *Biomaterials*. 2004;25:77-83.

[44] Dalby MJ, Riehle MO, Johnstone HJH, Affrossman S, Curtis ASG. Polymer-Demixed Nanotopography: Control of Fibroblast Spreading and Proliferation. *Tissue Engineering*. 2002;8(6):1099-107.

[45] Liliensiek SJ, Campbell S, Nealey PF, Murphy CJ. The scale of substratum topographic features modulates proliferation of corneal epithelial cells and corneal fibroblasts. *Journal of Biomedical Materials Research Part A*. 2006 Oct;79A(1):185-92.

[46] Dalby MJ, Berry CC, Riehle MO, Sutherland DS, Agheli H, Curtis ASG. Attempted endocytosis of nano-environment produced by colloidal lithography by human fibroblasts. *Experimental Cell Research*. 2004 May 1;295(2):387-94.

[47] Lim JY, Hansen JC, Siedlecki CA, Hengstebeck RW, Cheng J, Winograd N, et al. Osteob;ast adhesion on poly(L-lactic acid)/polystyrene demixed thin film blends. Effect of nanotopography, surface chemistry, and wettability. *Biomacromolecules*. 2005;6:3319-27.

[48] Lim JY, Hansen JC, Siedlecki CA, Runt J, Donahue HJ. Human foetal osteoblastic cell response to polymer-demixed nanotopographic interfaces. *Journal of the Royal Society Interface*. 2005 Mar;2(2):97-108.

[49] Dalby MJ, McCloy D, Robertson M, Wilkinson CDW, Oreffo ROC. Osteoprogenitor response to defined topographies with nanoscale depths. *Biomaterials*. 2006;27(8):1306-15.

[50] Dalby MJ, Gadegaard N, Tare R, Andar A, Riehle MO, Herzyk P, et al. The control of human mesenchymal cell differentiation using nanoscale symmetry and disorder. *Nature Materials*. 2007;6:997-1003.

[51] Hart A, Gadegaard N, Wilkinson CDW, Oreffo ROC, Dalby MJ. Osteoprogenitor response to low-adhesion nanotopographies originally fabricated by electron beam lithography. *Journal of Materials Science-Materials in Medicine*. 2007;18(6):1211-8.

[52] de Oliveira PT, Nanci A. Nanotexturing of titanium-based surfaces upregulates expression of bone sialoprotein and osteopontin by cultured osteogenic cells. *Biomaterials*. 2004 Feb;25(3):403-13.

[53] de Oliveira PT, Zalzal SF, Beloti MM, Rosa AL, Nanci A. Enhancement of in vitro osteogenesis on titanium by chemically produced nanotopography. *Journal of Biomedical Materials Research Part A*. 2007 Mar;80A(3):554-64.

[54] Osvaldo H, Fo S, Novaes AB, de Castro LMS, Rosa AL, de Oliveira PT. In vitro osteogenesis on a microstructured titanium surface with additional submicron-scale topography. *Clinical Oral Implants Research*. 2007 Jun;18(3):333-44.

[55] Choi CH, Hagvall SH, Wu BM, Dunn JCY, Beygui RE, Kim CJ. Cell interaction with three-dimensional sharp-tip nanotopography. *Biomaterials*. 2007 Mar;28(9):1672-9.

[56] Li WJ, Jiang YJ, Tuan RS. Chondrocyte phenotype in engineered fibrous matrix is regulated by fiber size. *Tissue Engineering*. 2006 Jul;12(7):1775-85.

[57] Schindler M, Ahmed I, Kamal J, Nur-E-Kamal A, Grafe TH, Chung HY, et al. A synthetic nanofibrillar matrix promotes in vivo-like organization and morphogenesis for cells in culture. *Biomaterials*. 2005 Oct;26(28):5624-31.

[58] Curtis ASG, Wilkinson CDW. Nanotechniques and approaches in biotechnology. *Trends in Biotechnology*. 2001;19:97-101.

[59] Kriparamanan R, Aswath P, Zhou A, Tang LP, Nguyen KT. Nanotopography: Cellular responses to nanostructured materials'. *J Nanoscience Nanotechnology*. 2006;6:1905-19.

[60] Schindler M, Nur-E-Kamal A, Ahmed I, Kamal J, Liu HY, Amor N, et al. Living in three dimensions - 3D nanostructured environments for cell culture and regenerative medicine. *Cell Biochemistry and Biophysics*. 2006;45(2):215-27.

[61] Brodie C. Sensitive cells - Stem cells take their cue from their physical surroundings. *American Scientist*. 2007 Mar-Apr;95(2):123-.

[62] Lo CM, Wang HB, Dembo M, Wang YL. Cell movement is guided by the rigidity of the substrate. *Biophys J*. 2000;79(1):144-52.

[63] Balaban NQ, Schwarz US, Riveline D, Goichberg P, Tzur G, Sabanay U, et al. Force and focal adhesion assembly: a close relationship studied using elastic micropatterned substrates. *Nature Cell Biology*. 2001;3:466-72.

[64] Bershadsky AD, Balaban NQ, Geiger B. Adhesion-dependent cell mechanosensitivity. *Annual Review of Cell and Developmental Biology*. 2003;19:677-95.

[65] Riveline D, Zamir E, Balaban NQ, Kam Z, Geiger B, Bershadsky AD. Focal contact as mechanosensor: Directional growth in response to local strain. *Molecular Biology of the Cell*. 1999 Nov;10:1971.

[66] Schwarz US, Balaban NQ, Riveline D, Addadi L, Bershadsky A, Safran SA, et al. Measurement of cellular forces at focal adhesions using elastic micro-patterned substrates. *Materials Science & Engineering C-Biomimetic and Supramolecular Systems*. 2003 Mar 3;23(3):387-94.

[67] Curtis A. Cell forces in tissues. *MedEngPhysics*. 2005;27:773-9.

[68] Curtis A, Sikolikova-Csaderova L, Aitchison G. Measuring cell forces by a photoelastic method. *Biophys J*. 2007;92:2255-61.

[69] Gadegaard N, Martines E, Riehle MO, Seunarine K, Wilkinson CDW. Applications of nano-patterning to tissue engineering. *Microelectronic Engineering*. 2006 Apr-Sep;83(4-9):1577-81.

[70] Martines E, Seunarine K, Morgan H, Gadegaard N, Wilinson CDW, Riehle MO. Superhydrophobicity and superhydrophilicity of regular nanopatterns. *Nanoletters*. 2005;5:2097-103.

[71] Martines E, Seunarine K, Morgan H, Gadegaard N, Wilkinson CDW, Riehle MO. Air-trapping on biocompatible nanopatterns. *Langmuir*. 2006 Dec;22(26):11230-3.

[72] Arnold M, Cavalcanti-Adam EA, Glass R, Blummel J, Eck W, Kantlehner M, et al. Activation of integrin function by nanopatterned adhesive interfaces. *Chemphyschem*. 2004;5(3):383-8.

[73] Cavalcanti-Adam EA, Micoulet A, Blummel J, Auernheimer J, Kessler H, Spatz JP. Lateral spacing of integrin ligands influences cell spreading and focal adhesion assembly. *European J.Cell Biology*. 2006;85(3-4):219-24.

[74] Cavalcanti-Adam EA, Satz J. Regulation of adhesion dynamics, motility and survival by spacing of extracellular matrix peptides. *J. Bone and Mineral Research*. 2006;21:S134-S.

[75] Cavalcanti-Adam EA, Shapiro IM, Composto RJ, Macarak EJ, Adams CS. RGD peptides immobilized on a mechanically deformable surface promote osteoblast differentiation. *J. Bone & Mineral Research*. 2002;17(12):2130-40.

[76] Cavalcanti-Adam EA, Volberg T, Micoulet A, Kessler H, Geiger B, Spatz JP. Cell spreading and focal adhesion dynamics are regulated by spacing of integrin ligands. *Biophysical Journal*. 2007;92(8):2964-74.

[77] Halegan A, SenguptamK., Kahn S, Sackmann E, Discher D. Topograpical pattern dynamics in passive adhesin of cell membranes. *BiophysJ*. 2004;87:3547-60.

[78] Pathak A, Deshpande VS, McMeeking RM, Evans AG. The simulation of stress fibre and focal adhesion development in cells on patterned substrates. *Journal of the Royal Society Interface*. 2008;5(22):507-24.

[79] Nicolas A, Geiger.B., Safran.S.A. Cell mechanosensitivity controls the anisotropy of focal adhesions. *PNAS*. 2004;101:12520-5.

[80] Pierres A, Benoliel AM, Touchard D, Bongrand P. How cells tiptoe on adhesive surfaces before sticking. *Biophysical Journal*. 2008;94(10):4114-22.

[81] Curtis ASG. The measurement of cell adhesiveness by an absolute method. *JEmbryol exp Morph*. 1969;22:305-25.

[82] Curtis ASG, Hocking LM. Collision efficiency of equal spherical particles in a shear flow: The influence of London-Van der Waals Forces Trans of Faraday Soc 1970;57066Pt. 6:1381-90.

[83] Tzvetkova-Chevolleau T, Stephanou A, Fuard D, Ohayon J, Schiavone P, Tracqui P. The motility of normal and cancer cells in response to the combined influence of the substrate rigidity and anisotropic microstructure. *Biomaterials*. 2008;29(10):1541-51.

[84] Chou L, Firth JD, Uitto V-J, Brunette DMD. Substratum surface topography alters cell shape and regulates fibronectin mRNA level, mRNA stability, secretion and assembly in human fibroblasts. *Journal of Cell Science*. 1995;108(4):1563-73.

[85] Dalby MJ, Hart A, Yarwood SJ. The effect of the RACK1 signalling protein on the regulation of cell adhesion and cell contact guidance on nanometric grooves. *Biomaterials*. 2008 Jan;29(3):282-9.

[86] Curtis ASG. Mechanical tensing of cells and chromosome arrangement. Biomechanics and cells (*Soc for Exp Biol Seminar Series 54*). 1994;54:121-30.

[87] Dalby MJ, Gadegaard N, Herzyk P, Sutherland D, Agheli H, Wilkinson CDW, et al. Nanomechanotransduction and interphase nuclear organization influence on genomic control. *Journal of Cellular Biochemistry*. 2007 Dec;102(5):1234-44.

[88] Dalby MJ, Gadegaard N, Herzyk P, Agheli H, Sutherland DS, Wilkinson CDW. Group analysis of regulation of fibroblast genome on low-adhesion nanostructures. *Biomaterials*. 2007;28(10):1761-9.

[89] Berry CC, Cacou C, Lee DA, Bader DL, Shelton JC. Dermal fibroblasts respond to mechanical conditioning in a strain profile dependent manner. *Biorheology*. 2003;40(1-3):337-45.

[90] Carvalho RS, Scott JE, Yen EHK. The effects of mechanical stimulation on the distribution of beta-1 integ.in and expression of beta-1-integrin mRNA in TE-85 human osteosarcoma cells. *Archives of Oral Biology*. 1995;40(3):257-64.

[91] Leung DY, Glagov S, Mathews MB. Cyclic stretching stimulates synthesis of matrix components by arterial smooth muscle cells in vitro. *Science*. 1976;191(4226):475-7.

In: New Cell Adhesion Research
Editors: Patrick Nott and Matthew Temple
ISBN 978-1-60692-378-8

Chapter VI

CADM1: A New Mast-Cell Adhesion Molecule That Mediates Interaction with Fibroblasts, Nerves and Smooth Muscles

Akihiko Ito* and Man Hagiyama
Division of Molecular Pathology, Department of Cancer Biology,
Institute of Medical Science, The University of Tokyo, Japan.

Abstract

Mast cells are a native composer of connective tissue of the skin dermis and intestinal and respiratory mucosa, and their interaction with fibroblasts is important for mast cell development and survival, because mast cells express c-kit receptor tyrosine kinase and fibroblasts express c-kit ligand. Part of mast cells are in close approximation with nerves in the skin and intestinal and respiratory mucosa. Independent lines of accumulated evidence indicate the existence of an intensive bidirectional crosstalk between mast cells and sensory nerves and suggest that mast cells and sensory nerves may be viewed as a functional unit, which could be of crucial importance in neuroimmunological pathways. Mast cells appear to have a property of influencing smooth muscle function via not only such nerve-mast cell effects, but also direct pathways. In bronchial asthma, mast cells infiltrate the airway smooth muscle layer, and interact directly with smooth muscle cells, suggesting pathogenic roles for mast cells in airway obstruction. Current studies on mast cell biology identified a novel adhesion molecule of mast cells, namely cell adhesion molecule-1, CADM1 [formerly referred to as Immunoglobulin Superfamily member 4 (IGSF4), Spermatogenic Immunoglobulin Superfamily (SgIGSF), Synaptic Cell Adhesion Molecule (SynCAM), Tumor Suppressor

* Correspondence concerning this article should be addressed to: Akihiko Ito, Division of Molecular Pathology, Department of Cancer Biology, Institute of Medical Science, The University of Tokyo, 4-6-1, Shirokanedai Minato-ku, Tokyo 108-8639, Japan, Fax: +81-3-5449-5407 E-mail: aito@ims.u-tokyo.ac.jp.

in Lung Cancer-1 (TSLC1), or Nectin-like Molecule-2 (Necl-2)]. This molecule is unique, because it serves as not only simple glue but also appears to promote functional communication between mast cells and fibroblasts, nerve or smooth muscle cells.

Keywords: neurogenic inflammation, substance P, asthma, SgIGSF, SynCAM, Nectin.

Introduction: Overview of Mast Cell Development

Mast cells are progeny of multipotential hematopoietic stem cells, but unlike other lineages of blood cells, they are a native composer of connective tissues, *i.e.*, the situation where mast cells exist is physiological in a variety of organs and tissues, such as the skin dermis, intestinal and respiratory mucosa, dura mater, thalamus and peritoneal cavity (Austen and Boyce, 2001; Kitamura *et al.*, 1977; Kitamura *et al.*, 1981; Kitamura *et al.*, 1989; Nakahata and Ogawa, 1982). The developmental process of mast cells is also unique. In contrast to the fact that most myeloid cells complete their differentiation within the bone marrow, mast cell-committed progenitors leave the bone marrow, invade connective or mucosal tissue, and then proliferate and differentiate to morphologically identifiable mast cells (Kitamura *et al.*, 1979; Chen *et al.*, 2005). The phenotype of mast cells, either connective tissue-type or mucosal, is determined by each site of final differentiation (Kitamura, 1989; Chen *et al.*, 2005). In addition, while most myeloid cells lose proliferation potential after maturation, peripherally resident mature mast cells still retain proliferation potential (Sonoda *et al.*, 1984). Even after they have executed degranulation, mature mast cells are able not only to proliferate but also to restore the original morphology by generating cytoplasmic granules (Kuriu *et al.*, 1989; Jozaki *et al.*, 1990).

Interaction of Mast Cells with Other Lineage Cells

Significance of Cell-Cell Adhesion in Mast Cell Survival

There are several murine loci that affect the development of mast cells (Kitamura *et al.*, 1978; Kitamura and Go., 1979). The well-known two are *W* and *Sl*. The *W* locus encodes c-*kit* receptor tyrosine kinase (KIT) (Chabot *et al.*, 1988; Geissler *et al.*, 1988), a growth-factor receptor that express on mast cells (Nocka *et al.*, 1989), and the *Sl* locus encodes stem cell factor (SCF), a cognate ligand of KIT that is produced by several cell types including fibroblasts (Flanagan *et al.*, 1990; Flanagan *et al.*, 1991; Huang *et al.*, 1990; Williams *et al.*, 1990; Zsebo *et al.*, 1990). SCF exerts its activities either as membrane-bound proteins or as proteolytically processed soluble proteins (Galli *et al.*, 1994). Systemic mast cell deficiency of mice with mutations at the *W* or *Sl* locus implies that the signals produced by binding of SCF to KIT are essential for development of mast cells. The *mi* locus also affects mast cell development (Kataoka *et al.*, 2002; Kim *et al.*, 1999; Morii *et al.*, 2001a; Morii *et al.*, 2001b). Mice with mutations at this locus lack mast cells in various tissues other than the dermis.

The *mi* locus encodes the microphthalmia transcription factor (MITF), a member of the basic-helix-loop-helix-luecine zipper family (Hodgkinson *et al.*, 1993; Hughes *et al.*, 1993). More than ten spontaneous (for example, *mi*, *mi-ce*, and *Mi-wh*) and transgene-insertional (i.e., *mi-vga9*) mutant alleles of the mi locus are known, and any homozygotes of each allele are partially mast cell-deficient mice. Molecular mechanisms of the mast cell deficiency remain precisely unknown, except that it does not seem to be attributable directly to SCF or KIT dysfunction.

Although SCF and KIT are essential for mast cell development, when bone marrow cells of *W* and *Sl* mutants are cultured in the presence of interleukin (IL)-3, primary mast cells (bone marrow-derived cultured mast cells, BMMCs) are established as easily as those from marrow cells of wild-type (+/+) mice (Crapper *et al.*, 1984). This indicates that the SCF-KIT signal is essential for mast cell development *in vivo*, but it can be substituted by IL-3 signals *in vitro*. Once established, however, BMMCs are useful to demonstrate the importance of SCF-KIT interaction in mast cell proliferation. When +/+ BMMCs are injected into the peritoneal cavity of W/W^v mice, they are capable of surviving. In contrast, they are incapable of surviving in the peritoneal cavity of Sl/Sl^d mice. These phenotypes are reproducible in vitro. When +/+ BMMCs are cocultured on NIH3T3 fibroblast monolayers, they are able to survive over the monolayer without any specific growth factors (Fujita *et al.*, 1988a; Fujita *et al.*, 1988b). However, when +/+ BMMCs are prevented from being in contact with NIH3T3 fibroblasts, they gradually disappear from the coculture (Fujita *et al.*, 1988b). In this coculture, adhesion of BMMCs to NIH3T3 fibroblasts appears to be important for BMMCs to proliferate, probably because the adhesion allows KIT on BMMCs to bind to membrane-bound SCF expressing on NIH3T3 fibroblasts. Not to mention, KIT-mutant BMMCs are incapable of surviving in this coculture. Interestingly, MITF-mutant BMMCs are also incapable of surviving, and in addition, they are incapable of attaching to NIH3T3 fibroblasts, suggesting that MITF mutations may result in poor adhesion causing mast cells not to receive growth-promoting signals through KIT evoked by membrane-bound SCF.

Isolation of CADM1 as a New Mast-Cell Adhesion Molecule

We applied a cDNA library subtraction method to probe molecular mechanisms of mast cell deficiency in MITF mutants. Repeated subtraction processes allowed us to isolate Cell Adhesion Molecule-1, CADM1, as a novel mast-cell adhesion molecule whose expression was critically regulated by MITF (Ito *et al.*, 2003a). Structurally, this molecule belongs to the immunoglobulin superfamily, and has 3 immunoglobulin-like motifs in the extracellular domain and a short cytoplasmic domain that interacts with other proteins via the protein 4.1-binding and PDZ-binding motifs (Wakayama *et al.*, 2003; Figure 1). To date, a variety of cells have been revealed to express CADM1, including epithelial cells, such as pulmonary and biliary epithelial cells (Ito *et al.*, 2003b; Ito *et al.*, 2007b) and pancreatic islet cells (Koma *et al.*, in press), and non-epithelial cells, such as neurons (Biederer *et al.*, 2002), spermatogenic cells (Wakayama *et al.*, 2003), and mast cells (Ito *et al.*, 2003a). CADM1 forms homodimers on the cell membrane through *cis*-interactions, and binds either *trans*-homophilically or heterophilically, depending on the cell type expressing the CADM1 and the binding partners available on adjacent cells. The *trans*-homophilic binding occurs among neurons (Biederer et al., 2002) and among epithelial cells of identical types (Ito *et al.*, 2003a;

Ito *et al.*, 2007b), and the *trans*-heterophilic binding occurs between mast cells and fibroblasts (Ito *et al.*, 2003a) and between spermatogenic and Sertoli cells (Wakayama *et al.*, 2007). Recently, one of the heterophilic binding partners was identified as Class-I-restricted T-cell-associated molecule (Boles *et al.*, 2005; Galibert *et al.*, 2005). In addition to the full-length form of approximately 100 kDa, CADM1 appears to have two shorter forms of about 35 and 15 kDa, which are probably generated by N-terminal truncation of the full-length form, as suggested in murine mast cells (Ito *et al.*, 2003a) and human mesothelial cells (Ito *et al.*, in press); however, the function of these truncated forms remains unclear.

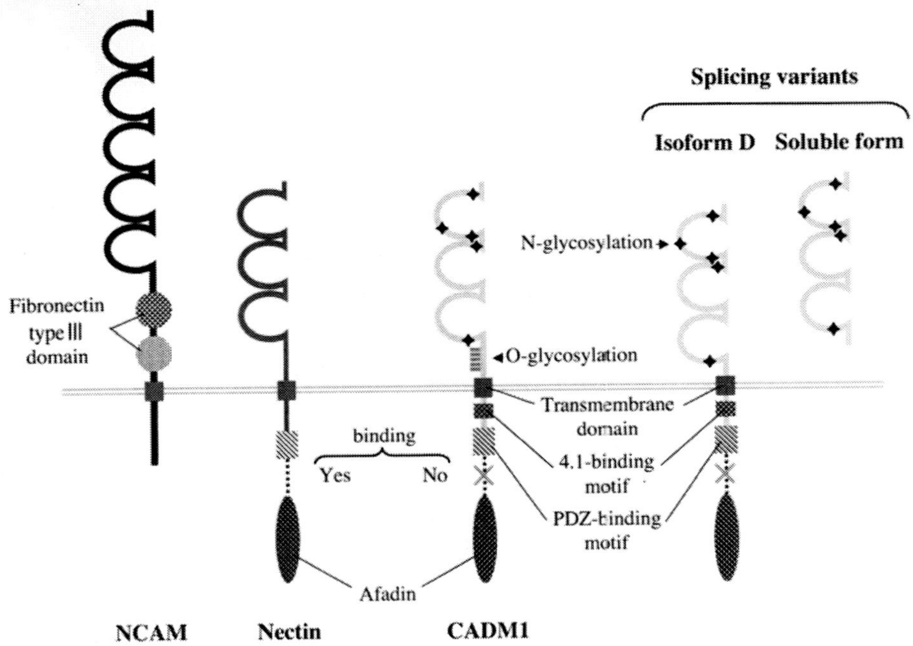

Figure 1. Putative structures of NCAM, Nectin, and CADM1. The three molecules belong to the immunoglobulin superfamily with one membrane-spanning region: NCAM has 5 immunoglobulin-like loops, and Nectin and CADM1 have 3 loops in the extracellular domains. Nectin and CADM1 are similar also in the cytoplasmic domains, but Nectin binds to Afadin, an F-actin binding protein, whereas CADM1 does not bind. CADM1 has several isoforms generated by alternative splicing. One lacks a heavily O-glycosylated threonine repeat located in the juxtamembrane region of the extracellular domain, while another is a soluble isoform which mRNA open reading frame ends just before the juxtamembrane region.

We generated CADM1-transgenic mice and mated them with MITF mutants to obtain MITF-mutant homozygotes that expressed CADM1 exogenously (Ito *et al.*, 2005). Although these mice still have the mast cell deficient phenotype similar to that of the MITF mutants, BMMCs from these mice expressed CADM1 as abundantly as +/+ BMMCs and were normal in adhesion to NIH3T3 fibroblasts in coculture. In addition, when these BMMCs were injected into the peritoneal cavity of W/W^v mice, their survival was significantly better than that of BMMCs from the non-transgenic, MITF-mutant homozygotes (Ito *et al.*, 2005). CADM1-mediated efficient attachment of mast cells to stromal cells appears to result in enhancement in growth-promoting signal transduction into mast cells mediated by binding between membrane-bound SCF on stromal cells and KIT on mast cells (Figure 2).

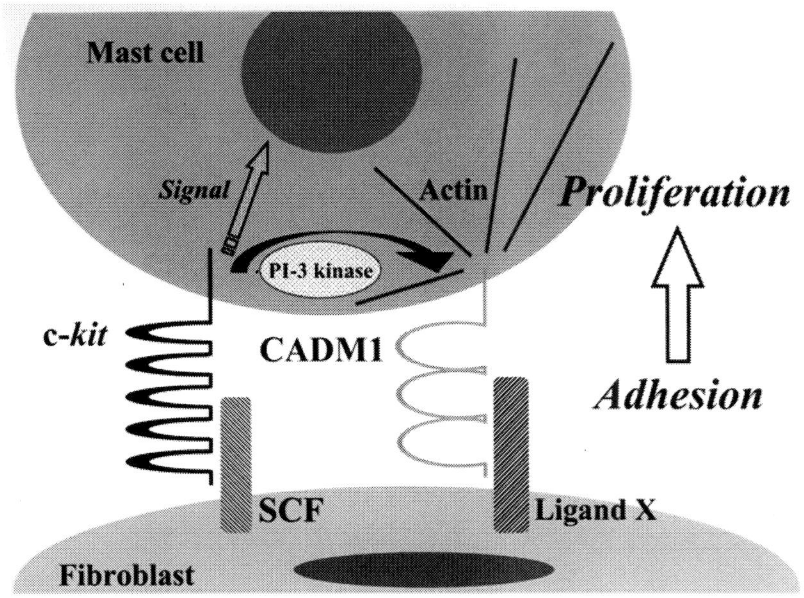

Figure 2. Mast cell-fibroblast attachment mediated by CADM1 and its involvement in mast cell survival. Mast cells exist physiologically in connective tissue of various organs, and are in contact with fibroblasts. Since fibroblasts and mast cells express membrane-bound SCF and KIT, respectively, direct contact between both results in KIT activation, leading to mast cell proliferation and survival. CADM1 is one of the molecules that sustain this cellular and molecular interaction. Interestingly, KIT-mediated intracellular signals through PI-3 kinase are necessary for CADM1 to function as an adhesion molecule (Koma et al, 2005). Although the binding partner(s) expressed on fibroblasts remain unknown, CADM1 appears to promote the attachment of mast cells to fibroblasts, and as a result, to enhance the efficiency of growth-promoting signal transduction in mast cells.

Nerve-Mast Cell Interaction Mediated by CADM1

As a composer of connective tissue, mast cells exist in various tissues and organs. Apparently, it seems that their distribution is at random in the skin dermis and intestinal and respiratory mucosa, but when one makes closer observation, he can notice that mast cells prefer to locate around vessels and nerve fibers (Selye, 1965). In fact, some mast cells are in a close approximation to nerves in connective tissues of various organs. According to electron microscopy, membrane-membrane apposition with a spatial gap of approximately 20 nm or less is detected between nerve and mast cells (Stead *et al.*, 1987; Stead *et al.*, 1989). These anatomical findings are suggestive of functional communication between nerve and mast cells.

Several lines of molecular events support this suggestion: (1) nerves in contact with mast cells often contain neuropeptides, such as substance P and calcitonin gene-related peptide, and release these neuropeptides on stimulation (Stead *et al.*, 1987; Keller and Marfurt, 1991), and (2) mast cells express receptors for many neuropeptides (Krumins and Broomfield, 1993; Cooke *et al.*, 1998; van der Kleij *et al.*, 2003; Bischoff *et al.*, 2004; De Jonge *et al.*, 2004).

Therefore, nerve activation results in mast cell activation, *i.e.*, degranulation or secretion of mediators (Dimitriadou *et al.*, 1991). On the other hand, mast cells synthesize and release a variety of molecules, which can, in turn, influence neuronal activity (Frieling *et al.*, 1991; Frieling *et al.*, 1993).

For example, tryptase directly activates proteinase-activated receptors on neurons (Corvera *et al.*, 1999), and TNFalpha and nerve growth factor cause changes in local nerves so as to lower their threshold to activation (Leon *et al.*, 1994; van Houwelingen *et al.*, 2002). These nerve-mast cell effects, together with the anatomical association between both, are assumed to be important for promotion and regulation of many inflammatory diseases, such as atopic dermatitis, alopecia, asthma, and irritable bowel syndrome (Theoharides, 1996; Barbara *et al.*, 2004; Theoharides and Cochrane, 2004). In spite of such clinical importance, the molecules that connect mast cells directly with nerves have not been studied intensively so far.

The time was 2003, when we isolated cell adhesion molecule-1, CADM1, as a novel mast cell adhesion molecule (Ito *et al.*, 2003a), and at the same time we realized that CADM1 is an adhesion molecule common to nerves and mast cells. Formerly, CADM1 was referred to as synaptic cell adhesion molecule (SynCAM), spermatogenic immunoglobulin superfamily (SgIGSF) (Wakayama *et al.*, 2001), tumor suppressor in lung cancer-1 (TSLC1) (Kuramochi *et al.*, 2001), nectin-like molecule-2 (Necl-2) (Shingai *et al.*, 2003), immunoglobulin superfamily 4A (IGSF4A), or retinoic acid-inducible gene 175 (RA175) (Urase *et al.*, 2001). As might be expected from the variety of names, CADM1 proved to be multifunctional. In a period from 2002 to 2007, several research groups reported CADM1 functions and roles independently. Murakami and coworkers (Kuramochi *et al.*, 2001) found the CADM1 gene in the chromosomal region where human lung cancers often exhibit loss of heterozygosity, and identified CADM1 as a tumor suppressor in human non-small lung cancer. Momoi and coworkers generated CADM1 knockout mice and demonstrated that loss of CADM1 results in male infertility (Fujita *et al.*, 2006). We found that biliary epithelial cells express CADM1 transiently and showed its possible role in bile duct formation (Ito *et al.*, 2007b). Of our great interest was the report by Biederer *et al.* (2002). They showed that CADM1 is localized to neuronal synapses in the brain, and it functions as a homophilic adhesion molecule that spans the synaptic cleft. More importantly, they demonstrated that CADM1 is an inducer of synaptic differentiation, because synapses are formed even in non-neuronal cells at the contact site with neuronal cells when the non-neuronal cells are transfected with CADM1 and glutamate receptor cDNAs. CADM1 appears to assemble machineries necessary for synapse formation around its cytoplasmic domain.

We examined whether CADM1 mediates the attachment between nerves and mast cells via *trans*-homophilic binding (Furuno *et al.*, 2005). BMMCs of various genotypes or IC-2 murine mast cells, which lack CADM1 expression, were seeded onto networks of neurites extending out from superior cervical ganglia (SCG), and the number of mast cells attached to the neurites per SCG cell was counted. Compared with +/+ BMMCs, mast cells lacking CADM1, such as MITF mutant-derived BMMCs and IC-2 cells, were defective in attachment to neurite, and transfection with CADM1 cDNA normalized this (Furuno *et al.*, 2005). On the other hand, the attachment of mast cells expressing CADM1, such as +/+ BMMCs and IC-2 cells transfected with CADM1 cDNA, was inhibited dose-dependently by a function-blocking antibody to CADM1. Immunocytochemistry revealed that CADM1 was locally concentrated at points of contact between neurites and mast cells (Furuno *et al.*, 2005). CADM1 appeared to mediate the adhesion between mast cells and neurites via *trans*-homophilic binding.

Next, we examined whether CADM1-mediated nerve-mast cell attachment results in efficient communication between both. In SCG neurite-mast cell cocultures, neurites were selectively activated with scorpion venom. Following neurite activation, some of IC-2 cells attendant to neurites became activated with a lag time of approximately 10 - 20 seconds (Furuno *et al.*, 2005). This event was considered to be a response of mast cells to neurite activation, because mast cells distant from neurites did not become activated and scorpion venom did not activate mast cells directly. One-quarter of IC-2 cells attendant to neurites responded, whereas CADM1 exogenously expressed in IC-2 cells doubled this proportion. The response of IC-2 cells was blocked by CP-99,994-1, an antagonist to neurokinin-1 receptor, which is receptive for substance P (Furuno *et al.*, 2005). Notably, a much higher concentration of the antagonist was required to inhibit the response of IC-2 cells expressing CADM1 exogenously (Furuno *et al.*, 2005). CADM1 appeared not only to function as simple glue in nerve-mast cell interaction, but also to promote the development of microenvironment where mast cells have an enhanced susceptibility to nerve activation. Homophilic binding of this adhesion molecule might result in an increase in the number of neurokinin-1 receptors on mast cells or in the amount of Substance P released from the neurite.

The role of CADM1 in nerve-mast cell interaction was assessed under the conditions nearer *in vivo* setting (Ito *et al.*, 2007a). In the rodent mesentery, a microanatomical contact has been demonstrated between mast cells and peptidergic nerves containing substance P (Stead *et al.*, 1989; Arizono *et al.*, 1990; Crivellato *et al.*, 1991). Therefore, we examined whether activation of mesenteric peptidergic nerves may result in degranulation of mast cells in contact with the nerves. The mesenteric nerve root was stimulated electrically, because it is shown to induce intestinal contraction by evoking efferent function of substance P-ergic sensory nerves in the presence of guanethidine and hexamethonium (Takaki et al., 1987; Sugumori *et al.*, 2000). In the steady-state mesentery of wild-type mice, the proportion of mast cells with morphologic features of degranulation was approximately 20%, and it increased nearly two-fold when the mesenteric nerve root was stimulated electrically (Ito *et al.*, 2007a). In contrast, there was no significant increase detectable in the mesentery of MITF-mutant mice. When CADM1 knockout mouse-derived BMMCs were transplanted to the mesentery of mast cell-deficient W/W^v mice, they did not degranulate in response to the mesenteric nerve stimulation, whereas transfection with CADM1 cDNA restored those responses (Ito *et al.*, 2007a). CADM1 appeared to promote communication between nerves and mast cells in the murine mesentery.

Possible Implication of CADM1-Mediated Nerve-Mast Cell Interaction in Pathophysiology of Human Diseases

Psychoemotional stressors have long been expected to exacerbate or trigger atopic and allergic diseases, such as atopic dermatitis and bronchial asthma. One of the most intriguing hypotheses for this effect is that stress induces neurogenic inflammation and cytokine production. Currently, Peters and coworkers demonstrated that stress worsens dermatitis via substance P-dependent neurogenic inflammation in mice (Pavlovic *et al.*, 2007). Maurer and coworkers demonstrated that sensory skin nerves augment mast cell-driven inflammatory responses by releasing neuropeptides that increases mast cell degranulation in a mouse model of passive cutaneous anaphylaxis (Siebenhaar *et al.*, 2008). Kurebayashi and coworkers attempted to generate 2,4,6-trinitrobenzene sulfonic acid-induced colitis in rats, and found that mast cell-deficient *Ws/Ws* rats did not manifest symptoms of visceral hypersensitivity in this model (Ohashi *et al.*, 2008). Many independent lines of evidence consistently indicate the existence of an intensive bidirectional crosstalk between mast cells and sensory nerves and suggest that mast cells and sensory nerves may be viewed as a functional unit and that the crosstalk between mast cells and sensory nerves could be of crucial importance in neurogenic inflammation pathways. Clinicians often experience that psychoemotional stress worsens patients' symptoms in a variety of non-infectious inflammatory diseases, such as migraines, atopic dermatitis, bronchial asthma, and irritable bowel syndrome (Theoharides and Cochrane, 2004). It is noticeable that mast cells are postulated to be involved in the pathophysiology of all these diseases. As these diseases gets worse, the magnitude of nerve-mast cell interaction may grow stronger. In fact, in the intestinal mucosa of patients with irritable bowel syndrome, the number of activated mast cells in close proximity to nerves correlates with severity and frequency of abdominal pain and/or discomfort (Barbara *et al.*, 2004). Neurogenic inflammation is likely to play a causative role in psychoemotional stress-evoked exacerbation of symptoms, and enhancement in nerve-mast cell interaction is assumed to underlie such pathophysiological conditions. Of our great interest is to examine how psychoemotional stress augments the extent of nerve-mast cell interaction. We are now intensively examining possible molecular linkage between CADM1 expression and psychoemotional stress or inflammatory response.

Smooth Muscle-Mast Cell Interaction Mediated by CADM1 in Asthma

Bronchial asthma is characterised by the presence of airflow obstruction, airway hyperresponsiveness and chronic airway inflammation. Mast cells play an important role in the pathophysiology of asthma through their chronic activation by IgE and allergen (Cruse *et al.*, 2005; Bradding *et al.*, 2006). To fulfil this role, mast cells not only secrete a plethora of bronchospastic and proinflammatory autacoid mediators, proteases and cytokines (Bradding *et al.*, 2006), but also appear to infiltrate the airway smooth muscle layer (Brightling *et al.*, 2002), because mast cells neither exist in this layer under physiologic conditions, nor infiltrate the layer of patients with eosinophilic bronchitis (Brightling *et al.*, 2002), who have high levels of mast cell mediators in their sputum and airway eosinophilia, but no bronchial hyperresponsiveness or symptoms of airflow obstruction. Noticeably, the specific infiltration of airway smooth muscle by mast cells is one of the critical determinants of the asthmatic phenotype: (1) mast cell density in the airway smooth muscle layer correlates with the severity of airway hypersenresponsiveness (Brightling *et al.*, 2002), and (2) the mast cells within the smooth muscle bundles in asthma exhibit ultrastructural features of activation (Begueret *et al.*, 2007). We recently demonstrated an important role for CADM1 in the specific recruitment of mast cells to airway smooth muscle. Human lung mast cells appeared to adhere to human bronchial smooth muscle cells, in part via CADM1 (Yang *et al.*, 2006). This action of CADM1 is probably dependent upon its *trans*-heterophilic binding manner, because smooth muscle cells do not express CADM1. Previously, we showed that *trans*-heterophilic binding of CADM1 mediates mast cell-fibroblast attachment and this attachment contributes to mast cell survival (Ito *et al.*, 2003a; Ito *et al.*, 2004). Although it remains unknown what molecule(s) are the *trans*-heterophilic binding partner expressed on smooth muscle cells and fibroblasts, it is intriguing to speculate that specific cellular crosstalk occurs between human lung mast cells and airway smooth muscle cells (Figure 3).

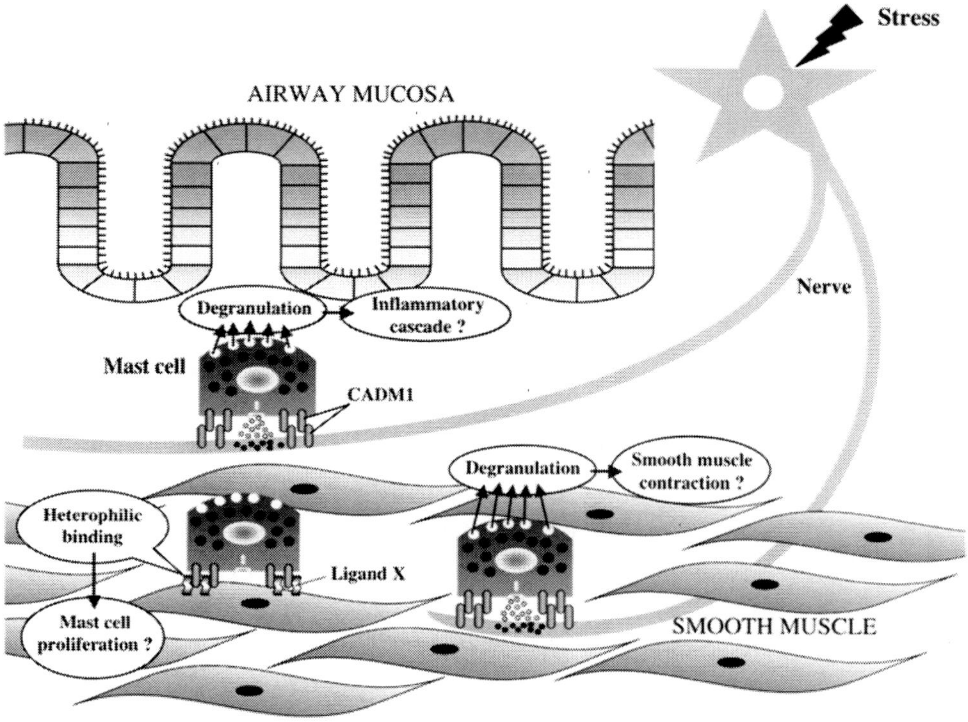

Figure 3. Possible involvement of nerve-mast cell and smooth muscle-mast cell interactions in bronchial asthma pathophysiology. Mast cells resident in the airway mucosa and smooth muscle layer may release cytoplasmic granules and cytokines in response to nerve activation, as a result of pathologic manifestation of neurogenic inflammation based on CADM1 homophilic binding. In addition, CADM1 appears to promote mast cell proliferation in the smooth muscle layer via its heterophilic binding to unknown ligand(s) expressed on smooth muscle cells. These dual actions of CADM1 are likely to contribute to exacerbation of asthmatic symptoms induced by psychoemotional stress.

CONCLUSION

Since mast cells are present physiologically in connective tissue of various organs, it is reasonable to speculate that mast cells are in contact with fibroblasts more often than with any other lineage cells. Past works clearly demonstrated inevitable importance of mast cell-fibroblast interaction in mast cell development and survival, based on the fact that fibroblasts express membrane-bound and soluble SCF, while mast cells express KIT. Direct contact of mast cells to fibroblasts results in membrane-bound SCF-evoked KIT activation, leading to mast cell proliferation. CADM1 appears to be one of the molecules that sustain this cellular and molecular mechanism.

During the last decade there has been a marked accumulation in the evidence indicating that nerve and mast cells communicate bi-directionally with each other. As we reviewed here, CADM1 is, to our knowledge, the first molecule that has proved to sustain and promote this communication. Importantly, CADM1 is not just simple glue but an enhancer of the communication. Therefore, CADM1 could play a causative role in exacerbation of neurogenic inflammation, if its expression or function is enhanced in response to psychoemotional stress and inflammatory cytokines. As we highlighted in asthma, CADM1 may also be an important mediator between smooth muscle and mast cells. In the coronary artery, mast cells are known to exist in the adventitia, and are assumed to be one of the inducers of medial smooth muscle spasm in response to physiological stress, such as coldness and fatigue, causing transient chest pain (Laine *et al.*, 1999; Laine *et al.*, 2000). CADM1 may play particular roles in such conditions by mediating both nerve-mast cell and smooth muscle-mast cell interactions. Currently, we demonstrated that CADM1 mediates nerve-pancreatic islet cell interaction via *trans*-homophilic binding and is likely to promote nerve-evoked hormone secretion from islet cells (Koma et al, in press). Further characterization of CADM1 functions under physiological and pathological conditions will lead us to future development of novel therapeutic strategies against a variety of neuroimmune-related and mast cell-involved diseases.

References

Arizono, N., Matsuda, S., Hattori, T., Kojima, K., Maeda, T. and Galli, S.J. (1990). Anatomical variation in mast cell nerve associations in the rat small intestine, heart, lung and skin. *Lab. Invest. 62*: 626-634.

Austen, K.F. and Boyce, J.A. (2001). Mast cell lineage development and phenotypic regulation. *Leuk. Res. 25*: 511-518.

Barbara, G., Stanghellini, V., De, Giorgio, R., Cremon, C., Cottrell, G.S., Santini, D., Pasquinelli, G., Morselli-Labate, A.M., Grady, E.F., Bunnett, N.W., Collins, S.M. and Corinaldesi, R. (2004). Activated mast cells in proximity to colonic nerves correlate with abdominal pain in irritable bowel syndrome. *Gastroenterol. 126*: 693-702.

Begueret, H., Berger, P., Vernejoux, J.M., Dubuisson, L., Marthan, R. and Tunon-de-Lara, J.M. (2007). Inflammation of bronchial smooth muscle in allergic asthma. *Thorax 62*: 8-15.

Biederer, T., Sara, Y., Mozhayeva, M., Atasoy, D., Liu, X., Kavalali, E.T. and Sudhof, T.C. (2002). SynCAM, a synaptic adhesion molecule that drives synapse assembly. *Science 297*: 1525-1531.

Bischoff, S.C., Schwengberg, S., Lorentz, A., Manns, M.P., Bektas, H., Sann, H., Levi-Schaffer, F., Shanahan, F. and Schemann, M. (2004). Substance P and other neuropeptides do not induce mediator release in isolated human intestinal mast cells. *Neurogastroenterol. Motil. 16*: 185-193.

Boles, K.S., Barchet, W., Diacovo, T., Cella, M. and Colonna, M. (2005) The tumor suppressor TSLC1/NECL-2 triggers NK-cell and CD8+ T-cell responses through the cell-surface receptor CRTAM. *Blood 106*: 779-786.

Bradding, P., Walls, A.F. and Holgate, S.T. (2006). The role of the mast cell in the pathophysiology of asthma. *J. Allergy Clin. Immunol. 117*: 1277-1284.

Brightling, C.E., Bradding, P., Symon, F.A., Holgate, S.T., Wardlaw, A.J. and Pavord, I.D. (2002). Mast-cell infiltration of airway smooth muscle in asthma. *N. Engl. J. Med. 346*: 1699-1705.

Chabot, B., Stephenson, D.A., Chapman, V.M., Besmer, P. and Bernstein, A. (1988). The proto-oncogene *c-kit* encoding a transmembrane tyrosine kinase receptor maps to the mouse W locus. *Nature 335*: 88-89.

Chen, C.C., Grimbaldeston, M.A., Trai, M., Weissman, I.L. and Galli, S.J. (2005). Identification of mast cell progenitors in adult mice. *Proc. Natl. Acad. Sci. U.S.A. 102*: 11408-11413.

Cooke, H.J., Fox, P., Alferes, L., Fox, C.C. and Wolfe, S.A. Jr. (1998). Presence of NK1 receptors on a mucosal-like mast cell line, RBL-2H3 cells. *Can. J. Physiol. Pharmacol. 76*: 188-193.

Corvera, C.U., Déry, O., McConalogue, K., Gamp, P., Thoma, M., Al-Ani, B., Caughey, G.H., Hollenberg, M.D. and Bunnett, N.W. (1999). Thrombin and mast cell tryptase regulate guinea-pig myenteric neurons through proteinase-activated receptors-1 and -2. *J. Physiol. (Lond.) 517*: 741-756.

Crapper, R.M., Thomas, W.R. and Schrader, J.W. (1984). In vivo transfer of persisting (P) cells; further evidence of their identity with T-dependent mast cells. *J. Immunol. 133*: 2174-2179.

Crivellato, E., Damiani, D., Mallardi, F. and Travan, L. (1991). Suggestive evidence for a microanatomical relationship between mast cells and nerve fibres containing substance P, calcitonin gene related peptide, vasoactive intestinal polypeptide, and somatostatin in the rat mesentery. *Acta. Anat. (Basel). 141*: 127-131.

Cruse, G., Kaur, D., Yang, W., Duffy, S.M., Brightling, C.E. and Bradding, P. (2005). Activation of human lung mast cells by monomeric immunoglobulin E. *Eur. Respir. J. 25*: 858-863.

De Jonge, F., De, Laet, A., van Nassauw, L., Brown, J.K., Miller, H.R., van Bogaert, P.P., Timmermans, J.P. and Kroese, A.B. (2004). In vitro activation of murine DRG neurons by CGRP-mediated mucosal mast cell degranulation. *Am. J. Physiol. 287*: G178-G191.

Dimitriadou, V., Buzzi, M.G., Moskowitz, M.A. and Theoharides, T.C. (1991). Trigeminal sensory fiber stimulation induces morphological changes reflecting secretion in rat dura mater mast cells. *Neuroscience. 44*: 97-112.

Echtenacher, B., Männel, D.N. and Hültner, L. (1996). Critical protective role of mast cells in a model of acute septic peritonitis. *Nature 381*: 21-22.

Flanagan, J.G. and Leder, P. (1990). The kit ligand: a cell surface molecule altered in steel mutant fibroblasts. *Cell 63*: 185-194.

Flanagan, J.G., Chan, D.C. and Leder, P. (1991). Transmembrane form of the kit ligand growth factor is determined by alternative splicing and is missing in the Sl^{d} mutant. *Cell 64*: 1025-1035.

Frieling, T., Cooke, H.J. and Wood, J.D. (1991). Serotonin receptors on submucous neurons in guinea pig colon. *Am. J. Physiol. 261*: G1017-G1023.

Frieling, T., Cooke, H.J. and Wood, J.D. (1993). Histamine receptors on submucous neurons in guinea pig colon. *Am. J. Physiol. 264*: G74-G80.

Fujita, E., Kouroku, Y., Ozeki, S., Tanabe, Y., Toyama, Y., Maekawa, M., Kojima, N., Senoo, H., Toshimori, K. and Momoi, T. (2006). Oligo-astheno-teratozoospermia in mice lacking RA175/TSLC1/SynCAM/IGSF4A, a cell adhesion molecule in the immunoglobulin superfamily. *Mol. Cell. Biol. 26*: 718-726.

Fujita, J., Nakayama, H., Onoue, H., Ebi, Y., Kanakura, Y., Kuriu, A. and Kitamura, Y. (1988a). Failure of *W/W^v^* mouse-derived cultured mast cells to enter S phase upon contact with NIH/3T3 fibroblasts. *Blood 72*: 463-468.

Fujita J, Nakayama H, Onoue H, Kanakura, Y., Nakano, T., Asai, H., Takeda, S., Honjo, T. and Kitamura, Y. (1988b). Fibroblastdependent growth of mouse mast cells in vitro: duplication of mast cell depletion in mutant mice of *W/W^v^* genotype. *J. Cell. Physiol. 134*: 78-84.

Furuno, T., Ito, A., Koma, Y., Watabe, K., Yokozaki, H., Bienenstock, J., Nakanishi, M. and Kitamura, Y. (2005). The spermatogenic Ig superfamily/synaptic cell adhesion molecule mast-cell adhesion molecule promotes interaction with nerves. *J. Immunol. 174*: 6934-6942.

Galibert, L., Diemer, G.S., Liu, Z., Johnson, R.S., Smith, J.L., Walzer, T., Comeau, M.R., Rauch, C.T., Wolfson, M.F., Sorensen, R.A., Van der Vuurst de Vries, A-R., Branstetter, D.G., Koelling, R.M., Scholler, J., Fanslow, W.C., Baum, P.R., Derry, J.M. and Yan, W. (2005) Nectin-like protein 2 defines a subset of T-cell zone dendritic cells and is a ligand for class-I-restricted T-cell-associated molecule. *J. Biol. Chem. 280*: 21955-21964.

Galli, S.J. and Wershil, B.K. (1996). The two faces of the mast cell. *Nature 381*: 21-26.

Galli, S.J., Zsebo, K.M. and Geissler, E.N. (1994). The kit ligand, stem cell factor. *Adv. Immunol. 55*: 1-96.

Geissler, E.N., Ryan, M.A. and Housman, D.E. (1988) The dominantwhite spotting (W) locus of the mouse encodes the *c-kit* proto-oncogene. *Cell 55*: 185-192.

Hart, S.P. and Dransfield, I. (2005). Activation of human lung mast cells by monomeric immunoglobulin E. *Eur. Respir. J. 26*: 744-745.

Hodgkinson, C.A., Moore, K.J., Nakayama, A., Steingrímsson, E., Copeland, N.G., Jenkins, N.A. and Arnheiter, H. (1993). Mutations at the mouse microphthalmia locus are associated with defects in a gene encoding a novel basic-helix-loop-helix-zipper protein. *Cell 74*: 395-404.

Huang, E., Nocka, K., Beier, D.R., Chu, T.Y., Buck, J., Lahm, H.W., Wellner, D., Leder, P. and Besmer, P. (1990). The hematopoietic growth factor KL is encoded by the *Sl* locus and is the ligand of the *c-kit* receptor, the gene product of the W locus. *Cell 63*: 225-233.

Hughes, J.J., Lingrel, J.B., Krakowsky, J.M. and Anderson, K.P. (1993). A helix-loop-helix transcription factor-like gene is located at the mi locus. *J. Biol. Chem.* 268: 20687-20690.

Ito, A., Hagiyama, M., Mimura, T., Matsumoto, M., Wakayama, T., Iseki, S., Yokozaki, H. and Okada, M. (2008). Expression of cell adhesion molecule 1 in malignant pleural mesothelioma as a cause of efficient adhesion and growth on mesothelium. *Lab. Invest.* (In press).

Ito, A., Hagiyama, M., Oonuma, J., Murakami, Y., Yokozaki, H. and Takaki, M. (2007a). Involvement of the SgIGSF/Necl-2 adhesion molecule in degranulation of mesenteric mast cells. *J. Neuroimmunol. 184*: 209-213.

Ito, A., Jippo, T., Wakayama, T., Morii, E., Koma, Y., Onda, H., Nojima, H., Iseki, S. and Kitamura, Y. (2003a). SgIGSF: a new mast-cell adhesion molecule used for attachment to fibroblasts and transcriptionally regulated by MITF. *Blood 101*: 2601-2608.

Ito, A., Koma, Y., Watabe, K., Jippo, T., Wakayama, T., Iseki, S. and Kitamura, Y. (2004). Contribution of the SgIGSF adhesion molecule to survival of cultured mast cells in vivo. *Biochem. Biophys. Res. Commun. 319*: 200-206.

Ito, A., Nishikawa, Y., Ohnuma, K., Ohnuma, I., Koma, Y., Sato, A., Enomoto, K., Tsujimura, T. and Yokozaki, H. (2007b). SgIGSF is a novel biliary-epithelial cell adhesion molecule mediating duct/ductule development. *Hepatol. 45*: 684-694.

Ito, A., Okada, M., Uchino, K., Wakayama, T., Koma, Y., Iseki, S., Tsubota, N., Okita, Y. and Kitamura, Y. (2003b). Expression of the TSLC1 adhesion molecule in pulmonary epithelium and its down-regulation in pulmonary adenocarcinoma other than bronchioloalveolar carcinoma. *Lab. Invest. 83*: 1175-1183.

Jozaki, K., Kuriu, A., Waki, N., Adachi, S., Yamatodani, A., Tarui, S.and Kitamura, Y. (1990). Proliferative potential of murine peritoneal mast cells after degranulation induced by compound 48/80, substance P, tetradecanoylphorbol acetate, or calcium ionophore A23187. *J. Immunol. 145*: 4252-4256.

Kataoka, T.R., Morii, E., Oboki, K., Jippo, T., Maeyama, K. and Kitamura, Y. (2002). Dual abnormal effects of mutant MITF encoded by Mi^{wh} allele on mouse mast cells: decreased but recognizable transactivation and inhibition of transactivation. *Biochem. Biophys. Res. Commun. 297*: 111-115.

Keller, J.T. and Marfurt, C.F. (1991). Peptidergic and serotoninergic innervation of the rat dura mater. *J. Comp. Neurol. 309*: 515-534.

Kim, D.K., Morii, E., Ogihara, H., Lee, Y.M., Jippo, T., Adachi, S., Maeyama, K., Kim, H.M. and Kitamura, Y. (1999). Different effect of various mutant MITF encoded by *mi*, Mi^{or}, or Mi^{wh} allele on phenotype of murine mast cells. *Blood 93*: 4179-4186.

Kitamura, Y. (1989). Heterogeneity of mast cells and phenotypic change between subpopulations. *Annu. Rev. Immunol. 7*: 59-76.

Kitamura, Y. and Go, S. (1979). Decreased production of mast cells in $S1/S1^{d}$ anemic mice. *Blood 53*: 492-497.

Kitamura, Y., Go, S. and Hatanaka, K. (1978). Decrease of mast cells in W/W^{v} mice and their increase by bone marrow transplantation. *Blood 52*: 447-452.

Kitamura, Y., Matsuda, H. and Hatanaka, K. (1979). Clonal nature of mast-cell clusters formed in W/W^{v} mice after bone marrow transplantation. *Nature 281*: 154-155.

Kitamura, Y., Shimada, M., Hatanaka, K.and Miyano, Y. (1977). Development of mast cells from grafted bone marrow cells in irradiated mice. *Nature 268*: 442-443.

Kitamura, Y., Yokoyama, M., Matsuda, H., Ohno, T. and Mori, K.J. (1981). Spleen colony-forming cell as common precursor for tissue mast cells and granulocytes. *Nature 291*: 159-160.

Kojima, S., Kitamura, Y. and Takasu, K. (1980). Prolonged infection of Nippostrogylus brasilensis in genetically mast cell depleted W/W^{v} mice. *Immunol. Let. 2*: 159-162.

Koma, Y., Furuno, T., Hagiyama, M., Hamaguchi, K., Nakanishi, M., Masuda, M., Hirota, S., Yokozaki, H. and Ito, A. (2008). CADM1 is a novel pancreatic-islet cell adhesion molecule that mediates nerve-islet cell interactions. *Gastroenterol.* (In press).

Koma, Y., Ito, A., Watabe, K., Hirata, T., Mizuki, M., Yokozaki, H., Kitamura, T., Kanakura, Y. and Kitamura Y. (2005). Distinct role for c-kit receptor tyrosine kinase and SgIGSF adhesion molecule in attachment of mast cells to fibroblasts. *Lab. Invest. 85*: 426-435.

Krumins, S.A. and Broomfield, C.A. (1993). C-terminal substance P fragments elicit histamine release from a murine mast cell line. *Neuropeptides 24*: 5-10.

Kuramochi, M., Fukuhara, H., Nobukuni, T., Kanbe, T., Maruyama, T., Ghosh, H.P., Pletcher, M., Isomura, M., Onizuka, M., Kitamura, T., Sekiya, T., Reeves, R.H. and Murakami, Y. (2001). TSLC1 is a tumor-suppressor gene in human non-small-cell lung cancer. *Nature Genet. 27*: 427-430.

Kuriu, A., Sonoda, S., Kanakura, Y., Jozaki, K., Yamatodani, A. and Kitamura, Y. (1989). Proliferative potential of degranulated murine peritoneal mast cells. *Blood 74*: 925-929.

Laine, P., Kaartinen, M., Penttilä, A., Panula, P., Paavonen, T. and Kovanen, P.T. (1999). Association between myocardial infarction and the mast cells in the adventitia of the infarct-related coronary artery. *Circulation 99*: 361-369. No citation in the text.

Laine, P., Naukkarinen, A., Heikkilä, L., Penttilä, A. and Kovanen, P.T. (2000). Adventitial mast cells connect with sensory nerve fibers in atherosclerotic coronary arteries. *Circulation 101*: 1665-1669.

Leon, A., Buriani, A., Dal, Toso, R., Fabris, M., Romanello, S., Aloe, L. and Levi-Montalcini, R. (1994). Mast cells synthesize, store, and release nerve growth factor. *Proc. Natl. Acad. Sci. USA. 91*: 3739-3743.

Malaviya, R., Ikeda, T., Ross, E. and Abraham, S.N. (1996). Mast cell modulation of neutrophil influx and bacterial clearance at sites of infection through TNF-alpha. *Nature 381*: 21-22.

Matsuda, H., Watanabe, N., Kiso, Y., Hirota, S., Ushio, H., Kannan, Y., Azuma, M., Koyama, H. and Kitamura, Y. (1990). Necessity of IgE antibodies and mast cells for manifestation of resistance against larval Haemaphysalis longicornis ticks in mice. *J. Immunol. 144*: 259-262.

Morii, E., Ogihara, H., Kim, D.K., Ito, A., Oboki, K., Lee, Y.M., Jippo, T., Nomura, S., Maeyama, K., Lamoreux, M.L. and Kitamura, Y. (2001a). Importance of leucine zipper domain of *mi* transcription factor (MITF) for differentiation of mast cells demonstrated using mi^{ce}/mi^{ce} mutant mice of which MITF lacks the zipper domain. *Blood 97*: 2038-2044.

Morii, E., Ogihara, H., Oboki, K., Kataoka, T.R., Maeyama, K., Fisher, D.E., Lamoreux, M.L. and Kitamura, Y. (2001b). Effect of a large deletion of the basic domain of mi transcription factor on differentiation of mast cells. *Blood 98*: 2577-2579.

Nakahata, T. and Ogawa, M. (1982). Identification in culuture of a class of hemopoietic colony-forming units with extensive capability to self-renew and generate multipotential hemopoietic colonies. *Proc. Natl. Acad. Sci. USA. 79*: 3843-3847.

Nocka, K., Majumder, S., Chabot, B., Ray, P., Cervone, M., Bernstein, A. and Besmer, P. (1989). Expression of *c-kit* gene products in known cellular targets of *W* mutations in normal and *W* mutant mice—evidence for an impaired *c-kit* kinase in mutant mice. *Genes. Dev. 3*: 816-826.

Ohashi, K., Sato, Y., Kawai, M. and Kurebayashi, Y. (2008). Abolishment of TNBS-induced visceral hypersensitivity in mast cell deficient rats. *Life. Sci.* (In press).

Pavlovic, S., Daniltchenko, M., Tobin, D.J., Hagen, E., Hunt, S.P., Klapp, B.F., Arck, P.C. and Peters, E.M. (2007). Further exploring the brain-skin connection: stress worsens dermatitis via substance P-dependent neurogenic inflammation in mice. *J. Invest. Dermatol. 128*: 434-446.

Ruitenberg, E.J. and Elgersma, A. (1976). Absence of intestinal mast cell response in congenitally athymic mice during *Trichinella spiralis* infection. *Nature 264*: 258-260.

Selye, H. (1965). The mast cells. Butterworths, Washington, D.C.

Shingai, T., Ikeda, W., Kakunaga, S., Morimoto, K., Takekuni, K., Itoh, S., Satoh, K., Takeuchi, M., Imai, T., Monden, M. and Takai, Y. (2003). Implications of nectin-like molecule-2/IGSF4/RA175/SgIGSF/TSLC1/SynCAM1 in cell-cell adhesion and transmembrane protein localization in epithelial cells. *J. Biol. Chem. 278*: 35421-35427.

Siebenhaar, F., Magerl, M., Peters, E.M., Hendrix, S., Metz, M. and Maurer, M. (2008). Mast cell-driven skin inflammation is impaired in the absence of sensory nerves. *J. Allergy. Clin. Immunol.* (In press).

Sonoda, T., Kanayama, Y., Hara, H., Hayashi, C., Tadokoro, M., Yonezawa, T. and Kitamura, Y. (1984). Proliferation of peritoneal mast cells in the skin of W/W^v mice that genetically lack mast cells. *J. Exp. Med. 160*: 138-151.

Stead, R.H., Dixon, M.F., Bramwell, N.H., Riddell, R.H. and Bienenstock, J. (1989). Mast cells are closely apposed to nerves in the human gastrointestinal mucosa. *Gastroenterol. 97*: 575-585.

Stead, R.H., Tomioka, M., Quinonez, G., Simon, G.T., Felten, S.Y. and Bienenstock, J. (1987). Intestinal mucosal mast cells in normal and nematode-infected rat intestines are in intimate contact with peptidergic nerves. *Proc. Natl. Acad. Sci. U.S.A. 84*: 2975-2979.

Sugimori, S., Kadowaki, M., Yoneda, S., Yamanouchi, M., Nakano, H. and Takaki, M. (2000). Nociceptin inhibits capsaicin-sensitive contraction to mesenteric nerve stimulation in the guinea-pig isolated ileum. *Auton. Neurosci. 86*: 65-69.

Takai, Y., Irie, K., Shimizu, K., Sakisaka, T. and Ikeda, W. (2003). Nectins and nectin-like molecules: roles in cell adhesion, migration, and polarization. *Cancer. Sci. 94*: 655-667.

Takaki, M., Jin, J.G., and Nakayama, S. (1987). Involvement of hexamethonium-sensitive components and capsaicin-sensitive components in cholinergic contractions induced by mesenteric nerve stimulation in guinea-pig ileum. *Biomed. Res. 8*: 195-203.

Theoharides, T.C. (1996). The mast cell: a neuroimmunoendocrine master player. *Int. J. Tissue. React. 18*: 1-21.

Theoharides, T.C. and Cochrane, D.E. (2004). Critical role of mast cells in inflammatory diseases and the effect of acute stress. *J. Neuroimmunol. 146*: 1-12.

Urase, K., Soyama, A., Fujita, E. and Momoi, T. (2001). Expression of RA175 mRNA, a new member of the immunoglobulin superfamily, in developing mouse brain. *Neuroreport 12*: 3217-3221.

van der Kleij, H.P., Ma, D., Redegeld, F.A., Kraneveld, A.D., Nijkamp, F.P. and Bienenstock, J. (2003). Functional expression of neurokinin 1 receptors on mast cells induced by IL-4 and stem cell factor. *J. Immunol. 171*: 2074-2079.

van Houwelingen, A.H., Kool, M., de Jager, S.C., Redegeld, F.A., van Heuven-Nolsen, D., Kraneveld, A.D. and Nijkamp, F.P. (2002). Mast cell-derived TNF-alpha primes sensory nerve endings in a pulmonary hypersensitivity reaction. *J. Immunol. 168*: 5297-5302.

Wakayama, T., Koami, H., Ariga, H., Kobayashi, D., Sai, Y., Tsuji, A., Yamamoto, M. and Iseki, S. (2003). Expression and functional characterization of the adhesion molecule spermatogenic immunoglobulin superfamily in the mouse testis. *Biol. Reprod. 68*: 1755-1763.

Wakayama, T., Ohashi, K., Mizuno, K. and Iseki, S. (2001). Cloning and characterization of a novel mouse immunoglobulin superfamily gene expressed in early spermatogenic cells. *Mol. Reprod. Dev. 60*: 158-164.

Wakayama, T., Sai, Y., Ito, A., Kato, Y., Kurobo, M., Murakami, Y., Nakashima, E., Tsuji, A., Kitamura, Y. and Iseki, S. (2007). Heterophilic binding of the adhesion molecules poliovirus receptor and immunoglobulin superfamily 4A in the interaction between mouse spermatogenic and Sertoli cells. *Biol. Reprod. 76*: 1081-1090.

Williams, D.E., Eisenman, J., Baird, A., Rauch, C., Van Ness, K., March, C.J., Park, L.S., Martin, U., Mochizuki, D.Y., Boswell, H.S., Burgess, G.M., Cosman, D. and Lyman, S.D. (1990). Identification of a ligand for the *c-kit* proto-oncogene. *Cell 63*: 167-174.

Yang, W., Kaur, D., Okayama, Y., Ito, A., Wardlaw, A.J., Brightling, C.E. and Bradding, P. (2006). Human lung mast cells adhere to human airway smooth muscle, in part, via tumor suppressor in lung cancer-1. *J. Immunol. 176*: 1238-1243.

Zsebo, K.M., Williams, D.A., Geissler, E.N., Broudy, V.C., Martin, F.H., Atkins, H.L., Hsu, R., Birkett, N.C., Okino, K.H., Murdock, D.C., Jacobsen, F.W., Langley, K.E., Smith, K.A., Takeish, T., Cattanach, B.M., Galli, S.J. and Suggs, S.V. (1990). Stem cell factor is encoded at the *Sl* locus of the mouse and is the ligand for the *c-kit* tyrosine kinase receptor. *Cell 63*: 213-224.

In: New Cell Adhesion Research
Editors: Patrick Nott and Matthew Temple

ISBN 978-1-60692-378-8

Chapter VII

CARBON NANOPARTICLES AS SUBSTRATES FOR CELL ADHESION AND GROWTH

Lucie Bacakova[1,*], Lubica Grausova[1], Marta Vandrovcova[1], Jiri Vacik[2,†], Aneta Frazcek[3,‡], Stanislaw Blazewicz[3], Alexander Kromka[4,§], Bohuslav Rezek[4], Milan Vanecek[4], Milos Nesladek[5,||], Vaclav Svorcik[6,#], Vladimir Vorlicek[7,&] and Milan Kopecek[1]

[1]Institute of Physiology, Academy of Sciences of the Czech Republic, Videnska 1083, 142 20 Prague 4-Krc, Czech Republic
[2]Nuclear Physics Institute, Academy of Sciences of the Czech Republic, 250 68 Rez near Prague, Czech Republic
[3]AGH University of Science and Technology, Faculty of Materials Science and Ceramics, Department of Biomaterials, Al. Mickiewicza 30, 30-059, Cracow, Poland
[4]Institute of Physics, Academy of Sciences of the Czech Republic, Cukrovarnicka 10, 162 53 Prague 6, Czech Republic
[5]CEA-LIST, Centre d'Etudes Saclay, Bat. 451, p. 84, 91191 Gif Sur Yvette, France
[6]Department of Solid State Engineering, Institute of Chemical Technology, Technicka 5, 166 28 Prague 6, Czech Republic
[7]Institute of Physics, Academy of Sciences of the Czech Republic, Na Slovance 2, 182 21 Prague 6, Czech Republic

[*] E-mail address: lucy@biomed.cas.cz; grausova@biomed.cas.cz; vandrovcova@biomed.cas.cz; kopecek@biomed.cas.cz.
[†] E-mail address: vacik@ujf.cas.cz.
[‡] E-mail address: afraczek@op.pl; blazew@agh.edu.pl.
[§] E-mail address: kromka@fzu.cz; rezek@fzu.cz; vanecek@fzu.cz.
[||] E-mail address: Milos.NESLADEK@cea.fr.
[#] E-mail address: vaclav.svorcik@vscht.cz.
[&] E-mail: vorlicek@fzu.cz.

ABSTRACT

Carbon nanoparticles, such as fullerenes, nanotubes and nanodiamonds, have been considered as promising building blocks in nanotechnology for broadscale nanostuctured devices and materials, e.g., microchips, nanorobots and biosensors, carriers for controlled drug and gene delivery or tracers for novel imaging technologies. Soon after macroscopic production was established, fullerenes, as donors and acceptors of electrons, started to be targeted for use in photodynamic therapy against tumor and microbial cells and also for quenching oxygen radicals. It has been discovered that nanodiamonds can also act as antioxidant and anti-inflammatory agents.

The interaction of carbon nanoparticles with cells and tissues has been investigated mainly using suspensions of these particles in cell culture media or other fluids. Relatively little is known about the influence of layers consisting of carbon nanoparticles on cell-substrate adhesion. Carbon nanoparticles can be advantageously used for surface modification of various artificial materials developed for the construction of tissue replacements and other body implants. In the form of films deposited on the surface of implants, carbon nanoparticles can not only improve the mechanical and other physical properties of a body implant, but can also enhance the attractiveness of the implant for cell colonization. The latter effect is probably mediated by the surface nanostructure of the films, which, at least to a certain degree, mimics the architecture of physiological cell adhesion substrates, such as extracellular matrix molecules. On nanostructured substrates, the cell adhesion-mediating molecules are adsorbed in the appropriate spectrum, amount and spatial conformation that make specific sites on these molecules (e.g. amino acid sequences like RGD) accessible to cell adhesion receptors. In addition, the nanostructured surfaces have been reported to enhance the adsorption of vitronectin, preferred by osteoblasts for their adhesion. The electrical conductivity of carbon nanotube films has also had beneficial effects on the growth and maturation of osteoblasts. Thus, the nanoparticles deposited on the bone-anchoring parts of bone, joint or dental replacements can improve the integration of these devices with the surrounding bone tissue. In addition, carbon nanoparticles admixed into polymers, designed for the fabrication of three-dimensional scaffolds for bone tissue engineering, could decorate the walls of the pores in these materials, and thus promote the ingrowth of bone cells. The interaction of cells with substrates modified with carbon nanoparticles could be further intensified by functionalizing these particles with various chemical functional groups or biomolecules, including KRSR-containing adhesion oligopeptides, recognized preferentially by osteoblasts.

This interdisciplinary review, based mainly on our own results, deals with the interaction of bone-derived cells with glass, polymeric and silicon substrates modified with fullerenes C_{60}, nanotubes, nanocrystalline diamond as well as other carbon allotropes, such as amorphous hydrogenated carbon or pyrolytic graphite. In general, these modifications have resulted in improved adhesion, growth, viability and osteogenic differentiation of these cells. The use of carbon and/or metal-carbon composite nanoparticles for the construction of patterned surfaces for regionally-selective adhesion and directed growth of cells is also discussed in this work.

1. INTRODUCTION

Carbon nanoparticles, such as fullerenes, nanotubes and nanodiamonds, are promising building blocks for the construction of novel nanomaterials for emerging industrial technologies, such as molecular electronics, advanced optics, and storage of hydrogen as a potential source of energy. In addition, they are considered as promising materials for a wide range of biomedical applications, including photodynamic therapy against tumors and infectious agents, quenching oxygen radicals, construction of microchips, nanorobots and biosensors, advanced imaging technologies, controlled drug or gene delivery, and also in the simulation of cellular components such as membrane pores or ion channels.

Fullerenes are spheroidal molecules made exclusively of carbon atoms (e.g., C_{60}, C_{70}), present in both terrestrial and extraterrestrial materials [1]. From these materials, fullerenes can be purified by chromatographic methods, selective complexation and fractional crystallization. For large-scale industrial production, various more efficient methods have been developed, such as filtration of a fullerene extract through a thin layer of activated carbon [2]. Fullerenes were discovered, prepared and systemized by Kroto *et al.* [3,4]. They exhibit a wide range of unique physical and chemical properties, e.g., extreme stability, ability to withstand high temperatures and pressures, a high proclivity to react with other species while maintaining their spherical geometry, the ability to entrap other smaller species such as helium, at the same time not reacting with the fullerene molecule. Potential technical applications for fullerenes include superconductors, lubricants, optical devices, chemical sensors, photovoltaics, polymer additives, polymer electronics such as Organic Field Effect Transistors (OFETS). Due to their high reactivity, fullerenes can serve as catalysts. They also represent an advantageous medium for hydrogen storage, because almost every carbon atom in C_{60} can absorb a hydrogen atom without disrupting the buckyball structure, which could lead to applications in fuel cells. Fullerenes have been also used as precursors to produce diamond films.

In addition, fullerene molecules display a diverse range of biological activity. Their unique hollow cage-like shape and structural analogy with clathrin-coated vesicles in cells support the idea of the potential use of fullerenes as drug or gene delivery agents [5-8]. Fullerenes are able to accept and release electrons. Thus, when irradiated with ultraviolet or visible light, they can convert molecular oxygen into highly reactive singlet (i.e., atomic) oxygen [8,9]. Thus, they have the potential to inflict photodynamic damage on biological systems, including damage to cellular membranes, inhibition of various enzymes or DNA cleavage [10-13]. This harmful effect can be advantageously exploited for photodynamic therapy against tumors, viruses and bacteria resistant to multiple drugs [14,15]. On the other hand, C_{60} is considered to be the world's most efficient radical scavenger. This is due to the relatively large number of conjugated double bonds in the fullerene molecule, which can be attacked by radical species. Thus, fullerenes would be suitable for applications in quenching oxygen radicals in medicine, as well as in cosmetics [9,16-22].

In their pristine unmodified state, fullerenes are highly hydrophobic and water-insoluble. On the other hand, they are relatively highly reactive, which enables them to be structurally modified. Fullerenes can form complexes with other atoms and molecules, e.g. metals, porphyrins, nucleic acids, proteins, as well as other carbon nanoparticles, e.g., nanotubes

[6,7,23-26]. In addition, fullerenes can be functionalized with various chemical groups, e.g., the hydroxyl, aldehydic, carbonyl, carboxyl, ester or amine groups, as well as amino acids and peptides. This usually renders them soluble in water and intensifies their interaction with biological systems [10,11,13,27,28]. For example, derivatized fullerenes decrease neuronal degeneration and death, protect the brain against ischemia [19,20], attenuate ischemia-reperfusion-induced lung injury [22,29], interact with receptors and inhibit various enzymes, such as HIV protease, HIV-reverse transcriptase, enzymes synthesizing cholesterol, redox enzymes or nitric oxide synthase [8,17,30]. Fullerenes C_{60} derivatized with monomalonic acid showed an inhibitory effect on nitric oxide-dependent relaxation of rabbit aortic smooth muscle [16].

Carbon nanotubes were discovered, synthesized and systemized by Iijima and Ichihashi [31-33]. These tubular structures are formed by a single cylindrically-shaped graphene sheet (single-wall carbon nanotubes, SWCNT or SWNT) or several graphene sheets arranged concentrically (multi-wall carbon nanotubes, MWCNT or MWNT). Their diameter is on the nanometer scale (e.g., from 0.4 nm to 2-3 nm in single-walled nanotubes), but their length can reach several micrometers. Related structures, with differences in shape but with a similar architectural principle, known as carbon nanorings, nanohorns, graphene nanoribbons or nanowires, have also been observed [23,34,35].

Carbon nanotubes can be prepared by arc-discharge and related methods [36], by laser ablation, laser evaporation or a laser beam pulse [37] or by chemical vapor deposition (CVD) methods [38]. The preparation of nanotubes usually requires the presence of metal catalysts, e.g., Fe, Co, Ni or Y, which can have adverse effects on cells and tissues [39-42]. In addition, in their pristine unmodified state, carbon nanotubes are hydrophobic, and therefore they are prone to form agglomerates in a water environment. To disperse these aggregates, various organic solvents or surfactants, such as Triton X and sodium dodecylbenzene sulfonate have been used [38,43,44], but these can alter biomolecules and damage cells. Thus, a more appropriate approach seems to be derivatization of nanotubes, especially with oxygen-containing chemical functional groups. Treatment of carbon nanotubes with acids (usually nitric and sulphuric, also hydrochloric or hydrofluoric acid), which attack the reactive ends of the carbon nanotubes, resulted in the creation of oxygen-containing groups, e.g. carboxylic. At the same time, this treatment purified the nanotubes from metal catalysts [36,42,45,46]. However, Bottini *et al.* [47] and Magrez *et al.* [48] reported increased toxicity of nanotubes after oxidation in acids. Oxidized nanotubes were shorter and straighter than their pristine counterparts, thus they could penetrate more easily inside the cells. In addition, they were better dispersed in an aqueous solution, and therefore reached higher concentrations of free nanotubes at similar weight per volume [47].

Carbon nanotubes have unique mechanical, physical and chemical properties in comparison with conventional carbon fibers [49]. Due to their sp^2 bonds, the tensile strength of SWCNT is about one hundred times higher than that of steel, while their density (i.e., specific weight) is about six times lower [50,33]. The excellent mechanical properties of carbon nanotubes could be utilized in hard tissue surgery, e.g., to reinforce bone implants, particularly scaffolds for bone tissue engineering made of synthetic polymers or chitosan [51,52]. In addition, nanotubes can resemble nanofibres of collagen and other extracellular proteins of the bone, and together with other carbon nanoparticles, they can also resemble hydroxyapatite and other inorganic crystals in the bone [53,54]. In addition, carbon nanotubes can be used effectively as substrates for growth and electrical stimulation of osteoblasts, which has been reported to promote their proliferation, differentiation, production of mineralized bone matrix and thus healing of the damaged bone [55-57]. Carbon nanotubes in the form of layers or as a component of 3D scaffolds can also be used as substrates for attachment, growth and electrical stimulation of neuronal cells [58]. For example, layers of multiwalled carbon nanotubes functionalized with 4-hydroxynonenal were successfully used as substrates for growing embryonic rat brain neurons *in vitro* [59].

Similarly as fullerenes, carbon nanotubes are relatively highly reactive, able to be functionalized, to form complexes with various atoms and molecules, and to emit photoluminescence. Due to these properties, both fullerenes and carbon nanotubes can be used in a wide range of advanced imaging, diagnostic and therapeutic strategies, such as transmission electron microscopy of DNA [60], labeling and tracer studies [1,61-63], magnetic resonance [25,26,64], radiotherapy of tumors, including advanced methods of boron neutron capture therapy [65,66], controlled drug and gene delivery [5-8,67] and biosensor technology [23,68-70]. The elongated shape of carbon nanotubes, resembling "tiny drinking straws", predisposes them to serve as tips for ultra-high-resolution atomic-force microscopy, and scanning tunneling microscopy [33,71,72]. In addition, carbon nanotubes and related materials, such as carbon nanohorns and nanotube-derived carbon foam, are promising substrates for storage of hydrogen as a potential source of energy [34,73-75], and also for storage of methane, a key technology for producing natural gas vehicles [76].

Carbon nanotubes also represent a novel class of antimicrobial agents [77], ion (e.g., K^+) channel blockers [78], artificial ion and water channels and components of nanoporous membranes [79-81].

On the other hand, the reactivity of nanotubes, their ability to emit electrons from their ends and to generate free radicals, can induce oxidative stress, inflammation and alterations of various biomolecules and cellular structures, such as genes, transcription factors, enzymes, extracellular matrix molecules, mitochondria and cell membranes [40,82-85], which should be taken into account in the design of nanotube-based materials for biomedical applications.

Nanocrystalline diamond (NCD) has also attracted considerable attention as a promising material for advanced biomedical applications. In the form of a powder, NCD has been prepared by detonation synthesis from a mixture of trinitrotoluene and hexogene [86]. Thin NCD films can also be deposited using microwave plasma-enhanced chemical vapor deposition (MW PECVD) from methane or fullerene C_{60} precursors [86,87]. Similarly as fullerenes and nanotubes, NCD can also be relatively easily functionalized with various chemical groups and biomolecules (e.g., nucleic acids, enzymes, antibodies and other proteins) with strong covalent bonding [88-90]. At the same time, in comparison with fullerenes and nanotubes, NCDs are practically non-toxic and thus they are the most advantageous among all carbon nanoparticles for biomedical applications [88,91]. In the form of a powder, NCD could serve as a carrier for controlled drug and gene delivery, for advanced imaging technologies or as an antioxidant and anti-inflammatory agent [91-93]. NCD films exhibit a wide range of unique physicochemical properties, such as mechanical hardness, chemical and thermal resistance, excellent optical transparency and controllable electrical properties (e.g., insulator/semiconductor/metal-like). Thus, biocompatible NCD films could markedly improve the mechanical and other physical properties of medical implants, particularly those designed for hard tissue surgery. Nanocrystalline diamond has already been used for coating the head and cup of artificial joint replacements, e.g., prostheses of temporomandibular joints [94]. In addition, NCD films could be applied in the bone-anchoring stems of articular prostheses or other permanent bone implants in order to improve their integration with the surrounding bone tissue. In studies *in vitro*, NCD provided excellent substrates for adhesion, growth, metabolic activity and maturation of human osteoblast-like MG 63 cells, human SAOS-2 osteoblasts, mouse MC3T3 osteoblasts, as well as other non-osteogenic cell types, such as human cervical carcinoma HeLa cells, rat pheochromocytoma PC12 cells or bovine pulmonary artery endothelial CPAE cells [57,88,95-97]. Type IIa diamond, microcontact-printed with laminin, supported adhesion of mouse cortical neurons and neurite outgrowth [98]. In studies *in vivo*, diamond layers, deposited by an MW CVD method on TiAl6V4 probes implanted into a rabbit femur, showed very high bonding strength to the metal base and also to the surrounding bone tissue, without any problems with corrosion [99]. Last but not least, similarly as fullerenes and nanotubes, NCDs are promising building blocks for constructing biomarkers, microchips, nanorobots or biosensors [88-91,100].

Layers of *other carbon allotropes*, e.g., a-C:H, graphite or hydrocarbon on pristine and modified polymers, are also characterized by nanostructure (i.e., the presence of particles less than 100 nm), in addition to their atomic structure. These layers are represented by carbon structures formed on the surface of synthetic polymers after irradiation, e.g., with plasma, by an ion beam or laser beam, or after modification of the material surface by magnetron sputtering, vacuum evaporation or chemical vapor deposition (CVD). These carbon allotropes could be used for surface modification of various materials designed for the construction of body implants or tissue engineering, e.g., synthetic polymers, metals or carbon-based composites [101-103]. These allotropes can be enriched with metals in order to improve their mechanical properties and to enhance their bioactivity [103-105].

Among all these exciting findings, we have concentrated on the influence of carbon nanoparticles, particularly when arranged into layers and used for biomaterial coating, on cell-substrate adhesion, subsequent growth, differentiation and viability of cells, especially bone-forming cells. A very interesting finding has been reported that nanostructured surfaces can promote preferential adhesion and growth of osteoblasts over other “competitive” cell types, including fibroblasts, and thus they can prevent fibrous encapsulation and loosening of bone implants. It is considered that the underlying mechanism is higher adsorption to the nanostructured surfaces of vitronectin, an extracellular matrix protein preferred by osteoblasts. Therefore, it can be expected that carbon nanoparticles may serve as novel building blocks for creating artificial bioinspired nanostructured surfaces for bone tissue engineering [53,54,106-111]. The preferential adhesion of osteoblasts could be further enhanced by the functionalization of carbon nanoparticles by the amino acid sequence Lys-Arg-Ser-Arg (KRSR), a ligand for adhesion receptors on osteoblasts [112-115]. In addition, carbon nanoparticles could be admixed to synthetic polymers, chitosan and other matrices designed for the fabrication of three-dimensional porous scaffolds for tissue engineering. The nanoparticles would form prominences and create a nanopattern on the pore walls, and thus promote the ingrowth of cells inside the scaffolds [109,116].

Therefore, in our interdisciplinary studies, the following types of materials modified with carbon nanoparticles were investigated:

(1) Continuous layers of fullerenes C_{60} deposited on microscopic glass coverslips or composites with a carbon matrix reinforced with carbon fibers, i.e., a material applicable in hard tissue surgery.
(2) Micropatterned layers created by deposition of fullerenes C_{60} through metallic grids on microscopic glass coverslips.
(3) Hybrid layers created by co-deposition of fullerenes C_{60} and metals on microscopic glass coverslips.
(4) Polymeric materials (i.e., a terpolymer of polytetrafluoroethylene, polyvinyldi fluoride and polypropylene, or polysulphone) mixed with single- or multi-walled carbon nanotubes.
(5) Nanostructured or hierarchically micro- and nanostructured nanocrystalline diamond layers deposited on silicon substrates.
(6) Layers of amorphous hydrogenated carbon (a-C:H), pyrolytic carbon, pyrolytic graphite as well as nanocomposite metal-hydrocarbon plasma polymer films.

The adhesion, growth, viability, metabolic activity and maturation of human osteoblast-like MG 63 cells were then studied in cultures on these materials. The cell behavior correlated with the physical and chemical properties of these materials, and also with results obtained by other authors.

2. Cell Growth on Fullerene C_{60} Layers

2.1. Continuous C_{60} Layers

In our experiments, fullerenes C_{60} (purity 99.5 %, SES Research, U.S.A.) were deposited on to microscopic glass coverslips (Menzel Glaser, Germany; diameter 12 mm) by evaporation of C_{60} in the Univex-300 vacuum system (Leybold, Germany) in the following conditions: room temperature of the substrates, C_{60} deposition rate $\leqslant$ 10 Å/s, temperature of C_{60} evaporation in the Knudsen cells about 450° C, time of deposition up to 50 minutes (1 A). The thickness was measured by atomic force microscopy (AFM). A scratch was made in the layer and its profile was measured in contact mode. The thickness of the layers increased proportionally to the temperature in the Knudsen cell and the time of deposition. We prepared two continuous films of different thicknesses, i.e., a thin layer of 505 ± 43 nm and a thick layer of 1090 ± 8 nm. Optical microscopy revealed that the fullerene layers were brownish in color (Figure 1 B), and the color intensity increased with layer thickness. Raman spectra of the deposited C_{60} films were measured in the back-scattering geometry at room temperature using a Ramascope 1000 Raman microscope (Renishaw, UK) with a 514.5 nm excitation wavelength Ar-ion laser. Immediately after deposition, the Raman spectra showed that the prepared fullerene films were of high quality, confirmed by a high peak $A_g(2)$ at wavenumber 1468 cm^{-1}, low peaks $H_g(7)$ and $H_g(8)$ and the absence of D (disorder, ~ 1350 cm^{-1}) and G (graphitic, ~ 1600 cm^{-1}) bands, which are signs of fragmentation and graphitization of C_{60}, respectively (Figure 2). The surface wettability of the films was estimated from the contact angle measured by a static method in a material-water droplet system using a reflection goniometer. The fullerene C_{60} films were relatively hydrophobic, having contact angles 100.6 ± 6.8° for thin and 96.7 ± 1.7° for thick continuous layers, respectively [117,118].

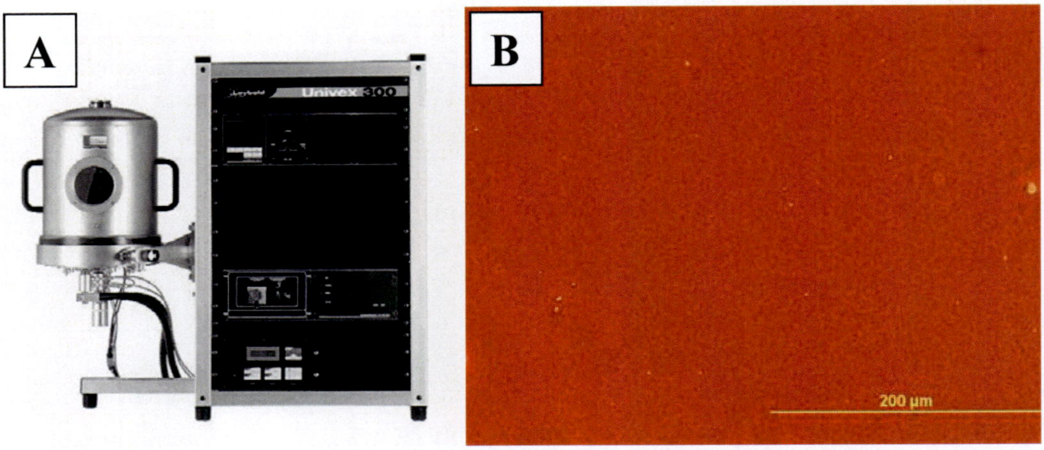

Figure 1. A: Univex-300 vacuum system (Leybold, Germany); B: continuous fullerene C_{60} layer (thickness of 1090 ± 8 nm) deposited on to a microscopic glass coverslip in this system [117,118].

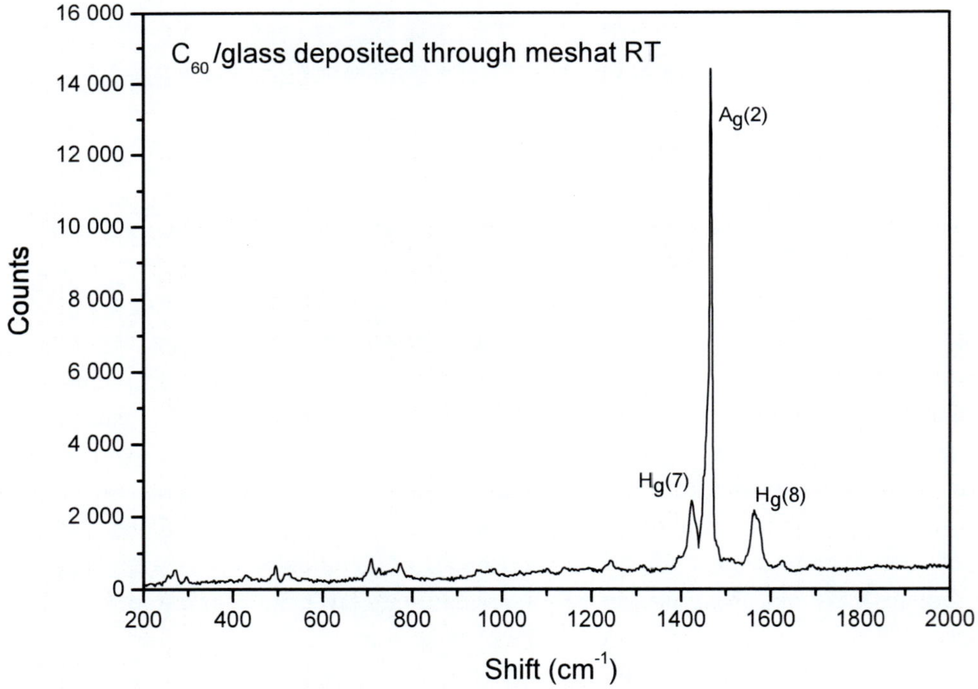

Figure 2. Raman analysis of a fullerene C_{60} layer immediately after deposition on glass coverslips. The quality of the fullerene layer is confirmed by a high peak $A_g(2)$ at wavenumber 1468 cm^{-1}, low peaks $H_g(7)$ and $H_g(8)$ and absence of D (disorder, ~ 1350 cm^{-1}) and G (graphitic, ~ 1600 cm^{-1}) bands, which are signs of fragmentation and graphitization of C_{60}, respectively [117,118].

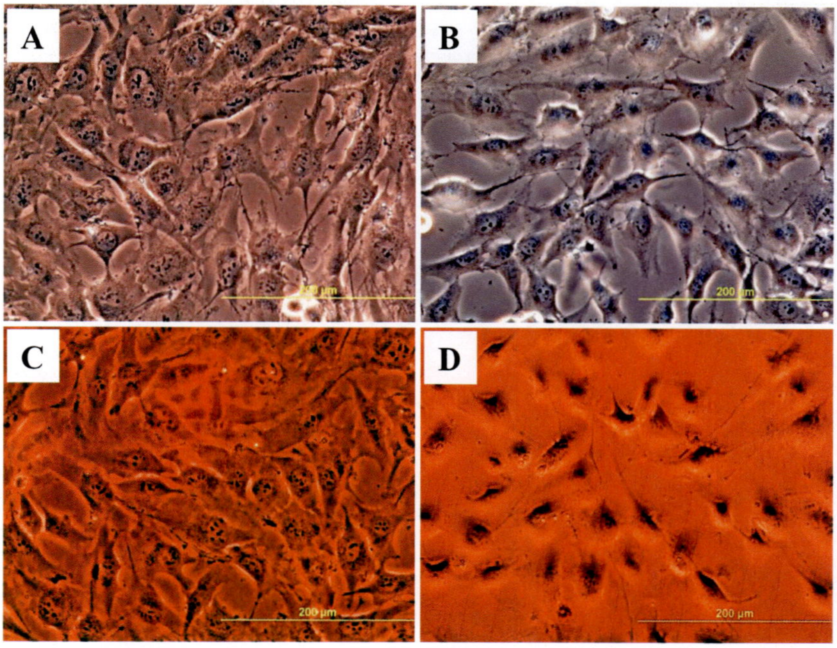

Figure 3. Morphology of human osteoblast-like MG 63 cells in 7-day-old cultures on a polystyrene dish (A) a microscopic glass coverslip (B), thin continuous (C) and thick continuous (D) fullerene C_{60} layers. Native cultures, Olympus IX 50 microscope, DP 70 digital camera, obj. 20x, bar = 200 μm [117].

To cultivate the cells, the fullerene-coated glass coverslips were sterilized with 70% ethanol for 1 hour, inserted into 24-well polystyrene multidishes (TPP, Switzerland; diameter 15 mm) and seeded with human osteoblast-like MG 63 cells (European Collection of Cell Cultures, Salisbury, UK). Each dish contained 5,000 cells (i.e., about 2,830 cells/cm^2) and 1.5 ml of Dulbecco's modified Eagle's Minimum Essential Medium supplemented with 10% fetal bovine serum and gentamicin (40 μg/ml). The cells were cultured for 1, 3 and 5 or 7 days at 37° C in a humidified air atmosphere containing 5% of CO_2. On day 1 after seeding, the cells on both continuous thin and continuous thick fullerene layers adhered in similar numbers (3,420 ± 420 cells/cm^2 and 2,880 ± 440 cells/cm^2, respectively), which was comparable to the values found on standard cell culture substrates, represented by the tissue culture polystyrene dish (3,080 ± 290 cells/cm^2) and the microscopic glass coverslip (2,560 ± 310 cells/cm^2). In addition, the cells on the fullerene layers and on the control substrates were of similar morphology, i.e., mostly polygonal and spread over a similar area (Figure 3). From days 1 to 5 after seeding, the cells on both continuous C_{60} films, polystyrene and glass also proliferated with similar cell population doubling times ranging from 18.6 ± 0.3 h to 19.6 ± 0.5 hours, and on day 5 they reached similar cell population densities (from 91 600 ± 9500 cells/cm^2 to 112 700 ± 9800 cells/cm^2; Figure 4). In addition, as revealed by staining the adhered cells with calcein and ethidium homodimer using the LIVE/DEAD kit, the percentage of viable cells on days 1 to 7 (after seeding on all fullerene layers) ranged from 80 ± 10 % to 100 ± 15 %, and was similar to the values obtained on the polystyrene culture dishes (99 ± 24 % to 100 ± 15 %), and also on the microscopic glass coverslips (88 ± 21 % to 99 ± 26 %). The trypan-blue exclusion test (performed on the trypsinized cells while they were being counted in the ViCell Analyser) also showed similar proportions of viable cells on the fullerene layers (82 ± 11 % to 89 ± 13 %), as well as on standard cell culture surfaces (80 ± 13 % to 95 ± 13 %). The average viability of the cells tended to increase with time of cultivation.

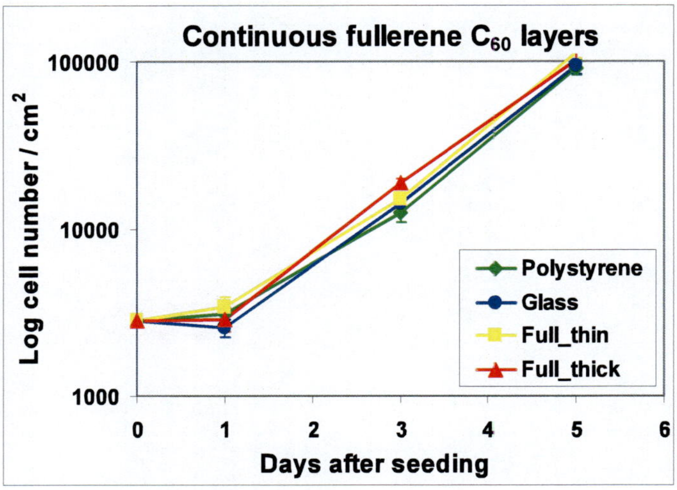

Figure 4. Growth curves of human osteoblast-like MG 63 cells in cultures on polystyrene dishes, microscopic glass coverslips or continuous thin and thick fullerene C_{60} layers. Mean ± S.E.M. (Standard Error of Mean) from 9 to 30 measurements. No statistically significant differences were found by ANOVA, Student-Newman-Keuls Method [117].

The supportive effects of fullerene C_{60} layers on cell colonization could be explained by their surface nanostructure, mimicking the nanoarchitecture of the natural extracellular matrix (ECM), basal lamina, as well as the cell membrane. The nanostructure of the cell adhesion substrate has been reported to enhance the adsorption of cell adhesion-mediating ECM molecules, such as fibronectin, collagen and especially vitronectin, provided by the serum of the culture medium or synthesized by the cells. The spatial conformation of the adsorbed molecules on our films may also be more appropriate for the accessibility of specific sites on these molecules by cell adhesion receptors, e.g., integrins [53,54,106-108,110,116,121,122]. As revealed by immunofluorescence staining of the MG 63 cells on our samples, β_1 integrins (i.e., receptors for collagen, fibronectin and vitronectin), and talin (an integrin-associated protein) were assembled in dot- or streak-like focal adhesion plaques, mainly visible on the cell periphery (Figure 5), and these processes were accompanied by the formation of a fine mesh-like β-actin cytoskeleton and by the presence of a considerable amount of osteopontin, a marker of osteogenic cell differentiation (Figure 6). All these events can be considered signs of active cell-substrate interaction, followed by signal transduction inside the cells. Moreover, β_1 integrins have been reported to enhance osteogenic cell differentiation, which is manifested by higher levels of osteocalcin, higher activity of alkaline phosphatase and a more pronounced response to 1,25-dihydroxyvitamin D_3 [111].

The proliferation and differentiation of chondrocytes (another cell type important in bone reconstruction) were also promoted by fullerenes. C_{60} added in cell culture enhanced the production of specific large proteoglycan ECM molecules, typical for cartilage, in rat embryonic limb bud cells in culture [123].

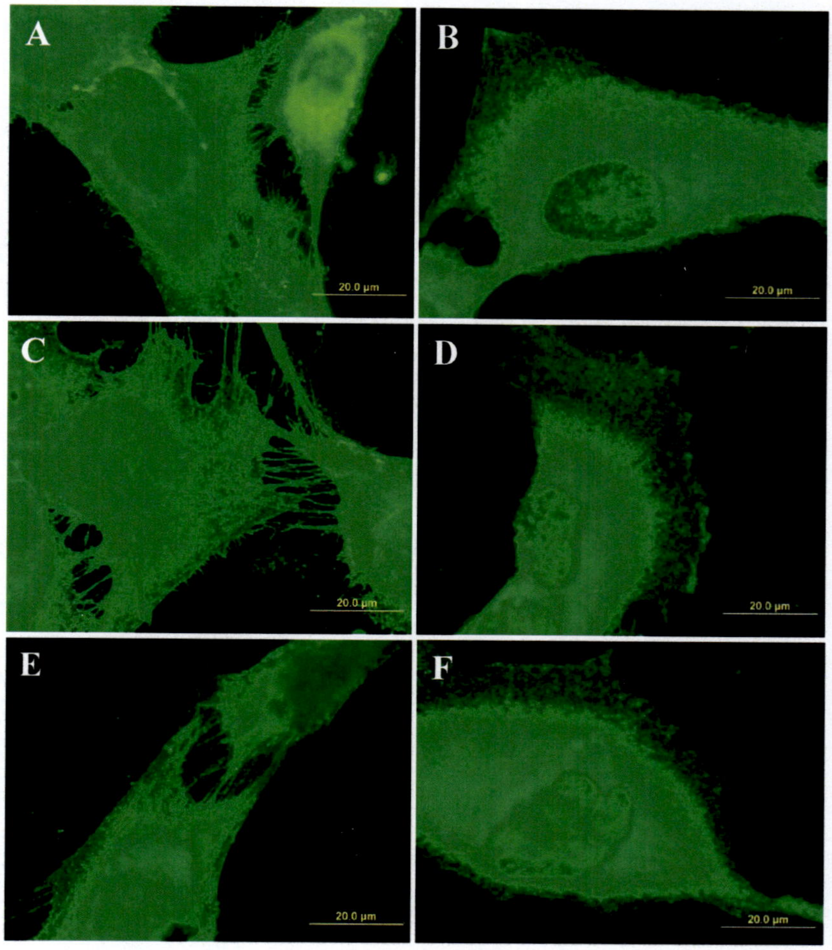

Figure 5. Immunofluorescence staining of β_1 integrins (A, C, E) and talin (B, D, F) in human osteoblast-like MG 63 cells on day 3 after seeding on microscopic glass coverslips (A, B) and thin (505 ± 43 nm; C, D) or thick (1090 ± 8 nm; E, F) continuous fullerene C_{60} layers. Olympus IX 50 epifluorescence microscope, DP 70 digital camera, obj. 100x, bar = 20 µm [117].

Similarly, C_3-fullero-*tris*-methanodicarboxyllic acid (a water-soluble fullerene derivative), supported the formation of focal adhesion plaques, an assembly of an actin cytoskeleton and cell spreading in human epidermoid carcinoma cells exposed to UV light, probably due to its antioxidative action [124]. The free radical scavenger activity of fullerenes and their derivatives (e.g., hydroxylated, carboxylated, butylated or sulphonated C_{60}) also protected keratinocytes from apoptosis induced by ultraviolet light B [27], minimized the oxidative damage of small bowel transplants [18], attenuated ischemia-reperfusion-induced lung injury [22], prevented focal cerebral ischemia and decreased neuronal degeneration and death [19,20]. However, hydroxylated fullerene derivatives protected cells against oxidative stress only at lower concentrations (10-50 µM in cell culture media), while at higher concentrations (1 mM and 1.5 mM), they caused cell death [22]. Hydroxylated fullerenes usually induced cell death characterized by signs of apoptosis, such as increased caspase activity, DNA fragmentation and annexin V binding to the phosphatidylserine, exposed on the outer side of the cell membrane, without increased membrane permeability [28]. On the other hand, a fullerene derivative $C_{60}(OH)_{24}$ (concentrations from 1 to 100 μg/ml) did not seem to induce apoptosis but caused the accumulation of polyubiquitinated proteins and facilitated autophagic cell death in human umbilical vein endothelial cells [125]. Fullerene colloids in water, also called nano-C_{60}, i.e. fullerene aggregates that readily form when pristine C_{60} is added to water, mainly induced cell necrosis characterized by oxidative damage to the cell membrane (e.g., peroxidation of the membrane lipids) and loss of membrane integrity [10], though the genotoxicity of these agents has also been demonstrated on human lymphocytes [11]. The harmful effects of fullerenes C_{60}, usually potentiated by the light, especially ultraviolet light, could be utilized for antimicrobial therapy or photodynamic therapy against tumor cells [126,127]. In addition, a nanocrystalline suspension of fullerenes C_{60} and C_{70} prepared by the solvent exchange method using tetrahydrofuran (THF/n$C_{60/70}$) augmented the cytotoxicity of the tumor necrosis factor [128,129].

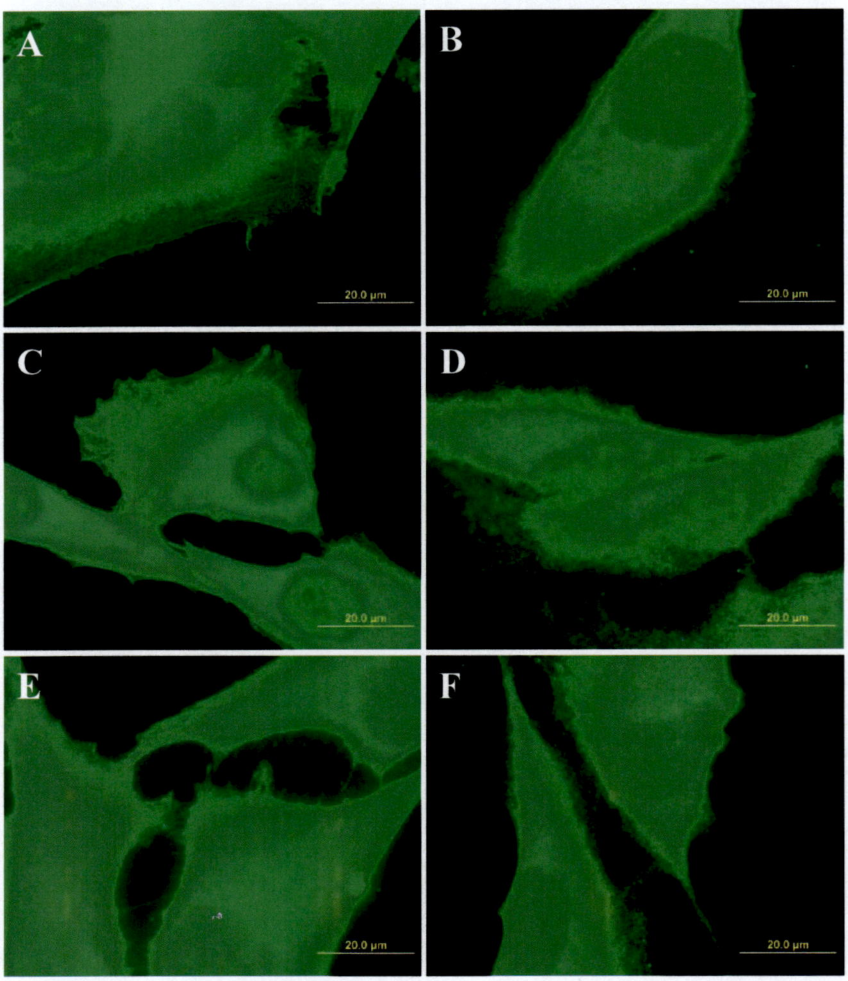

Figure 6. Immunofluorescence staining of β-actin (A, C, E) and osteopontin (B, D, F) in human osteoblast-like MG 63 cells on day 3 after seeding on microscopic glass coverslips (A, B) and thin (505 ± 43 nm; C, D) or thick (1090 ± 8 nm; E, F) continuous fullerene C_{60} layers. Olympus IX 50 epifluorescence microscope, DP 70 digital camera, obj. 100x, bar = 20 μm [117].

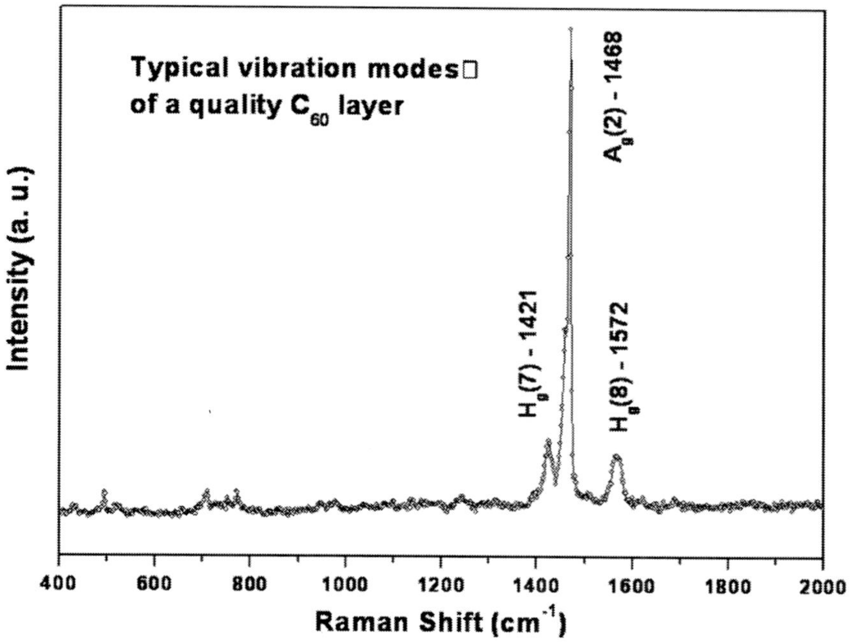

Figure 7. Raman analysis of a thin C_{60} layer deposited on a carbon-fibre reinforced composite (CFRC) substrate. The quality of the fullerene layer is confirmed by a high peak $A_g(2)$ at wavenumber 1468 cm^1, low peaks $H_g(7)$ and $H_g(8)$ and absence of D (disorder, ~ 1350 cm^{-1}) and G (graphitic, ~ 1600 cm^{-1}) bands, which are signs of fragmentation and graphitization of C_{60}, respectively [95].

Continuous layers of fullerenes C_{60}, experimentally developed on glass, were also tested on composites with the carbon matrix reinforced with carbon fibers (i.e., carbon fiber-reinforced carbon composites, CFRC), which have been considered as promising materials applicable in hard tissue surgery [104,130-133]. These composites were prepared at the Institute of Rock Structure and Mechanics, Acad. Sci. CR, Prague. Briefly, a commercially available woven fabric (made of Toray T 800 carbon fibers) was arranged in layers, infiltrated with a carbon matrix precursor (phenolic resin UMAFORM LE, Synpo Ltd., Pardubice, CR), pressed, cured, carbonized at 1000°C, and finally graphitized at 2200°C [104,130-133]. Prior to the deposition of fullerenes, the CFRCs were ground using a metallographic paper of 4000 grade in order to decrease the material surface micro-scale roughness, which has been shown to be less appropriate for spreading and subsequent growth of bone-derived and also vascular smooth muscle cells [104,130,131]. This grinding lowered the surface roughness of CFRC about twice. As measured by a profilometer (Rank Taylor Hobson Ltd., England), the departures of the roughness profile from the mean line (i.e., R_a parameter) decreased from 6.5 ± 1.8 μm to 3.5 ± 0.6 μm, and the mean spacing of the adjacent local peaks (parameter S) lengthened from 38 ± 11 μm to 96 ± 49 μm. The fullerene coating did not significantly change this surface microroughness, but created a nanostructured pattern on the pre-existing microarchitecture of the CFRC surfaces. The fullerene C_{60} layers were deposited on the CFRC by evaporation of C_{60} in the Leybold Univex-300 vacuum system (room temperature of the substrates, deposition rate ⩽ 1 Å/s, temperature of C_{60} evaporation in the Knudsen cells 450°C, time of deposition about 15 min, thickness of the fullerene layers < 100 nm). Similarly as in the continuous fullerene layers (deposited on microscopic glass coverslips), the Raman analysis confirmed again that the prepared fullerene films were of high quality and with no fragmentation or graphitization of C_{60} (Figure 7). One third of the CFRC samples (3 x 3 cm) were protected against fullerene coating by a mask in order to achieve regionally-selective growth of the fullerene layer. The other completely uncoated CFRC samples, as well as tissue culture polystyrene (TCPS) dishes were used as control materials [95].

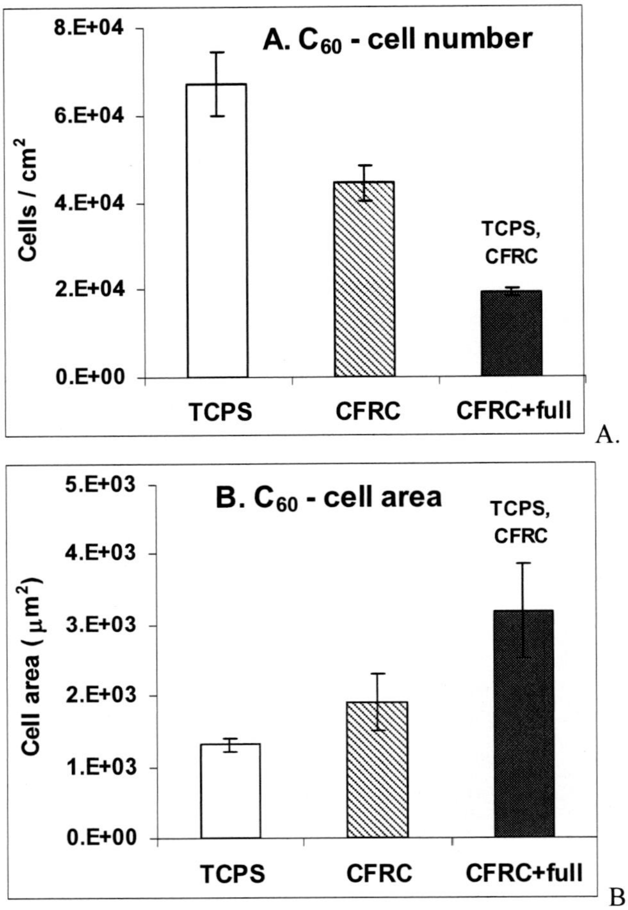

Figure 8. Population density (A) and adhesion area (B) of osteoblast-like MG 63 cells on day 2 after seeding on a tissue culture polystyrene dish (TCPS), carbon fibre-reinforced carbon composites (CFRC) and CFRC coated with a fullerene layer (CFRC+full). Mean ± S.E.M. (Standard Error of Mean) from 4-12 measurements, ANOVA, Student-Newman-Keuls method. Statistical significance: $^{TCPS, CFRC}$: $p \leq 0.05$ compared to the values on tissue culture polystyrene and pure CFRC [95].

The material was then seeded with human-osteoblast-like MG 63 cells. Unlike the fullerene layers on glass, the cell population density on day 2 after seeding (Figure 8 A) on C_{60}-coated CFRC was significantly lower (about 19,000 ± 700 cells/cm^2) than that on the control uncoated CFRC and polystyrene dishes (44,000 ± 4,000 cells/cm^2 and 67,000 ± 7,000 cells/cm^2, respectively). This finding could be attributed to a combination of at least two factors less favorable for cell adhesion, such as the remaining surface microroughness and relatively high hydrophobicity of the non-functionalized fullerenes (static water drop contact angle about 100°, see above). It is known that hydrophobic materials promote preferential adsorption of cell non-adhesive proteins from the serum of the culture media, such as albumin. In addition, the cell adhesion-mediating extracellular matrix proteins may be adsorbed in a relatively rigid state, thus their specific amino acid sequences were less accessible for cell adhesion receptors [121,122]. On the other hand, the water drop contact angle, measured by a static method in a material-water droplet system using a reflection goniometer, was not able to reveal any hydrophobicity of the fullerene layer deposited on CFRCs. The contact angle was non-measurable due to complete absorption of the water drop into the fullerene layer, which suggested a certain non-compactness or porosity of this layer on CFRC. The contact angle of the non-coated CFRC was 99.5 ± 1.0°.

The release of fullerenes into the culture media and their cytotoxic action seemed to be less probable in our experiments as an explanation of the lower number of cells adhered to C_{60}-coated CFRC. The cell attachment and spreading on the uncoated regions of the fullerene-modified CFRC, and also on the bottom of the polystyrene dishes containing fullerene-coated samples, were similar as in the control polystyrene dishes without fullerene samples (Figure 9). At the same time, the fullerene layer was resistant to mild wear, represented by swabbing with cotton, rinsing with liquids (water, phosphate-buffered saline, culture media) and exposure to cells and proteolytic enzymes (trypsin) used for cell harvesting. After these procedures and/or one-year storage at room temperature in a dark place, the Raman spectra did not change significantly. The peak $A_g(2)$ at wavenumber 1468 cm^{-1} still remained the highest, though some of the peaks $H_g(7)$ and $H_g(8)$ increased slightly, and additional small D and G bands appeared, which indicated some fragmentation, graphitization, polymerization or oxidation of C_{60} molecules.

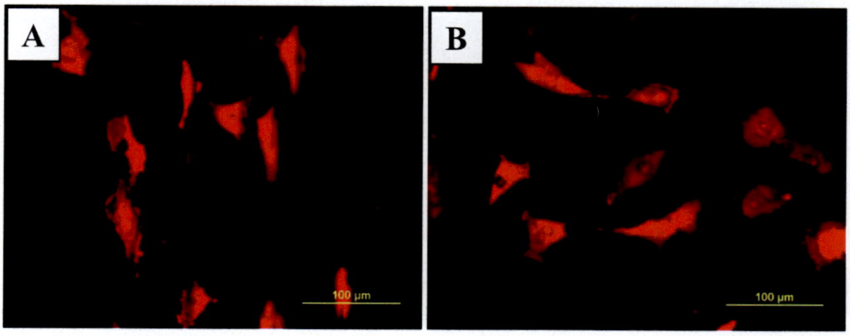

Figure 9. Human osteoblast-like MG 63 cells on day 1 after seeding on the uncoated region of CFRC partially coated by fullerenes (A) and the bottom of a polystyrene dish containing a fullerene-coated CFRC sample (B). Cells fixed with 70% ethanol and stained with propidium iodide. Olympus IX 50 microscope, DP 70 digital camera, obj. 20, bar=100 µm [95].

In addition, the fullerene-coated CFRC surfaces were stronger and less prone to release carbon particles, which is an important limitation to the potential biomedical use of CFRC [104,130,131]. Moreover, the spreading area of cells on the fullerene-coated samples amounted to 3,182 ± 670 μm^2, while on both control surfaces it was only 1,888 ± 400 and 1,300 ± 102 μm^2 (Figure 8 B). This could be explained by the low cell population density on the fullerene layer, which provided the cells with more space on which they could spread. On the other hand, the nanostructure of the fullerene layer might enhance the adsorption of vitronectin, i.e., an extracellular matrix protein promoting adhesion of osteoblasts [106-108]. The MG 63 cells on the fullerene-coated CFRC formed dot-like vinculin-containing focal adhesion plaques and a fine network of actin microfilaments (Figure 10), which is a sign of cell vitality and effective binding between cell adhesion receptors and extracellular matrix molecules adsorbed on the material surface [122].

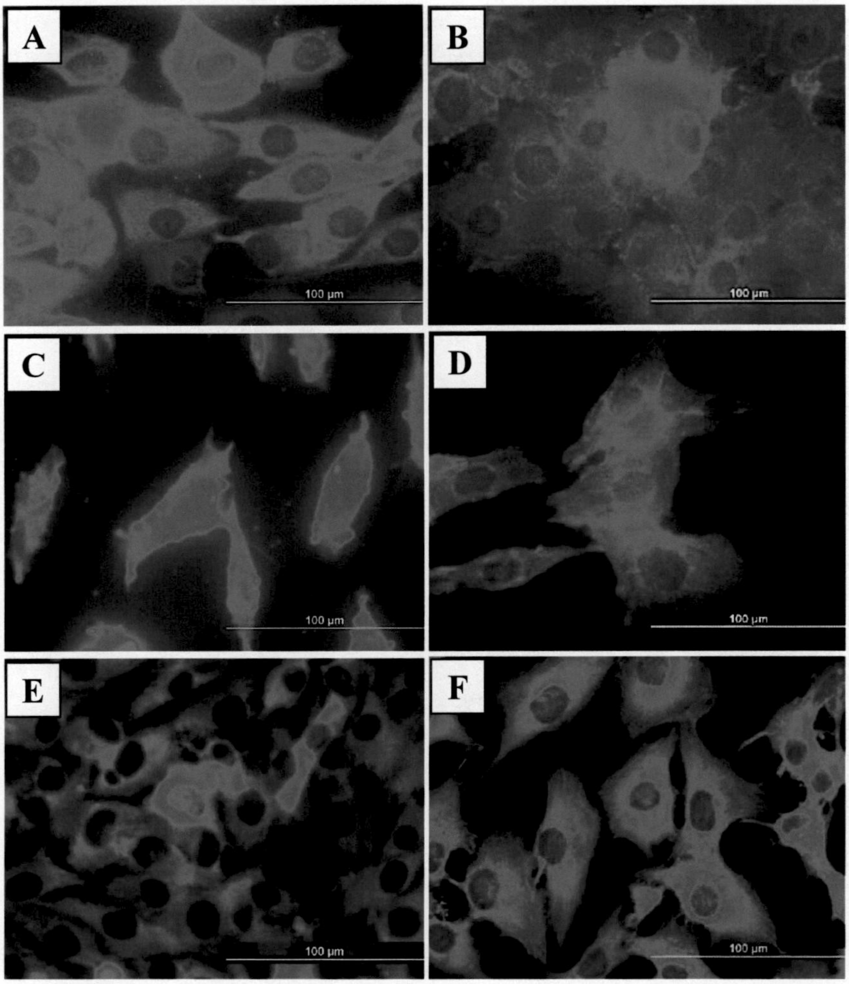

Figure 10. Immunofluorescence staining of vinculin (A, C, E) and β-actin (B, D, F) in osteoblast-like MG 63 cells on day 2 after seeding on CFRC (A, B), CFRC coated with a fullerene layer (C, D) and tissue culture polystyrene dish (E, F). Olympus IX 50 microscope, DP 70 digital camera, obj. 40, bar = 100 μm [95].

2.2. Micropatterned C_{60} Layers

An interesting idea is that fullerenes might be used for constructing patterned surfaces for regionally-selective adhesion and directed growth of cells. Domains adhesive for cells on the material surface have been created using a wide range of physico-chemical and biochemical approaches. These approaches involve attachment of moderately hydrophilic molecules (e.g., acrylic acid, [134]) or attachment of various chemical functional groups, particularly those containing oxygen. The latter can also be created by irradiation of the material surface, especially the surface of synthetic polymers, by plasma, ions or ultraviolet light [121,135-140]. In addition, the material surface can be functionalized by various biomolecules, particularly ligands for cell adhesion receptors, such as oligopeptides containing amino acid sequences recognized by all cell types (sequence RGD) or preferentially by a certain cell type (e.g., REDV by endothelial cells, VAPG by smooth muscle cells or KRSR recognized by osteoblasts) [112-114,122,141]. Regions repulsive to cells can be prepared using hydrophobic compounds, such as octadiene or polysiloxanes, or extremely hydrophilic molecules, e.g. polyethylene oxide [134,141,142].

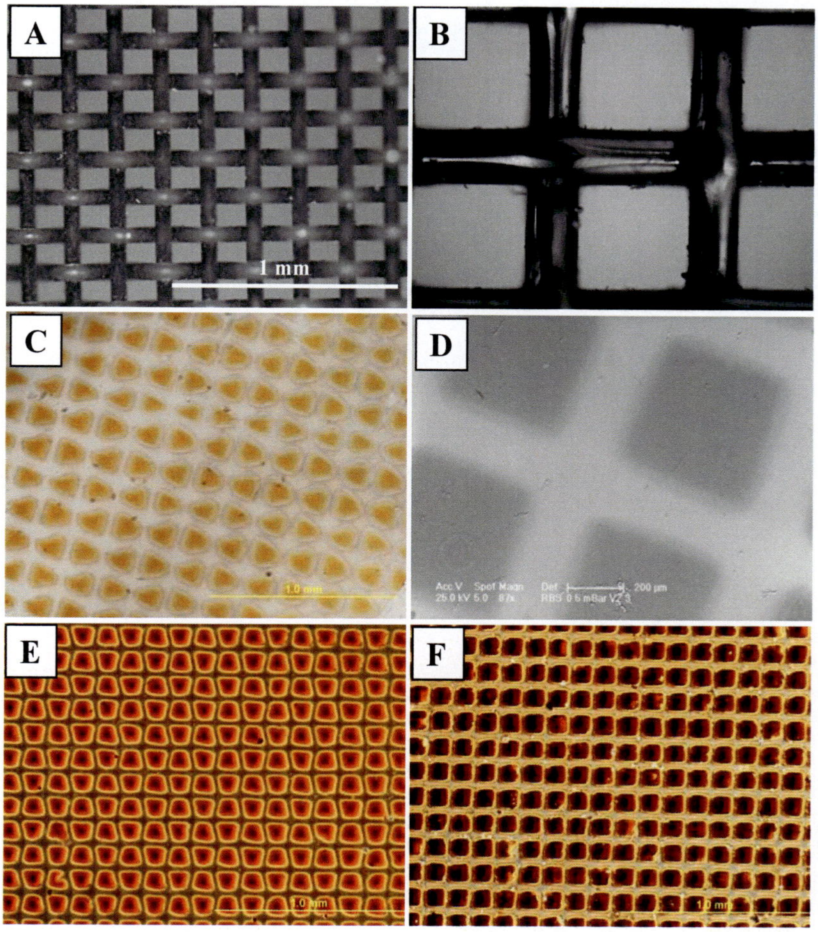

Figure 11. A, B: Metallic grids used for creating micropatterned fullerene C_{60} layers on microscopic glass coverslips. A: openings of 128 ± 3 μm per 98 ± 8 μm (about 12,500 μm^2) and 50 μm in distance, B: openings of 470 x 440 μm (about 207,000 μm^2), distance 200 μm. C-F: micropatterned layers created by the deposition of fullerenes C_{60} through a mesh with smaller (C, E, F) and larger (D) openings. The height of the fullerene protrusions was 128 ± 8 nm (C), 238 ± 3 nm (D), 484 ± 5 nm (E) and 1043 ± 57 nm (F). Olympus BX41 microscope (A, B), Olympus IX 50 microscope (C, E, F) and SEM-EDS scanning electron microscope (JOEL JSM-5600) (D). Objective 10x (A, B) or 4x (C, E, F); bar = 1mm (A, B, C, E, F) or 200 μm (D) [118].

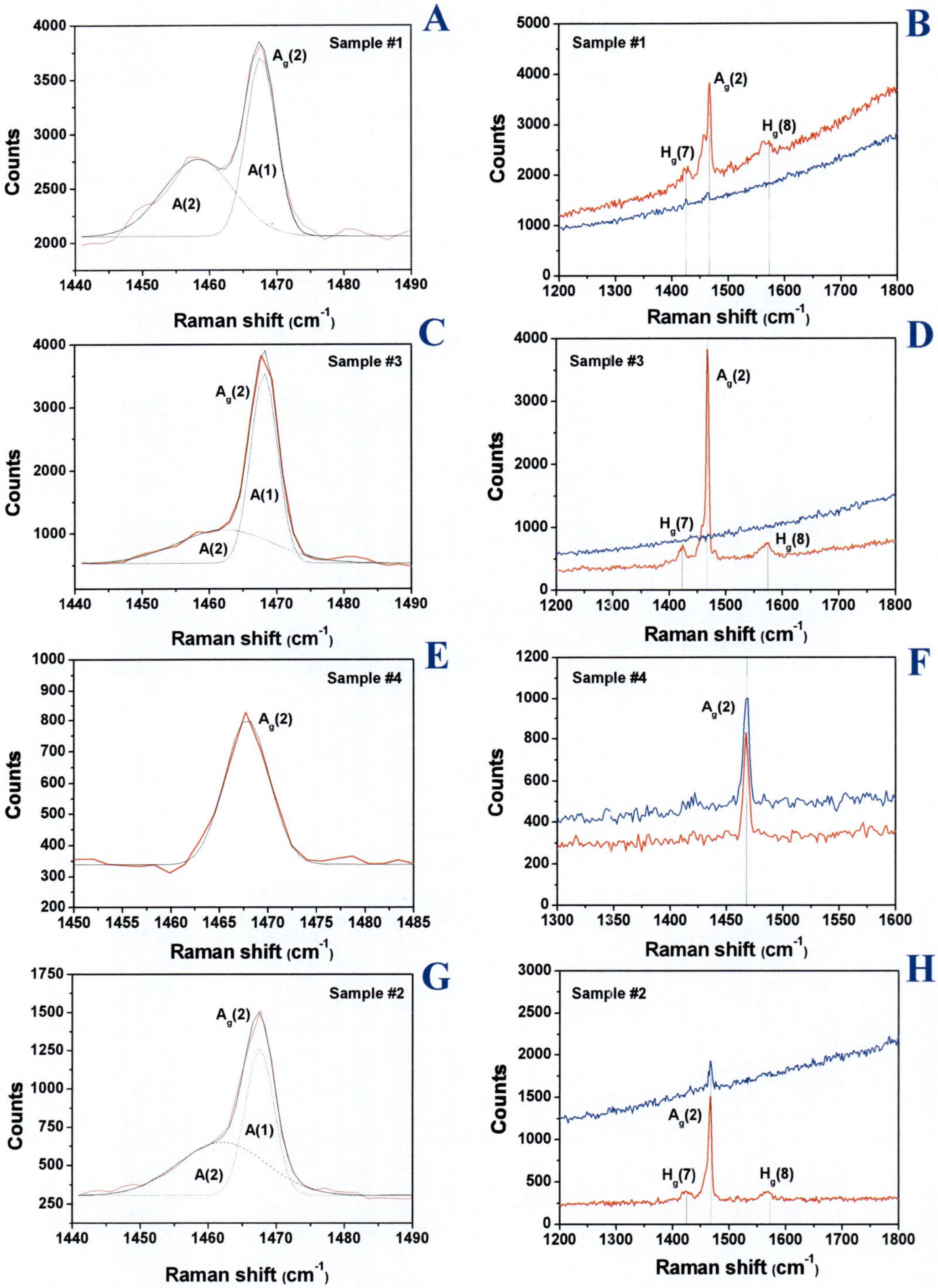

Figure 12. Raman analysis of microstructured C_{60} layers after 1-hour-sterilization with 70% ethanol. A, C, E, G: Gaussian peak fit analysis; B, D, F, H: Raman spectra below the openings (red line) or bars (blue line) of the metallic lattice. As revealed by AFM, the layer thickness below the openings and bars was 128 ± 8 and ≤ 8 nm, respectively, in the sample on B, 238 ± 3 nm and ≤ 3 nm on D, 484 ± 5 nm and 158 ± 5 nm on F, and 1043 ± 57 nm and ≤ 57 nm on H [117,118].

Patterned surfaces can also be created by changing the surface roughness and topography, i.e., by creating prominences and hollows of different size, shape and depth [116,143]. In our experiments, micropatterned layers were prepared by deposition of fullerenes C_{60} on microscopic glass coverslips through contact metallic masks with rectangular holes with an average size of 128 ± 3 μm per 98 ± 8 μm (about 12 430 μm^2) and 50 μm in distance, or 470 x 440 μm (about 207,000 μm^2), distance 200 μm (Figure 11). The fullerenes formed bulge-like prominences below the openings of the metallic grid. Similarly as in the continuous fullerene C_{60} films, the thickness of these prominences increased proportionally to the temperature in the Knudsen cell and the time of deposition. As revealed by AFM, the thickness of the layers on sites underlying the openings of the grid was 128 ± 8 nm, 238 ± 3 nm, 484 ± 5 nm or 1043 ± 57 (data presented as mean ± standard deviation). Fullerene layers were also formed below the metallic part of the grid, where their thickness was either within the size of the standard deviations (the 1st, 2nd and 4th layer) or it amounted to 158 ± 5 nm (the 3rd layer). Therefore, in the 3rd layer, the effective height of the fullerene prominences was 326 ± 5 nm [117,118].

For cell culture, the fullerene-modified samples were sterilized by 70% ethanol in water (1 hour, at room temperature). Analysis of the vibration mode $A_g(2)$ on the Raman spectra showed that, after sterilization, some of the micropatterned fullerene layers were almost intact (the 3rd layer, Figure 12 E), whereas in the other, some of the C_{60} molecules reacted with oxygen or polymerized, as indicated by the satellite peak A(2). The proportion of C_{60} molecules involved in these chemical changes was determined by the ratio A(2)/A(1), and reached about 30% in the 2nd layer (Figure 12 C) and 50% in the 1st and 4th layer (Figure 12 A, G). Similarly as in AFM, fullerenes were also found below the metallic bars of the grid, though in the thinner layers 1 and 2, the amount was very low and barely detectable (Figure 12 B, D, F, H).

Reflection goniometry showed that similarly to the continuous fullerene C_{60} layers, all of the micropatterned C_{60} layers were relatively highly hydrophobic. All micropatterned layers had similar water drop contact angles, i.e., 93.6 ± 3.6° (layer 1), 94.5 ± 5.5° (layer 2), 95.3 ± 3.1° (layer 3) and 95.6 ± 3.8° (layer 4).

Optical microscopy revealed that below the openings of the grids, the fullerene prominences were brownish in color, and the color intensity increased with layer thickness. Up to a thickness of 484 ± 5 nm, the layers were relatively well transparent in a conventional light microscope (Figure 11). Thus, the cells growing on the layers were well observable in native or hematoxylin- and eosin-stained cultures, and the fullerene bulges and grooves were also well distinguishable. However, the fullerene prominences 1043 ± 57 nm in thickness were less transparent, thus the cells were better observable after fluorescence staining (Figure 13).

The micropatterned layers were seeded with human osteoblast-like MG 63 cells. From day 1 to 7 after seeding, the cells on the layers with fullerene prominences up to 326 ± 5 nm were distributed almost homogeneously over the entire material surface (Figure 13), thus the cell population densities on the prominences and in the grooves were similar (Figure 14). At the same time, the cells on these layers adhered and grew to an extent comparable to that on tissue culture polystyrene dishes and microscopic glass coverslips. Similarly as in continuous fullerene C_{60} layers, these beneficial effects on cell colonization could be explained by their surface nanostructure, mimicking the nanoarchitecture of the natural extracellular matrix [53,54,106-110]. Moreover, due to the micropatterning, the fullerene layers in our study were constructed with a hierarchically-organized micro- and nanostructure, i.e., with a similar architectural principle as natural tissues [116]. The presence of oxygen, suggested in some of our fullerene layers by Raman spectroscopy, might also contribute to the supportive effects of these films on cell adhesion and growth. Oxygen-containing chemical functional groups have been repeatedly shown to enhance the colonization of various materials with cells [121,135-137,144]. In the case of fullerene-grafted polyurethane, the presence of oxygen-containing and amine functional groups (due to pre-treatment of the polymer with oxygen plasma and the use of aminosilanes as coupling agents) was also attributed to the enhanced adhesion and activation of platelets [120].

However, the cells on the layers with prominences of 1043 ± 57 nm were found preferentially in the grooves (Figure 13). These grooves contained from 75.0 ± 13.4 % to 99.6 ± 0.3 % of cells, though they occupied only 44 ± 3 % of the material surface. The cell population density in the grooves was about 5 to 192 times higher than on the bulges, and these differences increased proportionally to the time of cultivation (Figure 14). Preferential growth of bone cells in grooves, pits and other types of hollows has been observed on various artificially microfabricated or primary rough polymeric, metallic or carbon-based materials [104,116,130,143,145-148]. However, the cells were also able to colonize the prominences on these surfaces, though the prominences were usually much higher (i.e., several micrometers or even tens of µm) than those on the thick micropatterned fullerene layers in our present study (only about 1 µm). Surprisingly, on our fullerene C_{60} layers, the MG 63 cells were not able to "climb up" relatively low prominences only about 1 µm in height, even at a relatively late culture interval of 7 days after seeding. This may be due to a synergetic action of certain physical and chemical properties of the fullerene bulges less appropriate for cell adhesion, such as their hydrophobicity, a steep rise of the prominences, as well as the tendency of spherical ball-like fullerene C_{60} molecules to diffuse out of the prominences [149].

Fluorescence staining of cells using the LIVE/DEAD kit revealed that, on day 1 after seeding, the percentage of viable cells on all micropatterned fullerene layers ranged between 80 ± 10 % and 87 ± 20 %, and was similar to the values obtained on the control tissue culture polystyrene dishes (96 ± 21%) and microscopic glass coverslips (67 ± 16 %). It can be assumed that the presence of a certain number of dead cells was due to the stress to which the cells were submitted during the cell seeding procedure, particularly during trypsinization and keeping the anchorage-dependent cells in suspension, rather than due to a cytotoxic action of the fullerenes. Accordingly, the percentage of viable cells increased with time of cultivation, reaching 97 ± 25 % to 99 ± 15 % on day 7 after seeding on all tested surfaces.

Thus, micropatterned fullerene films could be used as templates for regionally-selective cell adhesion and growth. Moreover, osteogenic cells growing in hollows have been reported to be more active in the phosphorylation of various kinases and transcription factors, signal transduction and differentiation [143,145-148]. From this point of view, the microstructured fullerene layers (if they adhere strongly to the underlying materials and are resistant to the release of fullerene molecules) could be used as a bioactive coating for bone implants in order to improve their integration with the surrounding bone tissue.

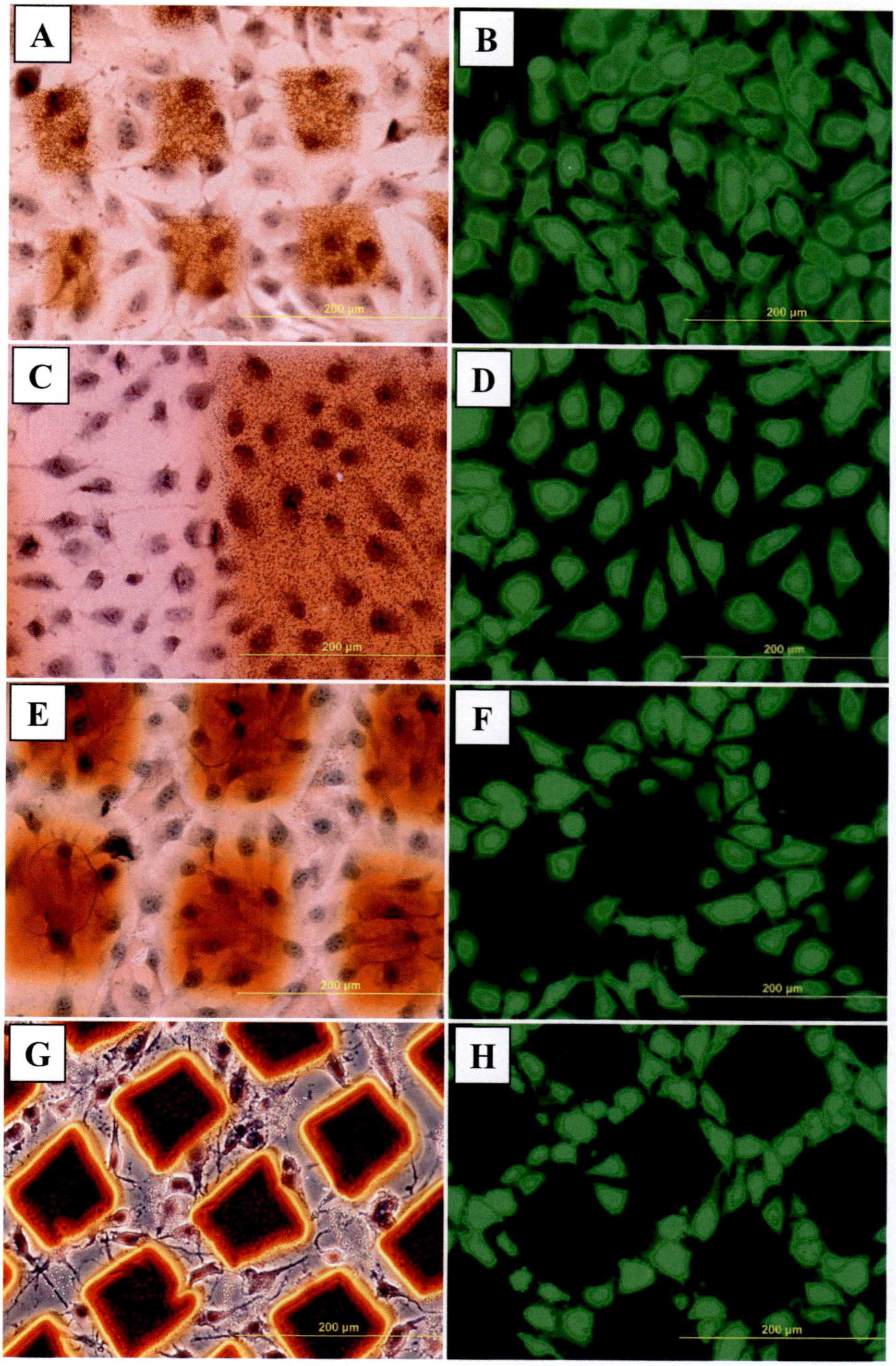

Figure 13. Human osteoblast-like MG 63 cells on day 7 after seeding on fullerene layers micropatterned with prominences 128 ± 8 nm in height (A, B), 238 ± 3 nm in height (C, D), 326 ± 5 nm in height (E, F) or 1043 ± 57 nm in height (G, H). Stained with hematoxylin and eosin (A, C, E, G) or LIVE/DEAD viability/cytotoxicity kit (B, D, F, H). Olympus IX 50 microscope, DP 70 digital camera, obj. 20x, bar = 200 µm [117, 118].

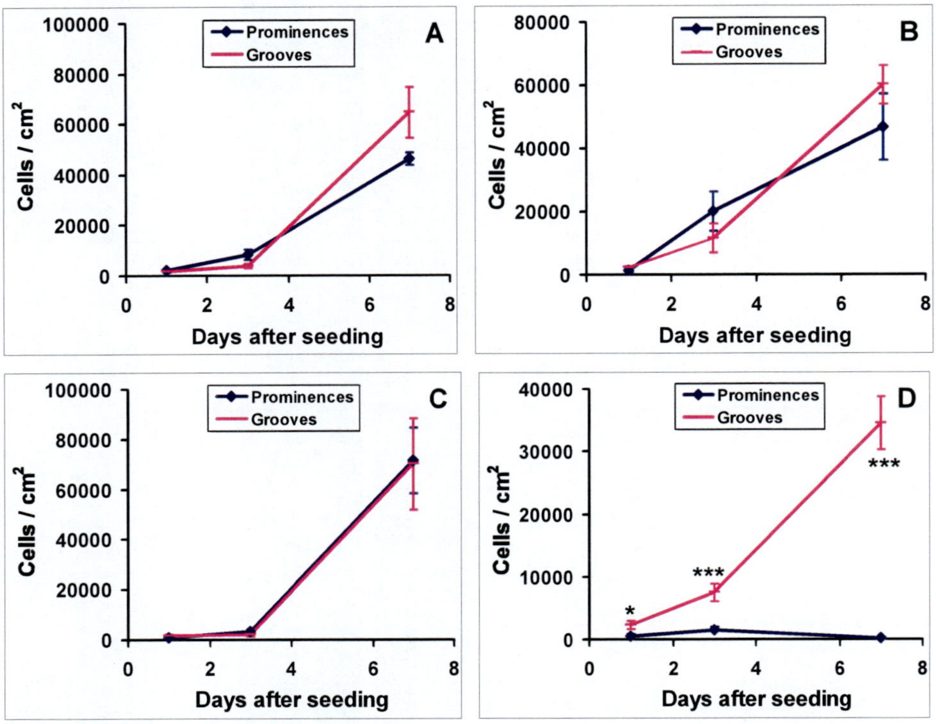

Figure 14. Population densities of MG 63 cells on prominences and grooves on fullerene layers micropatterned with prominences 128 ± 8 nm in height (A), 238 ± 3 nm in height (B), 326 ± 5 nm in height (C) or 1043 ± 57 nm in height (D). Mean ± Standard Error of Mean (S.E.M.) from 9 to 21 microphotographs, Student t-test, statistical significance: *p≤0.05, ***p≤0.001 [118].

2.3. Fullerene C_{60} Layers Modified by Ion- or Laser-irradiation

Both continuous and micropatterned fullerene C_{60} layers were irradiated with ion beams (Au^+ ions with energy of 1.8 MeV and fluence up to 2 x 10^{15} cm^{-2}) with the aim (i) to convert C_{60} into another biocompatible carbon allotrope, and/or (ii) to induce pattern formation in various scales and dimensions (in order to promote bio-compatibility and bio-attractivity of the micro- and/or nano-patterened carbon substrates with respect to the inspected MG 63 cells).

Figure 15 A shows that using ion irradiation (1.8 MeV Au^+, 2 x 10^{14} cm^{-2}) both goals can be achieved – i.e., conversion of C_{60} into bio-compatible amorphous carbon and selforganization of the as-deposited carbon allotrope layer (i.e., alteration of the original array of rectangular bulges into a system of periodically repeating islands of trapezoidal shape). The form of the trapezoids changes in one direction as a mirror-reflection recurrence; in the perpendicular direction the change is more complex, again combining a mirror reflection with a flip-flop turn. The bulge array form alteration may be due to a coordinated stress relaxation that is accumulated in the system (of a thin C_{60} backing layer supporting the bulge prominences) during ion irradiation. The continuous transformation of C_{60} into amorphous carbon (during progressive ion irradiation) results in the gradual inner stress growth that has (by contrast) a tendency to be relieved. This tendency grows with the accumulation of stress energy, and it results in a sudden stress relaxation. The consequence is a different sample configuration with a distinct topography that has lower area (stress) energy. The observed trapezoidal form and repeating spatial arrangement of the bulge prominences is an interesting self-shaping pattern formation that deserves further attention.

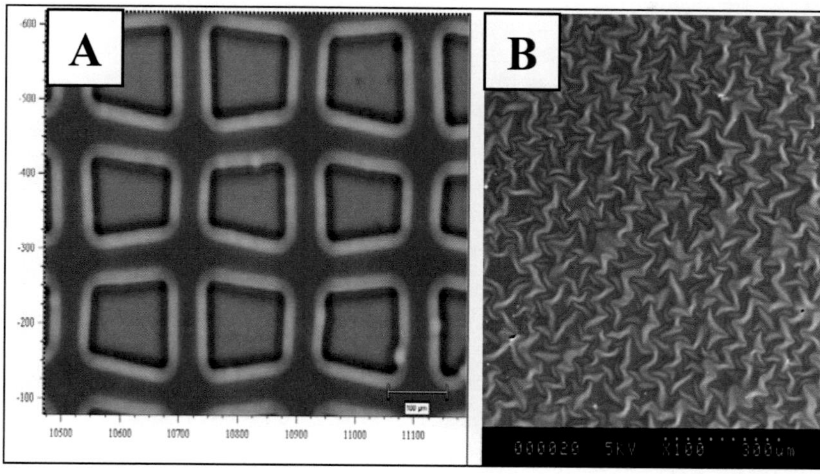

Figure 15. A. Fullerene C_{60} microdomains, created by deposition through a metallic mesh and irradiated with Au^+ ions (energy of 1.8 MeV and fluence of $2x10^{14}$ cm^{-2}). After irradiation, the original system of symmetrical (rectangular) C_{60} bulges was converted into a system of amorphous carbon bulges of trapezoid shapes varying (as a mirror image) periodically. B. A continuous fullerene C_{60} layer irradiated with Au^+ ions (1.8 MeV, $2x10^{15}$ cm^{-2}). The irradiation led to corrugation of the originally smooth fullerene layer.

A similar effect can be found when a continuous layer of fullerenes (a thin film fullerite) is irradiated with energetic ions. In Figure 15 B, a dramatic alteration of the originally smooth (as-deposited) C_{60} layer is demonstrated after 1.8 MeV Au^+ ion irradiation to the fluence 2 x 10^{14} cm^{-2}. It can be seen that ion irradiation leads to the formation of ridges and grooves in the surface morphology that could advantageously be controlled by a selected fluence of the ions. As mentioned above, C_{60} is converted during ion irradiation into amorphous carbon, which (due to a different molecular arrangement) requires expansion of the transformed layer into a larger area (than the original C_{60} film holds). This results in an interesting topography of a periodic-like corrugation of the amorphous carbon film. It might be expected that irradiation of the fullerite film through a metallic mask will create a periodic C_{60} - amorphous carbon system with original spatial patterns. Based on the effect of higher bio-compatibility of the micro- and nano-structured surface morphologies, micro-corrugated amorphous carbon films might become attractive for cell adhesion and growth.

The first experiments confirmed that the ion-irradiated fullerene layers enable homogeneous colonization with MG 63 cells to an extent similar to the non-irradiated layers, as well as reference cell culture substrates, represented by microscopic glass coverslips and standard polystyrene dishes. The vast majority of the cells were viable, of normal polygonal or spindle-shaped morphology, well-spread and with a well-assembled fine mesh-like beta-actin cytoskeleton (Figure 16).

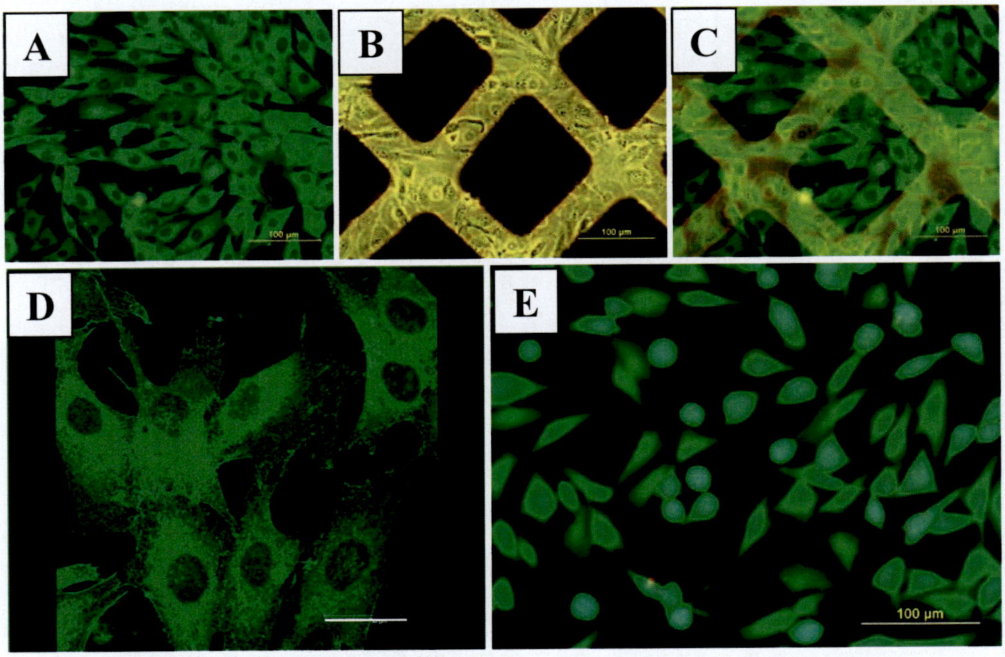

Figure 16. Human osteoblast-like MG 63 cells in cultures on a fullerene C_{60} layer micropatterned by deposition through metallic grid and then irradiated with Au ions of the energy of 1635 keV and fluence of $2x10^{14}$ cm^{-2}. A-D: immunofluorescence staining of beta-actin, day 3 after seeding; E: staining with LIVE/DEAD viability/cytotoxicity kit, day 1 after seeding. A, B, C, E: Olympus IX 50 microscope, DP 70 digital camera. A: fluorescence; B: conventional light; C: combination of both fluorescence and conventional light. Bar = 100 µm. D: Leica DM 2500 confocal microscope, obj. 100, bar = 30 µm.

2.4. Metal-fullerene C_{60} Composites

Another novel and interesting material with potential use in biotechnology and medicine is represented by binary metal/C_{60} composites. In our experiments, several different kinds of these hybrid (metallic-organic) systems, mainly thin films of Ni-C_{60} and Ti-C_{60}, have been inspected. These composites combine important properties of both phases and are assumed also to exhibit satisfying bio-compatible qualities.

In our earlier studies, nanocomposite Ti/hydrocarbon plasma polymer films, created by DC magnetron sputtering and containing amorphous hydrogenated carbon-like material with nanoclusters of Ti, supported the adhesion, growth and maturation of human bone-derived MG 63 cells to an extent comparable to that found on standard cell culture materials, represented by polystyrene dishes, microscopic glass slides or coverslips. In the case of vascular endothelial cells of the line CPAE, derived from the bovine pulmonary artery, the cell adhesion, growth and maturation was even improved in comparison with the standard culture materials and hydrocarbon plasma polymer films without Ti [103]. The beneficial effects of Ti on the colonization of carbon fibre-reinforced carbon composites (with MG 63 cells, as well as primary or low-passaged vascular smooth muscle cells, derived from the rat aorta) were also observed after the CFRC was coated with a carbon-titanium layer [104].

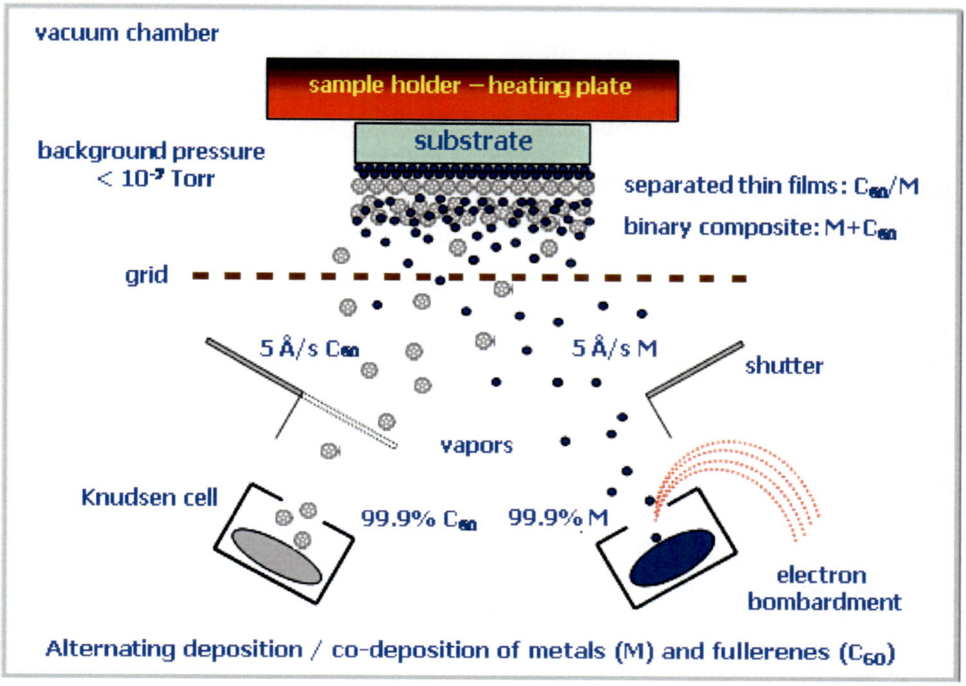

Figure 17. A scheme of hybridization of the metal/C_{60} composites in an Univex-300 vacuum system (Leybold, Germany; see Figure 1). Deposition rates: DR(M) = DR(C_{60}) ~ 5 Å/s; temperatures during deposition: 500°C (M), 120°C (C_{60}); total thickness of the hybrid composites 100 - 1000 nm.

In the case of metal-C_{60} hybrids, both inspected systems (Ni-C_{60} and Ti-C_{60}) were deposited on various supports (e.g., Si and MgO; Figure 17). They exhibited a number of interesting properties, mainly a tendency toward spontaneous self-organization, either already during hybrid synthesis (i.e., simultaneous co-deposition of both metal and organic phases) or during post-deposition processing (e.g., thermal annealing). A typical example of pattern formation of an as-deposited Ni-C_{60} hybrid is shown in Figure 18 A. Under certain deposition kinetics a periodic system of stripe domains was formed. The domains consist of sub-half-micrometer large droplets encompassed with a thin C_{60}-based rind with a polymeric structure. The stripes are embedded in the double-layer platform falling into the epitaxial Ni and amorphous carbon thin films. It has been clarified that during co-deposition the C_{60} molecules partly disintegrate and transform toward amorphous carbon. Due to the accumulation of the amorphous carbon (limited solubility in Ni), thermodynamic instability arises in the system, and then after it reaches a certain level it triggers spontaneous partitioning and self-organization of deposited material. A sequential drift and coordinate release of the thermodynamic instability has been proposed as a principal pattern of formation mechanism [150].

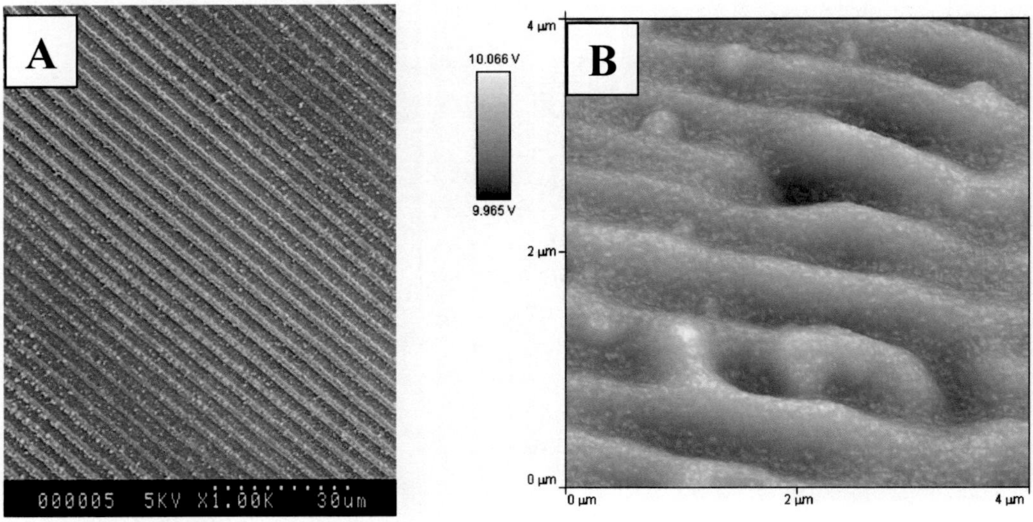

Figure 18. A. SEM micrograph of a stripe system spontaneously formed during co-deposition of Ni and C_{60} on the MgO (001) substrate. SEM-EDS (JOEL JSM-5600), bar = 30 µm. B. MFM (Magnetic Force Microscopy) analysis of the Ni+C_{60} composite synthesized at room temperature on the Si (001) substrate. Veeco CP-II microscope, scan area of 5 x 5 µm.

Another interesting manifestation of the pattern formation proclivity of the hybrid system is shown in Figure 18 B. Here, we observe an interesting property of Ni-C_{60} hybrids deposited on Si at RT - i.e., a buried self-organized morphology, consisting of periodic stripe-like magnetic domains, that has been revealed using the MFM (Magnetic Force Microscopy) method. The 'hidden' system of magnetic domains indicates that even at RT (when only a very small additional thermal energy may enter the system) a certain coordinated rearrangement (relaxation) of the stressed microstructure may appear [151].

An interesting issue is how these hybrid composites and their micro- and nano-pattern forms may affect the bio-activity of the cells. Preliminary results (obtained only recently) have demonstrated that the composites are bio-compatible and support adhesion and growth of the human osteoblast-like MG 63 cells (Figure 19 A-F).

A study of ion-irradiated composites has shown that, similarly as in the case of plain fullerite films, also in the case of hybrids the fullerene molecules are converted to amorphous carbon. Because of the metallic phase (which keeps the integrity of the composites firm) no manifestation of the spatial geometry alteration has been observed during the $C_{60} \rightarrow$ amorphous carbon transformation (as it is in the case of the plain fullerites). A thorough analysis of the systems showed that the hybrids are (after ion irradiation) more stable than the original structures (which are highly stressed due mainly to low miscibility of the C_{60} and metallic phases). More significant activity of the phase components (e.g., phase separation) can be observed only when the system is exposed to an energy flow, e.g., during thermal annealing or laser illumination. The preliminary results of our study showed that colonization of the human osteoblast-like MG 63 cells may also be possible on ion-irradiated and/or thermally processed hybrids, especially of the Ti-C_{60} type (Figure 19 G-I).

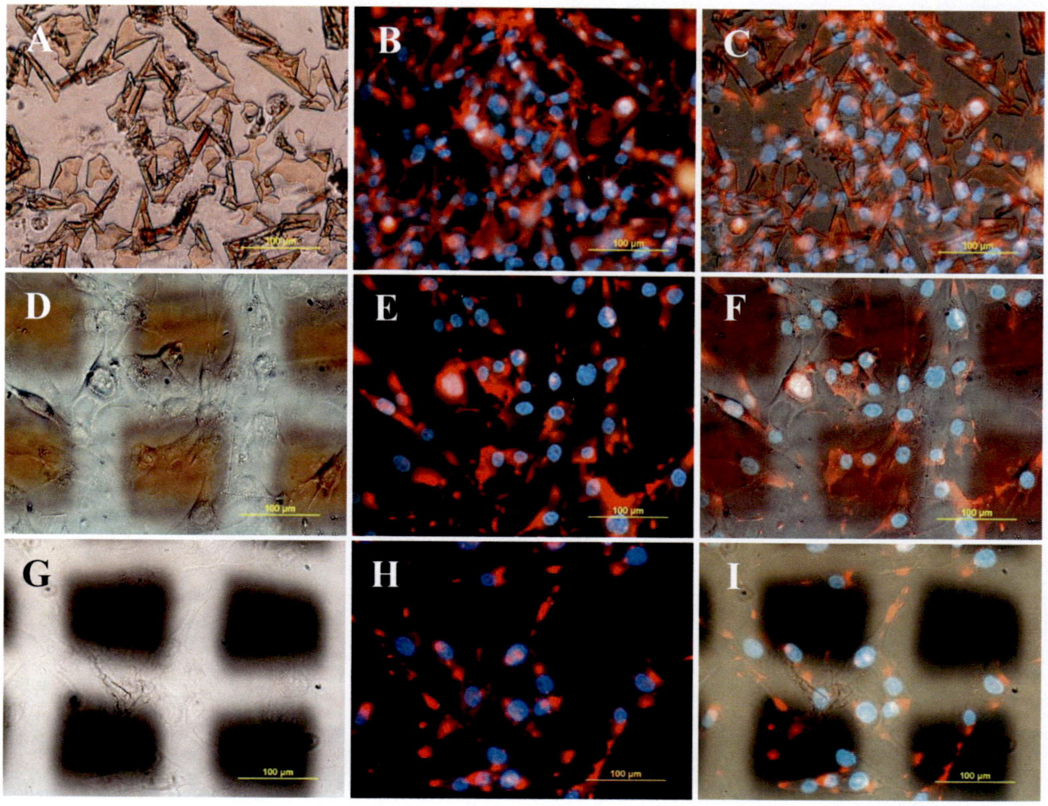

Figure 19. Human osteoblast-like MG 63 cells on day 4 after seeding on hybrid Ti-C_{60} composites deposited on microscopic glass coverslips in the form of continuous films (A-C) or films micropatterned by deposition through a metallic grid (D-F) and irradiated with Au^{+} ions of the energy of 1635 keV and fluence of $2x10^{14}$ cm^{-2} (G-I). The cells were fixed with paraformaldehyde and stained with Hoechst 33342 and Texas Red C_2-maleimide fluorescence dyes. Olympus IX 50 microscope, DP 70 digital camera. A,D,G: conventional light; B,E,H: fluorescence; C,F,I: combination of both conventional light and fluorescence. Bar = 100 µm.

3. Cell Growth on Polymer-Carbon Nanotube Composites

Our studies on cell interaction with nanotube-modified materials were carried out on polymer-nanotube composites, i.e., synthetic polymers mixed with single- or multi-walled carbon nanotubes. As the polymeric matrix, we used a terpolymer of polytetrafluoroethylene, polyvinyldifluoride and polypropylene (PTFE/PVDF/PP) or polysulphone (PSU), i.e., polymers which were considered to be promising for the construction of bone implants. The nanotubes were expected to reinforce the relatively soft and elastic polymeric material and thus to improve its mechanical properties for hard tissue surgery. The other desired function of nanotubes was the formation of prominences in nanoscale on the material surface, which could promote its colonization with bone cells.

3.1. Composites of Carbon Nanotubes and PTFE/PVDF/PP

For the preparation of composites consisting of carbon nanotubes and a PTFE/PVDF/PP terpolymer matrix, a commercially available terpolymer of polytetrafluoroethylene, polyvinyldifluoride and polypropylene (density of 1600 g/dm^3, Aldrich Chemical Co., U.S.A.) was dissolved in acetone at a concentration of 0.1 g/ml. Single-wall carbon nanohorns (SWNH) or high crystalline electric arc multi-wall nanotubes (MWNT-A; both from NanoCraft Inc., Renton, U.S.A.) were mixed with acetone in a sonicator for 5 minutes, and then with the terpolymer solution for 15 minutes to a concentration of 2, 4, 6 or 8 wt. %. The mixtures were then poured on to Petri dishes and left to evaporate the solvent [95].

Despite sonication, both SWNH and MWNT-A formed aggregates in the polymeric matrix, which created microscale irregularities on the material surface (Figure 20). The surface microscale roughness, measured by surface profilometry, increased proportionally to the nanoparticle concentration. In the pure terpolymer, the R_a parameter, i.e., the arithmetic mean of the departures of the roughness profile from the mean line, was only 0.20 ± 0.04 μm, whereas in the SWNH- and MWNT-A-modified material, the R_a parameter ranged from 0.41 ± 0.01 μm to 2.22 ± 0.36 μm in terpolymer samples with 2 to 8 wt. % of nanoparticles (Figure 21 A). The R_a parameter was usually similar in materials containing the same concentrations of SWNH or MWNT-A.

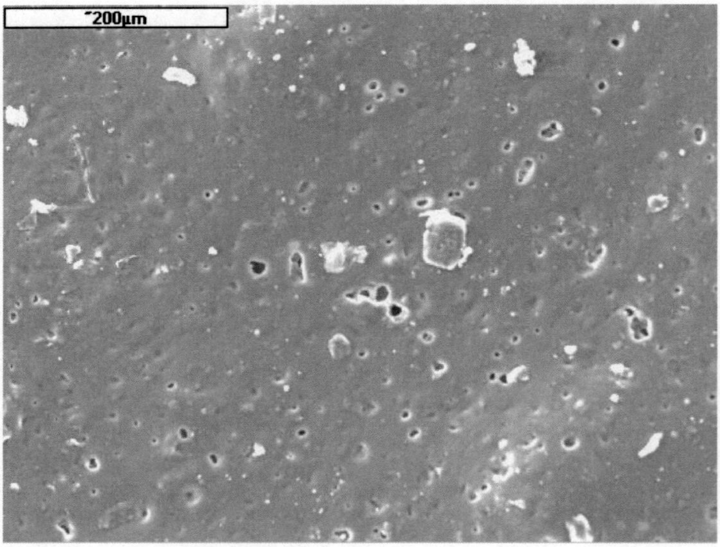

Figure 20. An example of the surface morphology of the terpolymer of polytetrafluoroethylene, polyvinyldifluoride and polypropylene mixed with 4 wt.% of single-wall carbon nanohorns. Jeol JSM 5400 Scanning Electron Microscope, obj. 2000x. Bar = 200μm [95].

At the same time, the surfaces of the terpolymer-nanotube composites contained irregularities in nanoscale, as revealed by AFM (Figure 21 B). Therefore, in fact, similarly to the fullerene-coated CFRC and micropatterned surfaces (created by deposition of C_{60} through metallic grids), the surface of the carbon nanotube-polymer composites was of a hierarchically organized micro- and nanostructure.

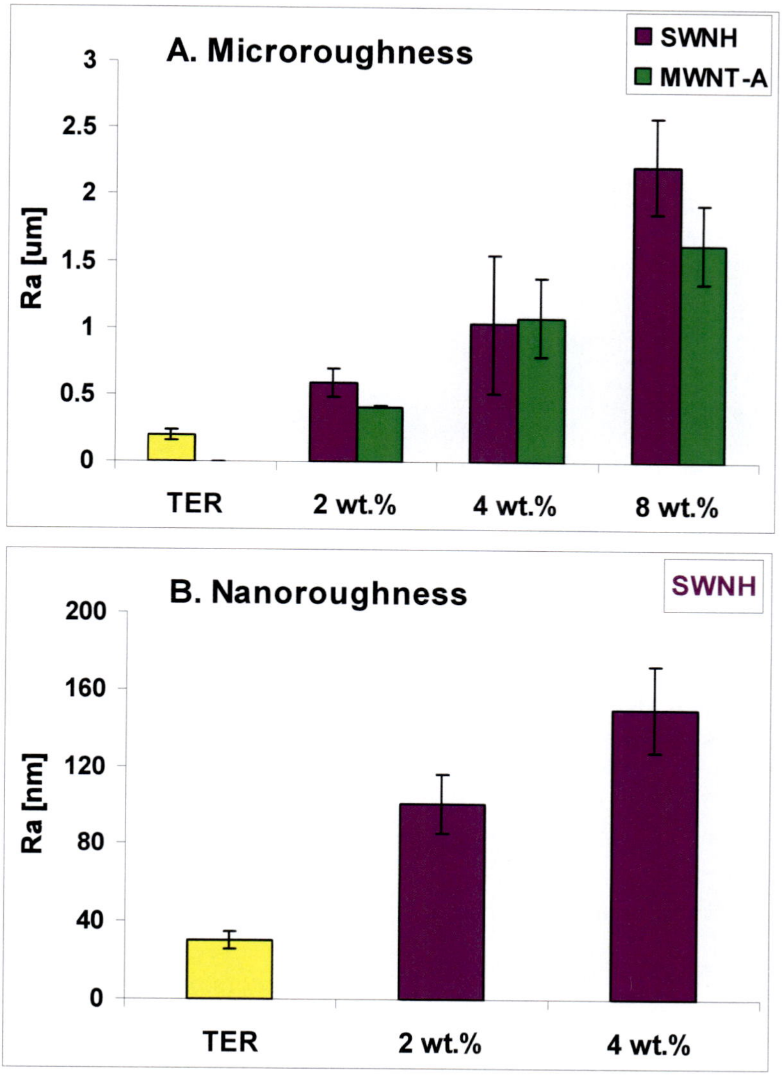

Figure 21. Surface roughness in microscale (A) and nanoscale (B) of unmodified terpolymer of polytetrafluoroethylene, polyvinyldifluoride and polypropylene (TER) and terpolymer samples mixed with 2, 4, 6 and 8 wt.% of single-wall carbon nanohorns (SWNH) or multi-wall carbon nanotubes (MWNT-A). R_a is an average deviation of the roughness profile from the mean line. Measured with a T500 Hommel Tester (Hommelwerke Co., Germany; A) or by atomic force microscopy using SPM LAB 5.01 software (B). Mean ± S.D. (Standard Deviation) from 3 measurements [95].

The composites of PTFE/PVDF/PP and carbon nanotubes were relatively highly hydrophobic. The water drop contact angle, measured in a static water drop-material contact system, was in the range from 99.3 ± 6.56° to 105.2 ± 2.12°, and was similar in the unmodified and all carbon-nanoparticle-modified terpolymer samples (Figure 22). The surface energy, measured in terpolymers modified with 4 wt. % of SWNH or MWNT-A, was also similar in pure terpolymer and both types of composite samples. Only in the sample with 4 wt. % of SWNH was the polar component of surface energy significantly lower (Table 1).

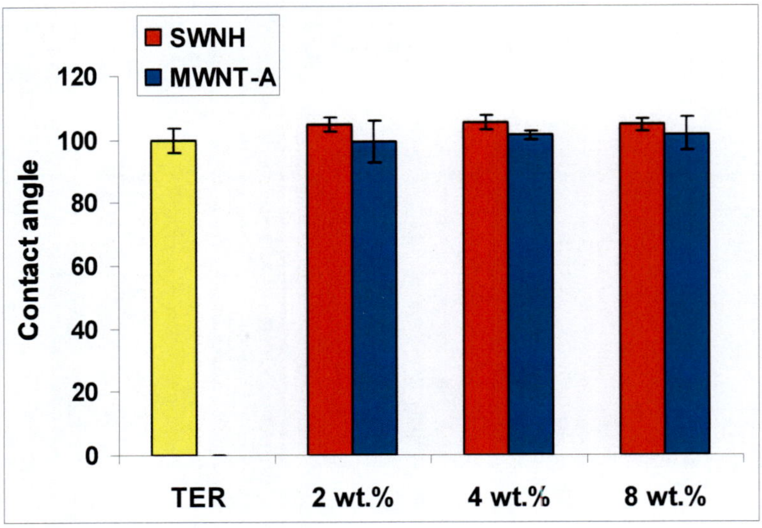

Figure 22. Static water drop contact angle measured on unmodified surface terpolymer of polytetrafluoroethylene, polyvinyldifluoride and polypropylene (TER) and terpolymer samples mixed with 2, 4, 6 and 8 wt.% of single-wall carbon nanohorns (SWNH) or multi-wall carbon nanotubes (MWNT-A). Measurement by the sessile drop method using the DSA 10 Mk2 automatic drop shape analysis system (Kruss, Germany). Mean ± S.D. (Standard Deviation) from 6-10 measurements [95].

Table 1. The surface energy and its dispersion and polar components in pure terpolymer and terpolymer modified with 4 wt. % of SWNH or MWNT-A, measured by contact angle for polar liquid (water) and for non-polar liquid (diodomethane)

Sample name	Surface energy [mN/m]	Dispersion component [mN/m]	Polar component [mN/m]
Terpolymer	22.63 ± 0.61	21.18 ± 0.47	1.45 ± 0.13
4 wt. % SWNH	21.12 ± 0.43	20.47 ± 0.38	0.65 ± 0.04
4 wt. % MWNT-A	22.61 ± 1.08	21.50 ± 0.94	1.11 ± 0.14

Mean ± S.D. (Standard Deviation) from 12 measurements.

The addition of carbon nanotubes to the PTFE/PVDF/PP terpolymer markedly improved several parameters of cell adhesion, growth and maturation. The human osteoblast-like MG 63 cells in cultures on PTFE/PVDF/PP mixed with SWNH or MWNT-A were well-spread, polygonal, and contained distinct beta-actin filament bundles, whereas most cells on the pure terpolymer were less spread or even round and clustered into aggregates (Figure 23). The enzyme-linked immunosorbent assay (ELISA) revealed that the cells on the material with 4 wt. % SWNH contained a higher concentration of vinculin, a component of focal adhesion plaques, in comparison with the values in cells on the pure terpolymer and also on the tissue culture polystyrene (Figure 24 A). The concentration of talin, another important integrin-associated focal adhesion protein, was also higher in cells grown on terpolymer samples with 4 and 8 wt.% of SWNH (Figure 24 C). However, the presence of MWNT-A did not affect the concentration of vinculin and talin in cells cultured on these samples (Figure 24 B, D). The concentration of osteocalcin, a marker of osteogenic cell differentiation, was similar in cells on all SWNH-modified and pure terpolymer samples, and also TCPS (Figure 24 E). However, in cells on samples with 4 and 6 wt. % of MWNT-A, the concentration of osteocalcin was significantly lower than the concentration in cells on the pure terpolymers and TCPS (Figure 24 F). This could be explained by a higher proliferation activity of these cells (Figure 25), which may delay the process of cell differentiation [122]. Another factor affecting the osteogenic maturation of cells is the softness of the adhesion substrate. On softer substrates, human mesenchymal stem cells differentiated towards muscle or neuronal cells, whereas osteogenic differentiation required matrices of relatively high rigidity [152]. The terpolymer used in our studies, though reinforced with carbon nanoparticles, could still be considered as a relatively soft material in comparison with the metals and ceramics that are also used in bone tissue engineering. Nevertheless, the concentration of intercellular adhesion molecule-1 (ICAM-1), an adhesion molecule of the immunoglobulin superfamily and an important marker of cell immune activation, was not increased in cells on terpolymers modified with SWNH or MWNT-A (Figure 24 G, H), which suggests that the material could be tolerated by the host tissue.

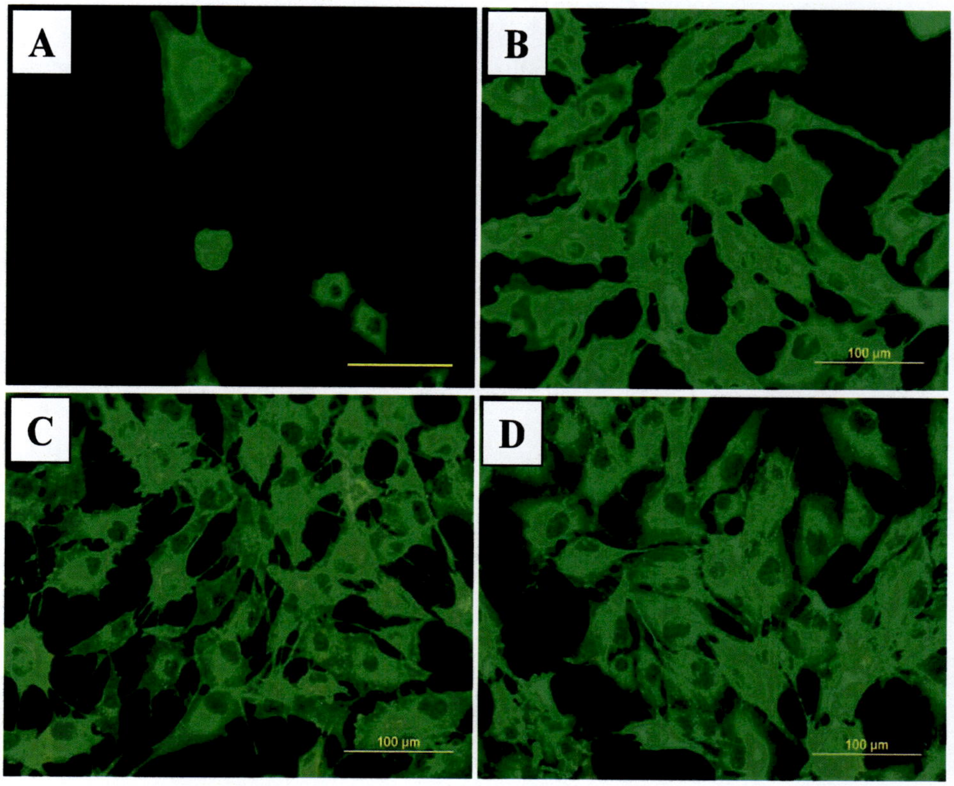

Figure 23. Immunofluorescence staining of β-actin in osteoblast-like MG 63 cells on day 3 after seeding on a terpolymer of polytetrafluoroethylene, polyvinyldifluoride and polypropylene (A), terpolymer mixed with 4 wt. % of single-wall carbon nanohorns (B) or 4 wt. % of multi-wall carbon nanotubes (C) and tissue culture polystyrene dish (D). Olympus IX 50 microscope, DP 70 digital camera, obj. 20, bar = 100 µm [95].

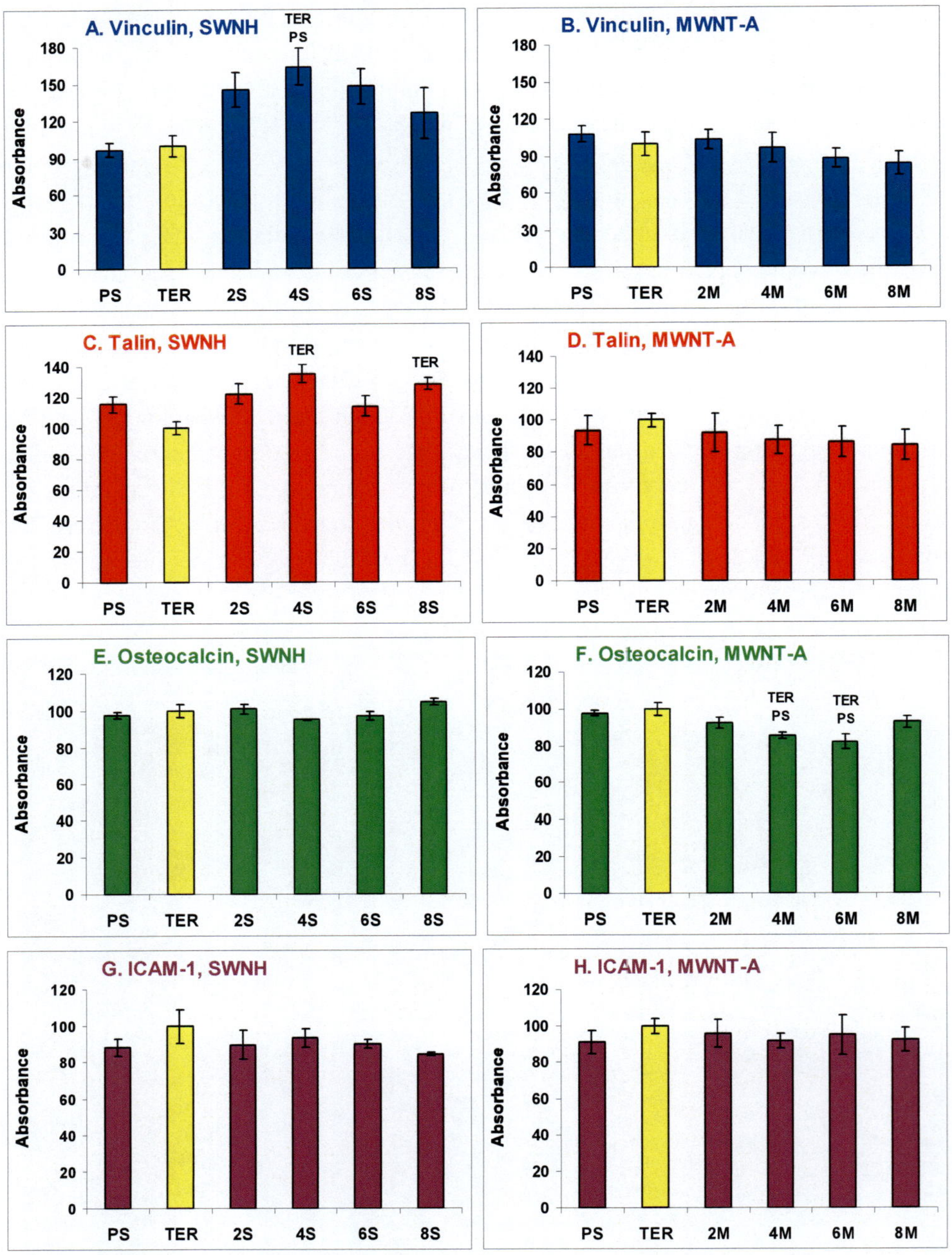

Figure 24. Concentration of vinculin (A, B), talin (C, D), and osteocalcin (E, F) in osteoblast-like MG 63 cells on day 8 after seeding on the pure terpolymer of polytetrafluoroethylene, polyvinyldifluoride and polypropylene (TER), terpolymer mixed with 2, 4, 6 or 8 wt.% (2S, 4S, 6S, 8S) of single-wall carbon nanohorns (SWNH) or 2, 4, 6 or 8 wt.% (2M, 4M, 6M, 8M) of multi-wall carbon nanotubes (MWNT-A) and tissue culture polystyrene (PS). Measured by ELISA per mg of protein; the absorbance values of cells from the modified terpolymers were expressed in % of the values obtained from the control cells grown on the unmodified terpolymer. Mean ± S.E.M. (Standard Error of Mean) from 4 measurements, ANOVA, Student-Newman-Keuls method. Statistical significance: $^{TER, PS}$: $p \leq 0.05$ compared to the values on the pure terpolymer and tissue culture polystyrene, respectively [95].

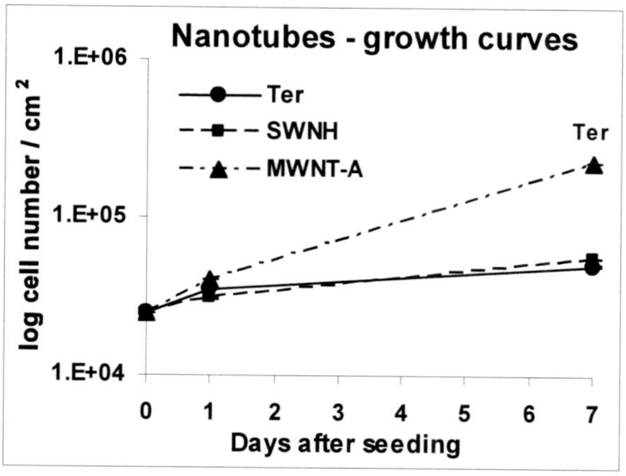

Figure 25. Growth curves of MG 63 cells on a terpolymer of polytetrafluoroethylene, polyvinyldifluoride and polypropylene (Ter), terpolymer mixed with 4 wt.% of single-wall carbon nanohorns (SWNH) or 4 wt.% of multi-wall carbon nanotubes (MWNT-A) [95].

In summary, the improved colonization of the PTFE/PVDF/PP terpolymer with human osteoblast-like MG 63 cells after it has been modified with carbon nanotubes could be attributed to changes in its surface roughness and topography, especially the creation of the surface nanostructure, rather than to its surface wettability, which remained unchanged and relatively low. Only the increased proliferation activity of cells on samples with 4 wt. % of MWNT-A (Figure 25) could be attributed to a favorable combination of the material surface nanostructure and a relatively high polar component of the free surface energy (Table 1) [153-156]. Such a combination was not observed on the other samples, which were either nanostructured (terpolymer with SWNH) or only had a higher polar component of the surface energy (pure terpolymer). On the other hand, the terpolymer modified with 4 wt. % of SWNH seemed to be the most appropriate substrate for cell adhesion, spreading and formation of focal adhesion plaques.

3.2. Composites of Carbon Nanotubes and PSU

The composites of polysulphone (PSU) and carbon nanotubes were prepared by a similar approach as the constructs with a PTFE/PVDF/PP matrix. The PSU (Aldrich Chemical Co., U.S.A.) was dissolved in dichloromethane at a concentration of 0.1 g/ml, mixed with 0.5, 1.0 or 2.0 wt.% of SWNH or MWNT-A, sonicated, poured on to Petri dishes and left to evaporate freely. The PSU-carbon nanotube composites also had a combined micro- and nanostructure of the material surface. The microroughness was similar in all nanotube-containing PSU samples, but was higher in nanotube-modified PSU than in unmodified polymer. The surface nanoroughness, measured by AFM, increased proportionally with the concentration of carbon nanotubes in the polymer matrix (Figure 26) [157,158].

In comparison with the terpolymer PTFE/PVDF/PP, the PSU used in our studies was more wettable, having a sessile water drop contact angle of $84.8 \pm 5.4^{\circ}$. Unlike the terpolymer, PSU in its pristine unmodified state gave good support for the adhesion and growth of bone-derived MG 63 cells. PSU proved to be an appropriate growth substrate for human periodontal ligament fibroblasts [159], freshly isolated rat hepatocytes [160], as well as a suitable component of three-dimensional scaffolds for the construction of bioartificial liver assist devices [161]. After the addition of MWNT-A, the surface wettability of the composite remained similar (contact angle in the range from 85.0 ± 4.6° to 90.1 ± 2.7°), while after addition of SWNH, the wettability decreased slightly (contact angle from 93.2 ± 4.3° to 91.2 ± 3.0 °).

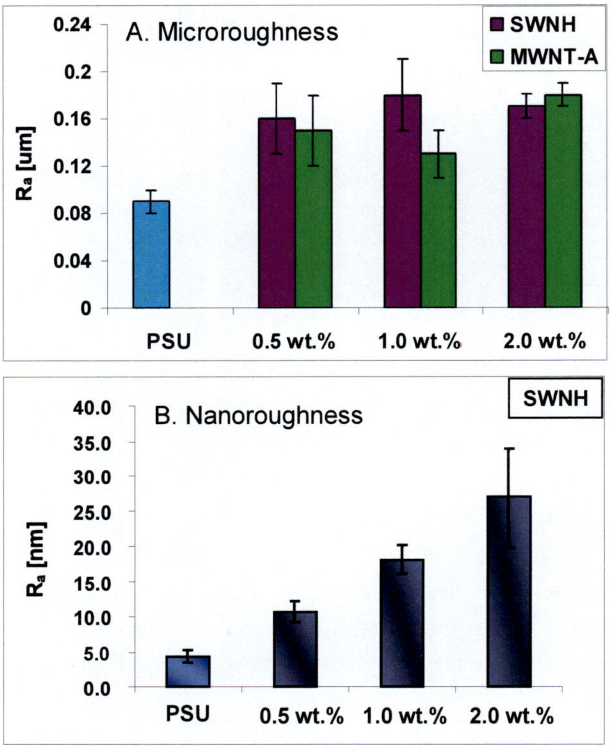

Figure 26. Surface roughness in microscale (A) and nanoscale (B) of unmodified polysulphone (PSU) and PSU mixed with 0.5, 1.0 or 2.0 wt.% of single-wall carbon nanohorns (SWNH) or multi-wall carbon nanotubes (MWNT-A). R_a is an average deviation of the roughness profile from the mean line. Measured using the T500 Hommel Tester (Hommelwerke Co., Germany; A) or by or by atomic force microscopy using SPM LAB 5.01 software (B). (B). Mean ± S.D. (Standard Deviation) from 6 measurements [157,158].

As for the colonization with cells, the numbers of MG 63 cells achieved on PSU-nanotube composites were usually similar to those obtained on unmodified PSU and standard polystyrene cell culture dishes. Only on day 3 after seeding on PSU with 1 wt. % of MWNT-A was the cell population density significantly higher (Figure 27). However, the addition of carbon nanotubes improved the cell spreading. The cell adhesion area, i.e., the cell area projected on the material surface, and measured in cells stained against beta-actin using the immunofluorescence technique, was the largest in cells grown on PSU samples with 0.5, 1 and especially 2 wt. % of MWNT-A (Figure 28). Moreover, the cells on PSU-carbon nanotube composites developed thick beta-actin cables oriented in parallel and resembling those observed in cells of muscle type (Figure 29). However, similarly as in cells on the composites PTFE/PVDF/PP and MWNT-A, this phenotype was often associated with a loss of markers of osteogenic differentiation, particularly osteocalcin [157,158].

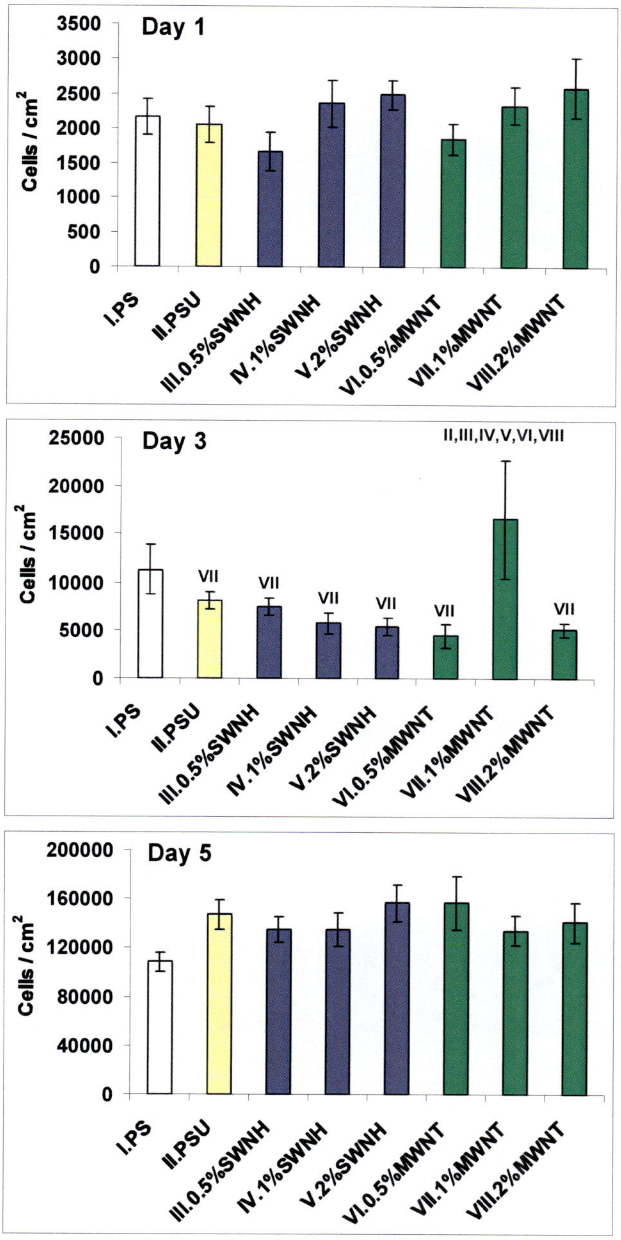

Figure 27. Cell population density (cells/cm^2) of human osteoblast-like MG 63 cells on day 1, 3 and 5 after seeding on cell culture polystyrene dishes (PS), polysulphone (PSU) and polysulphone mixed with 0.5, 1 or 2 wt.% of single-wall carbon nanohorns (SWNH) or multi-wall carbon nanotubes (MWNT). Mean ± S.E.M. (Standard Error of Mean) from 15-50 measurements. ANOVA, Student-Newman-Keuls method. Statistical significance: $^{I, II, III, IV, V, VI, VII, VIII}$: $p \leq 0.05$ compared to the values on the sample labeled with the same number [157,158].

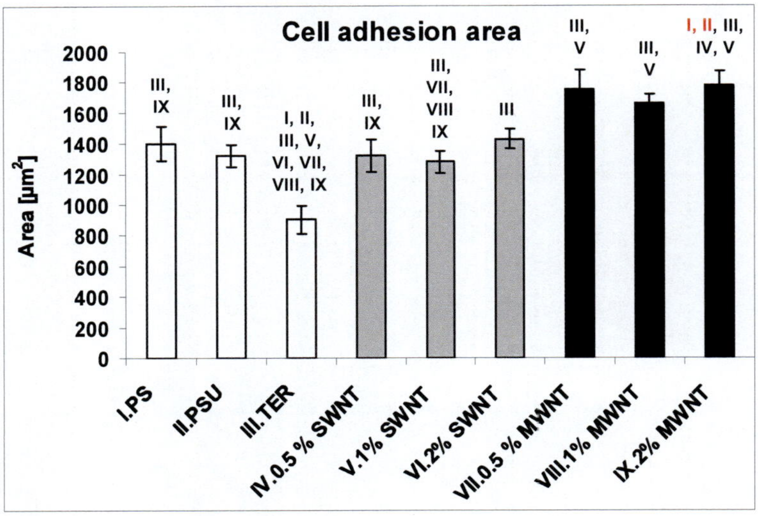

Figure 28. Cell spreading area (µm^2) of human osteoblast-like MG 63 cells on day 1 after seeding on cell culture polystyrene dishes (PS), polysulphone (PSU), terpolymer of polytetrafluoroethylene, polyvinyldifluoride and polypropylene (TER) and polysulphone mixed with 0.5, 1 or 2 wt.% of single-wall carbon nanohorns (SWNH) or multi-wall carbon nanotubes (MWNT). Mean ± S.E.M. (Standard Error of Mean) from 26-124 cells for each experimental group. ANOVA, Student-Newman-Keuls method. Statistical significance: I, II, III, IV, V, VI, VII, IX: $p \leq 0.05$ compared to the values on the sample labeled with the same number [157,158].

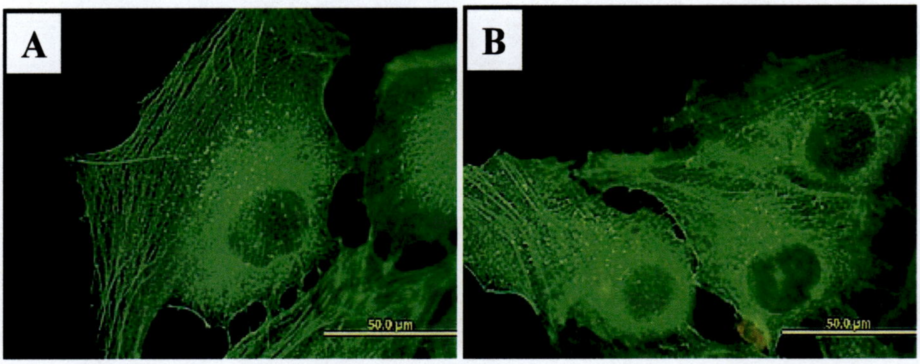

Figure 29. Immunofluorescence of beta-actin in human osteoblast-like MG 63 cells on day 3 after seeding on polysulphone mixed with 2 wt.% of single-wall carbon nanohorns (A) or 2 wt.% of multi-wall carbon nanotubes (B). Olympus IX 50 microscope, DP 70 digital camera, obj. 100, bar = 50 µm [157,158].

3.3. Interaction of Cells with Carbon Nanotubes Suspended in the Culture Media

Carbon nanotubes can be added to polymers designed for the construction of three-dimensional scaffolds for tissue engineering. These nanoparticles would form nanoscale prominences on the pore walls and thus improve the ingrowth of cells inside the scaffolds. In addition, they would reinforce the scaffolds and improve their mechanical properties in order to be more appropriate for bone tissue engineering, and they would also enable electrical stimulation of cells, which has been reported to promote their growth and maturation, manifested by collagen synthesis and the formation of mineralized bone tissue matrix [55,56]. An important question is what would happen with the nanotubes if the scaffolds were made of degradable polymers, such as polylactide, polyglycolide, polycaprolactone and their copolymers, which are often used in tissue engineering [162]. While in stable terpolymer PTFE/PVDF/PP, as well as PSU, the nanotubes were permanently bound in the polymeric matrix, they would be released from degradable materials. The nanotubes could then accumulate in adjacent and remote tissues, penetrate inside the cells and have adverse effects on the cells, e.g., a cytotoxic, immunogenic or even genotoxic, mutagenic and carcinogenic action [83-85]. On the other hand, a recent paper showed that there is a real chance of clearing the nanotubes from the organism relatively quickly by glomerular filtration [163].

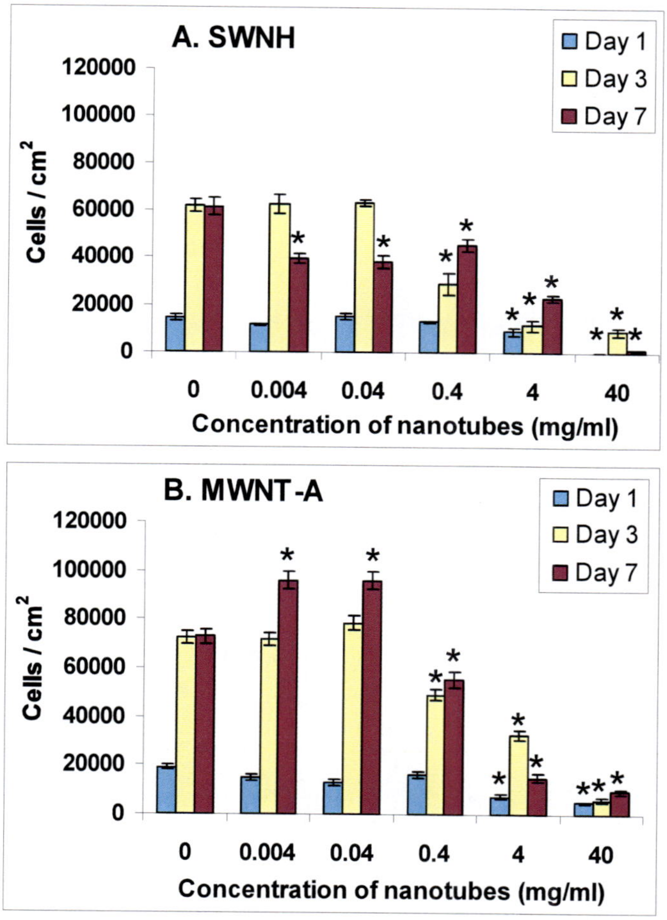

Figure 30. Cell population density (cells/cm^2) of human osteoblast-like MG 63 cells on day 1, 3 and 7 after seeding in the culture medium containing 0-40 mg/ml of single-wall carbon nanohorns (SWNH) or multi-wall carbon nanotubes (MWNT-A). Mean ± S.E.M. (Standard Error of Mean) from 3 samples for each experimental group. ANOVA, Student-Newman-Keuls method. Statistical significance: *: $p \leq 0.05$ compared to the corresponding values obtained in the medium without nanotubes [164].

In order to simulate less or more massive release of carbon nanotubes from the scaffolds and to evaluate their direct effects on cells, we suspended SWNH or MWNT-A in the culture media (DMEM, Sigma, Cat. No. D5648, supplemented with 10% of fetal bovine serum) at concentrations of 0.004, 0.04, 0.4, 4 and 40 mg/ml. This concentration range was chosen because it involves similar concentrations of nanotubes present in our polymer-nanotube composites, as well as concentrations lower or higher by several orders. The nanotube-containing media were then used for seeding human osteoblast-like MG 63 cells into standard polystyrene 24-well multidishes (TPP, Switzerland). Each well contained 30,000 cells (about 17,000 cells/cm^2 and 1.5 ml of the nanotube-containing culture medium). The medium without nanotubes served as a control sample [164]. The cell number and morphology on days 1, 3 and 7 after seeding were then evaluated.

On day 1 after seeding, the cells in both media adhered in numbers similar to the control value obtained in the medium without nanotubes up to a concentration of 0.4 mg/ml, where the numbers of initially adhering cells were significantly lower (Figure 30). This may be due not only to a cytotoxic action of carbon nanotubes but also to the fact that these nanoparticles, at higher concentrations (4 and 40 mg/ml), covered most of the bottom of the culture well and left only a limited space for cell attachment and spreading (Figures 31 and 32). On day 3 after seeding, the cell number in the media with nanotubes in concentrations of 0.004 and 0.04 mg/ml was similar (SWNH) or even higher (MWNT-A) than in the control nanotube-free medium (Figure 30). A significant decrease in the cell number was apparent from the concentration of 0.4 mg/ml of both types of nanotubes. However, on day 7 after seeding, the cell numbers in the media with SWNH were already significantly decreased from the lowest nanotube concentration of 0.004 mg/ml; whereas in the MWNT-A-containing media, the reduction was only from 0.4 mg/ml, i.e., a concentration 100 times higher (Figure 30). Therefore, it can be concluded from these experiments that the cell growth was inhibited more in the media containing single-walled nanotubes than in media containing multi-walled nanotubes. Similar results were also obtained in cultures of human fibroblasts, where single-wall carbon nanotubes induced stronger toxic effects, manifested by cellular apoptosis or necrosis, than other carbon nanoparticles represented by active carbon, carbon black, multi-wall carbon nanotubes and graphite [165]. In alveolar macrophages, the single-wall carbon nanotubes also impaired phagocytosis and induced cell degeneration and death to an extent more profound than multi-wall carbon nanotubes and fullerenes C_{60} [166]. In addition, single-wall carbon nanotubes stimulated the aggregation of human platelets *in vitro* and vascular thrombosis in rat carotid arteries *in vivo* more than their multi-wall counterparts [167].

The more pronounced adverse effects of single-wall carbon nanotubes on cells could be attributed to their relatively small size, which allows them to penetrate inside the cells more easily than bigger particles. In our experiments, the SWNH were only 2 to 3 nm in diameter and 30 to 50 nm in length, while the diameters of MWNT-A were about 5-20 nm and they were from 300 to 2000 nm in length. In addition, the SWNH had a relatively sharp closed horn-like end (19 degrees), which might further facilitate their penetration inside the cells. Recently, the more pronounced cytotoxicity of single-wall carbon nanotubes has been attributed to their relatively small surface area [165]. The cytotoxicity of both single- and multi-wall carbon nanotubes can be enhanced by the presence of impurities, particularly metal catalysts used during their preparation [39-42], the formation of big, stiff and solid agglomerates resembling asbestos [42,84,168] or some types of nanotube derivatization, e.g., oxidation by treatment in acids, as mentioned in the Introduction [47,48]. On the other hand, other types of nanotube functionalization, such as by 1,3-dipolar cycloaddition reaction, oxidation/amidation treatment or side-wall fuctionalization with phenyl-SO_3H, SO_3Na or phenyl-$(COOH)_2$ has been reported to attenuate the toxicity of carbon nanotubes [44,169].

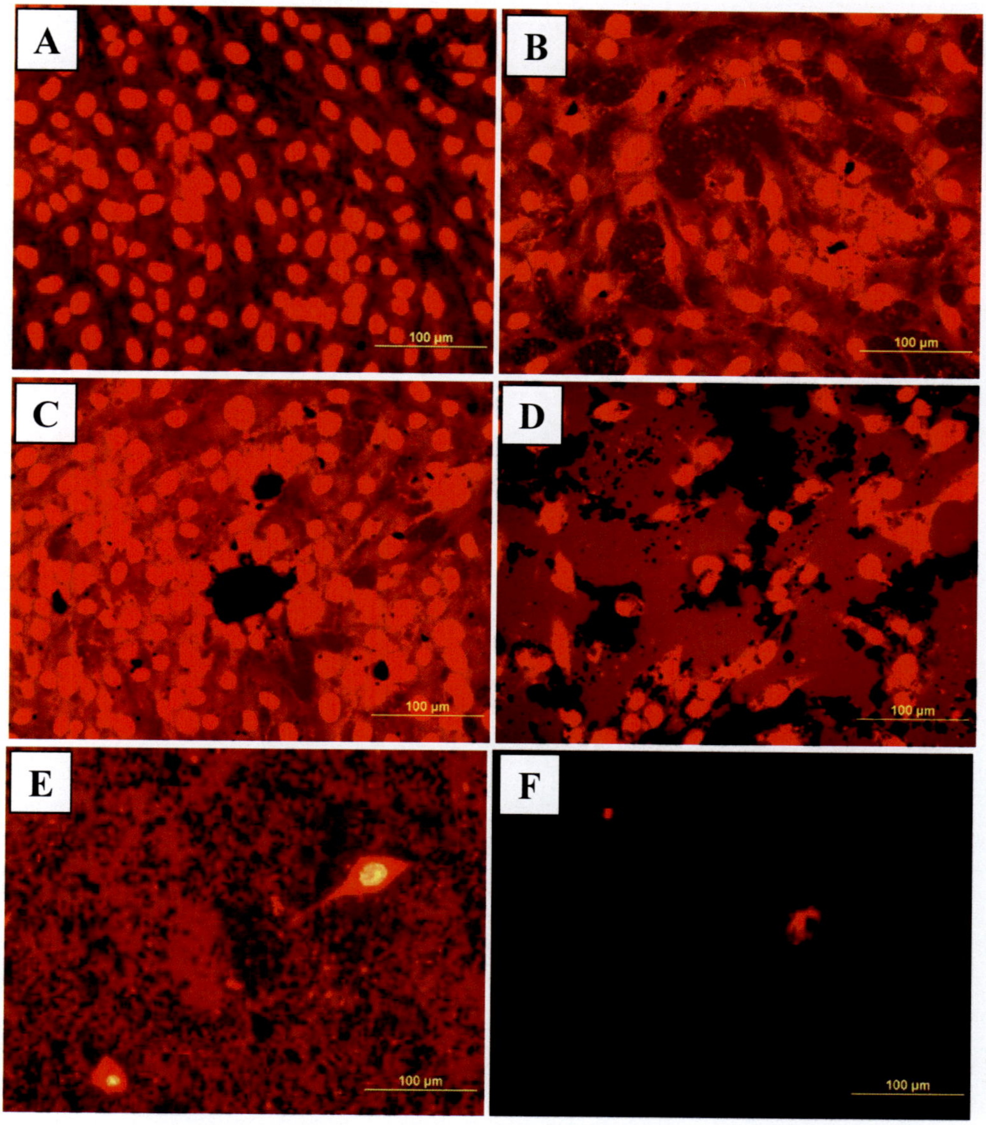

Figure 31. Morphology of human osteoblast-like MG 63 cells on day 3 after seeding in the culture medium containing single-wall carbon nanohorns in concentrations of 0 (A), 0.004 (B), 0.04 (C), 0.4 (D), 4 (E) or 40 (F) mg/ml. Cells fixed with 70% ethanol, stained with propidium iodide. Olympus IX 50 microscope, DP 70 digital camera, obj. 20, bar = 100 μm [164].

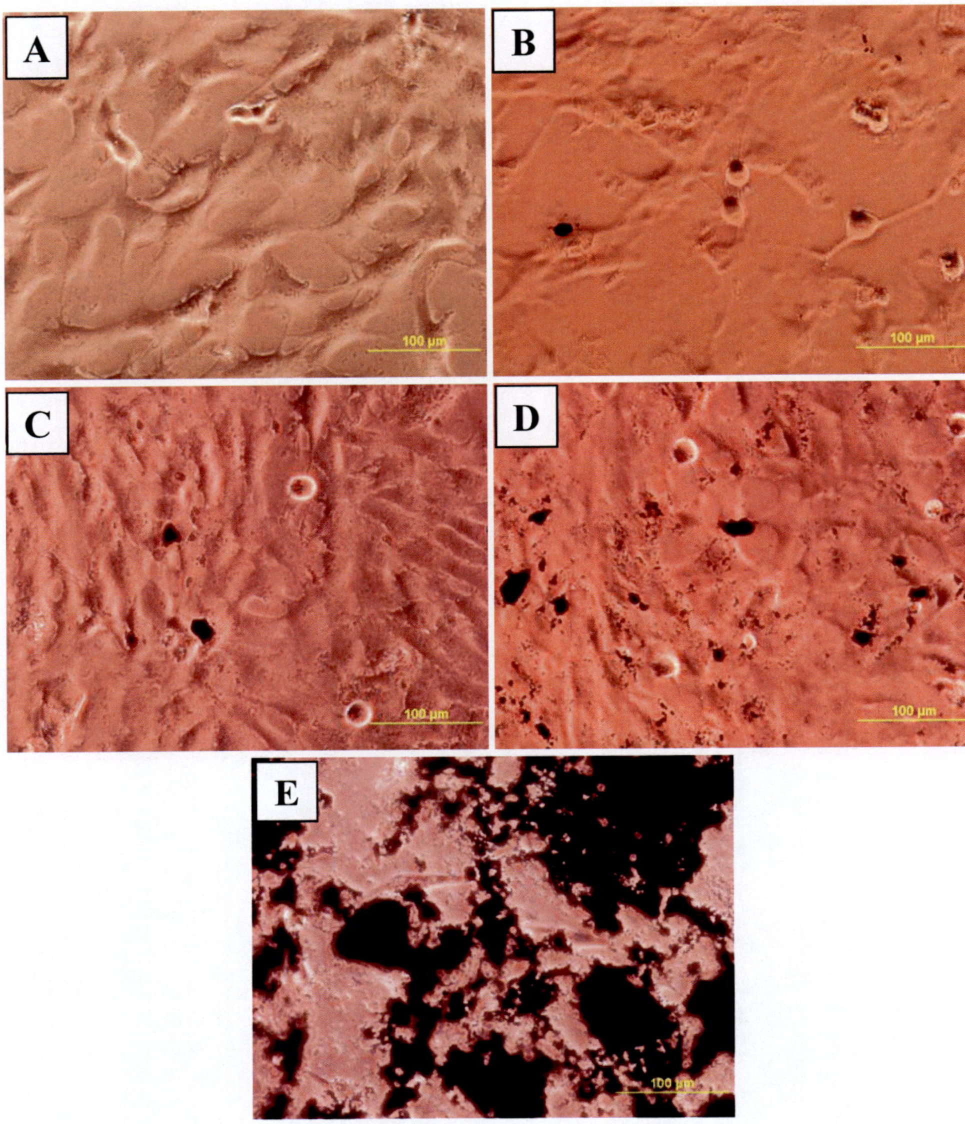

Figure 32. Morphology of human osteoblast-like MG 63 cells on day 3 after seeding in the culture medium containing multi-wall carbon nanotubes in concentrations of 0 (A), 0.004 (B), 0.04 (C), 0.4 (D) or 4 (E) mg/ml. Living cells in native cultures, Olympus IX 50 microscope, DP 70 digital camera, obj. 20, bar = 100 µm [164].

In cultures of human embryo kidney HEK 293 cells, single-wall carbon nanotubes in concentrations from 0.001 mg to 0.25 mg/ml, i.e., a concentration range corresponding, at least partly, to that used in our studies, inhibited cell growth by alterations in the expression of cell-cycle associated genes, such as cdk2, cdk4, cdk6 and cyclin D3, leading to an arrest in the G1 phase of the cell cycle, down-regulation of several signal transduction-associated genes (mad2, jak1, ttk, pcdha9 and erk), as well as a lower expression of cell adhesion-associated proteins such as laminin, fibronectin, collagen IV, cadherin and focal adhesion kinase. High nanotube concentration of 0.25 mg/ml induced cell death within 24 hours [82]. In a concentration of 0.1 mg/ml, single-walled carbon nanotubes significantly decreased the number of rat aortic smooth muscle cells relative to the cultures in the control pure medium [168]. In human epidermal keratinocytes, 6-aminohexanoic acid (AHA)-derivatized single-wall carbon nanotubes, applied to the culture media in concentrations from 0.00005 to 0.05 mg/ml, decreased the activity of mitochondrial enzymes, measured by MTT assay. At the same time, the cells treated with 0.05 mg/ml of AHA-SWNTs increased their production of interleukins IL-6 and IL-8 [84]. On the other hand, in human lung A549 cells, the EC_{50} value after 24 hours of exposure to a medium with single-wall carbon nanotubes exceeded the concentration of 8 mg/ml [83]. Relatively safe concentrations of single-walled carbon nanotubes, recommended by Bottini *et al.* [47], did not exceed 40 μg (0.04 mg) per ml of cell culture media. As for the multi-wall carbon nanotubes, when these nanoparticles of high purity were applied to the culture media in concentrations of 16 μg/ml and 20 μg/ml, the growth rate, vitality and shape of Crandell feline kidney fibroblasts remained unchanged, though the nanotube aggregates showed high adhesion to the cell membrane, and were also observed beneath the cell membrane [41].

4. Nanocrystalline Diamond Layers

In our studies, nanocrystalline diamond (NCD) films were grown on silicon substrates by a microwave plasma enhanced chemical vapor deposition (MW PECVD) method in an ellipsoidal cavity reactor (Figure 33) [95-97,170]. The silicon substrates were either mechanically lapped to root mean square (*rms*) roughness up to 300 nm or polished to atomic flatness (*rms* roughness about 1 nm). Prior to the deposition process, the substrates were mechanically seeded in an ultrasonic bath, using a 5 nm diamond powder for 40 minutes. The nucleation procedure was followed by PECVD growth using H_2/CH_4 gas mixtures. A constant methane concentration (1 % CH_4 in H_2), at a total gas pressure of 30 mbar was used. The substrate temperature was 860 °C, as measured by a two-color pyrometer working at wavelengths of 1.35 and 1.55 μm. The silicon substrates were overcoated with an NCD film on both sides, i.e., on the top and bottom side, respectively. Thus, hermetic sealing of the Si substrate minimized any unwanted biochemical reaction, as an opened Si area could result in pollution and disturbances of the subsequent cell experiments. As calculated from optical measurements [171], the film thickness was 330 nm on the atomically polished side.

Prior to the biological test, all NCD films were wet-cleaned in a solution of H_2SO_4 + KNO_3 for 30 min at 200 °C to remove any non-carbon phases from the surface of the deposited NCD films. Finally, the wet-cleaned NCD films were treated in oxygen plasma for 3 minutes at a total power of 300 W and gas pressure 10 mbar to enhance the hydrophilic character of the diamond surface. As mentioned above, moderately wettable surfaces have repeatedly been shown to promote adhesion and growth of cells [57,121,134-140]. The wetting angle, measured by a static method in a material-water droplet system using a reflection goniometer, was 35° for both types of NCD surfaces deposited on flat or rough Si substrates. Generally, the NCD surfaces were stable without significant changes of the wetting angle. Virgin silicon substrates treated in oxygen plasma also showed a low contact angle value (~17°), but this value was significantly less stable than for the NCD surface. A Si sample, standing in air for several hours showed local non-stabilities, and the wetting angle varied up to 42° [96,97].

Figure 33. An ellipsoidal cavity reactor (AIXTRON - P6, Germany), characterized by the following parameters: microwave frequency 2.45 GHz, maximum power 6 kW, substrate size 2" or 3", rotating substrate stage, stable plasma at high pressure, fully PC controlled.

The Raman spectra of the deposited NCD films were measured using a 514.5 nm excitation wavelength laser, which enabled us to determine the diamond character of the deposited films (i.e., the sp^3 hybridization). The typical Raman spectrum of an NCD film displayed one dominant peak centered at wavenumber 1333 cm^{-1} (optical phonon in diamond, Figure 34) [172]. In addition, X-ray photoelectron spectroscopy (XPS) determined that the ratio of carbon in sp^3 hybridization, characteristic for diamond, to the sp^2 C was more than 95 % [173].

As confirmed by AFM, two types of NCD layers were prepared. The first was nanostructured with a relatively flat surface (Figure 35A). On a detailed AFM scan of 1 x 1 µm, the development of crystal faceting was observable. This NCD film consisted of small crystals up to 30 nm in size. The calculated *rms* roughness was 8.21 nm. The surface features were similar and independent of the scanned area. The second type of layer was hierarchically micro- and nanostructured (Figure 35B). A scan over a large area (50 x 50 µm) exhibited features of relatively high, i.e. submicron, roughness up to ~300 nm. As shown by the detailed scan of 1 x 1 µm, the surface of these irregularities contained nano-scale features with *rms* roughness of 7.6 nm. Both nanostructured and hierarchically micro-nanostructured films thus differ mainly in surface roughness on a large scale [96,97].

The reason for preparing the hierarchically micro- and nanostructured NCD films was that not only the nanoroughness of the material surface, but also its microstructure, is important for its colonization with osteoblasts and for regeneration of the well functioning bone tissue. Surface roughness on the submicron or micrometer scale (R_a parameter, i.e., departures of the roughness profile from the mean line, in the range approx. from hundreds of nanometers to several micrometers) has repeatedly been shown to enhance the strength of osteoblast adhesion, their spreading and differentiation [111,154,174-176], including the deposition of mineralized bone extracellular matrix [177], and the acquisition of osteoblast phenotype in mesenchymal and osteoprogenitor cells [178,179]. At the same time, increasing surface microroughness reduced the activity of osteoblasts, and thus bone resorption [180]. Microstructured surfaces also modulated the number of attached cells, cell shape, presence and activity of cell adhesion receptors, the assembly of focal adhesion plaques and cytoskeleton, locomotion, proliferation, reactivity to hormones, vitamins and growth factors, as well as the production of various bioactive molecules in cells [111,154,174,175,177-182]. In addition, as mentioned several times, hierarchically-organized structures of graduated sizes represent the main architectural principle of natural tissues and organs [116].

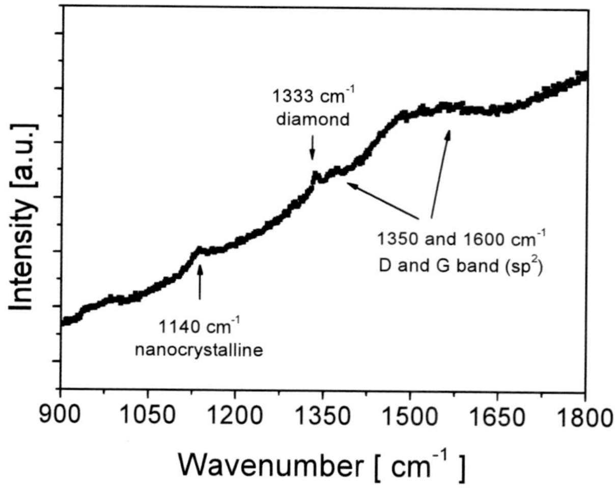

A

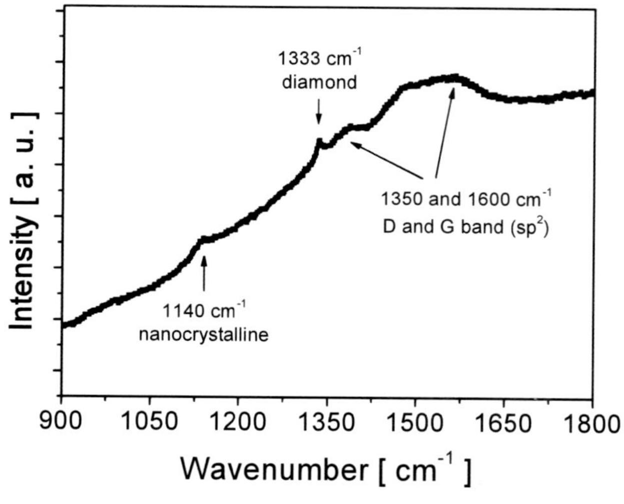

B

Figure 34. Raman spectroscopy of nanostructured (A) and hierarchically micro- and nanostructured (B) nanocrystalline diamond (NCD) films. The peak 1333 cm^{-1}, characteristic for diamond, and 1140 cm^{-1}, often attributed to NCD, are well visible in both types of films [96].

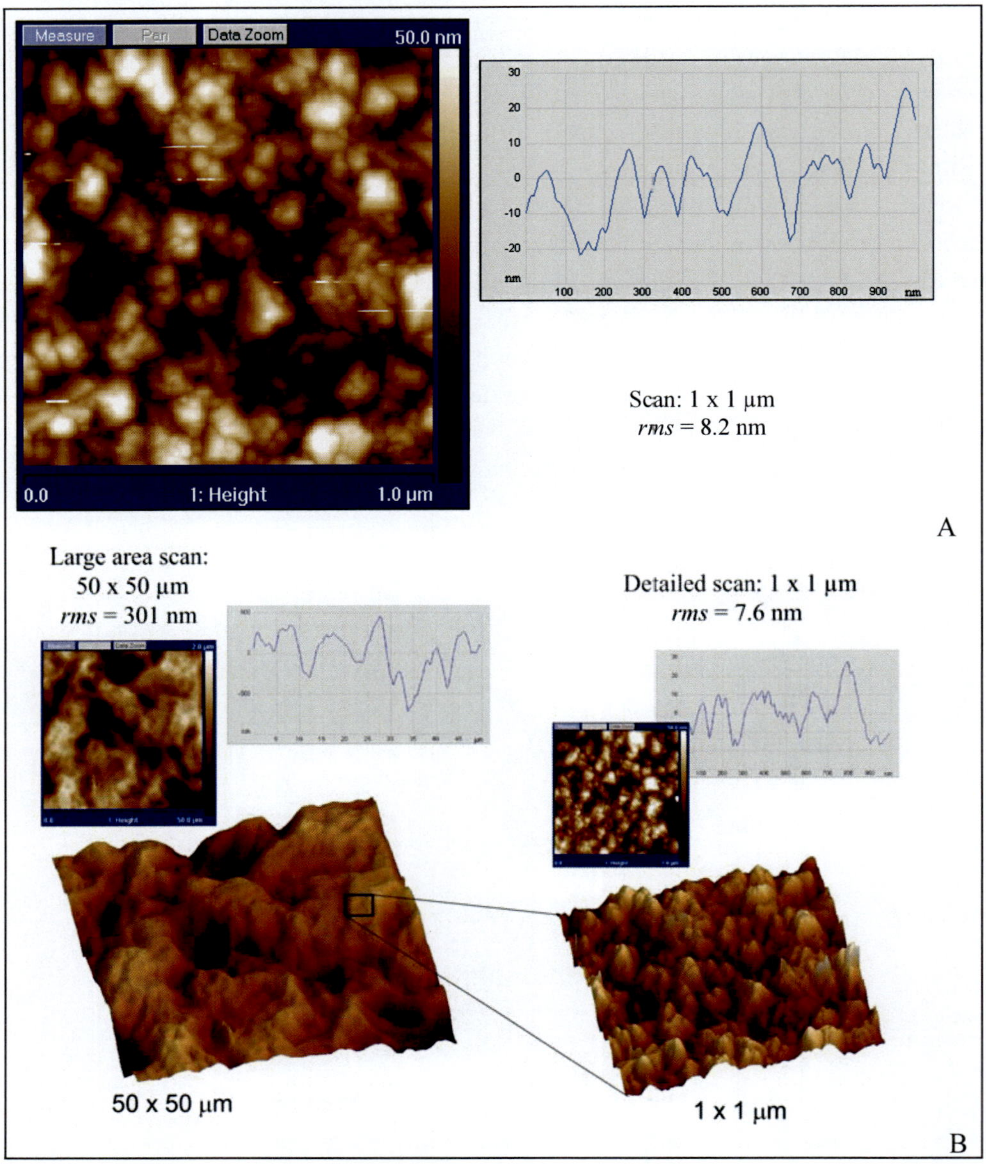

Figure 35. Atomic force microscopy of nanostructured (A) and hierarchically micro-nanostructured (B) NCD films. Root mean square (*rms*) roughness is 8.2 nm (A) or 301 nm and 7.6 nm (B). AFM Microscope Dimension 3100, Veeco [96].

NCD films were also deposited on microscopic glass slides using the MW PECVD method (26.7 mbar, 710 °C, 0.8-1% CH_4). After the growth, the films were chemically oxidized and subsequently hydrogenated. Some NCD films were also doped with boron (NCD-B; 3 000 ppm B:C; $\rho \sim 1\ 10^{-1}\ \Omega$ cm). The reason was that boron-doped NCD can be used in active sensing applications *in vivo* [175,176,183], and the presence of charged states of the ionized B acceptors on the diamond film surface can also influence cell colonization. The wetting angle of both types of films was 85-90°, and the root mean square roughness (*rms*), measured by AFM, was 18 nm.

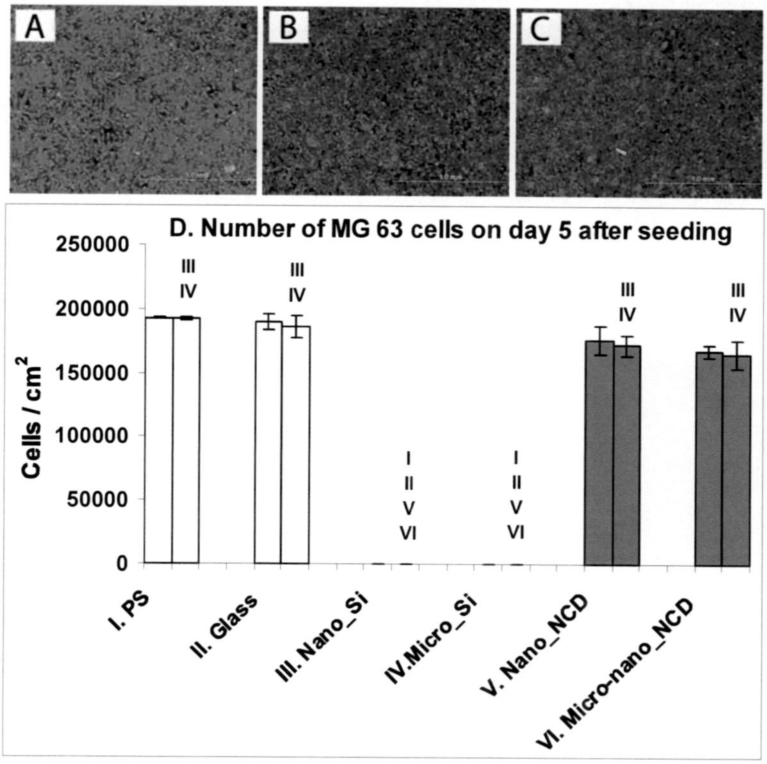

Figure 36. A-C: Fluorescence staining of MG 63 cells with a LIVE/DEAD kit on day 5 after seeding on a control tissue culture polystyrene dish (A), nanostructured NCD films (B) and hierarchically organized micro- and nanostructured NCD (C). D: Number of MG 63 cells on polystyrene culture dishes (PS), microscopic glass coverslips (Glass), nanostructured silicon substrates (Nano_Si), microstructured silicon substrates (Micro_Si), nanostructured diamond films (Nano_NCD) and diamond films with hierarchically organized micro- and nanostructure (Micro-nano_NCD). First column: total cell number, second column: number of viable cells. Mean ± S.E.M. (Standard Error of Mean) from 3 samples. ANOVA, Student-Newman-Keuls Method. Statistical significance: I, II, III, IV, V, VI: $p \leq 0.05$ in comparison with the sample of the same number [97].

All samples were sterilized in 70% ethanol, inserted in culture wells in polystyrene multidish plates and seeded with human osteoblast-like MG 63 cells. The number of initially adhered cells, evaluated 24 hours after seeding, was usually similar on all tested NCD layers, microscopic glass slides and cell culture polystyrene dishes. Only on silicon substrates was the number of initially adhering viable cells markedly lower, and this effect was more pronounced in CPAE cells than in MG 63 cells. On the other hand, the percentage of viable CPAE cells on silicon substrates improved with time of cultivation, though their absolute number still remained lower than on NCD-coated surfaces, glass slides and polystyrene wells. In MG 63 cells, the number of viable cells decreased with time of cultivation, and on day 5 after seeding, no cells were detected on Si substrates (Figure 36). It has been reported that the cytotoxicity of silica nanoparticles is dependent on the metabolic and growth activity of a given cell type [184]. For example, fibroblasts with relatively long doubling times were more susceptible to injury induced by silica exposure than tumor cells with short doubling times [184]. Silicon-containing materials can damage cells by the induction of reactive oxygen species, damage to cell membrane and mitochondrial metabolism, as well as inflammatory reactions [184,185]. Lower adhesion and spreading on silicon substrates was also observed in human cervical carcinoma HeLa cells, rat pheochromocytoma PC12 cells or mouse MC3T3 osteoblasts [88]. Silicon substrates implanted into rabbit eyes behaved in a highly corrosive manner [186].

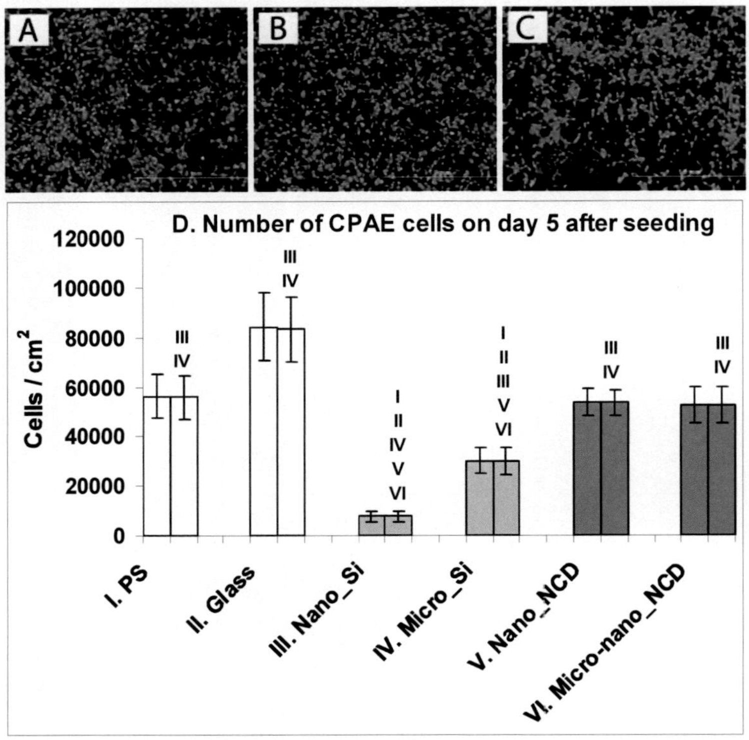

Figure 37. A-C: Fluorescence staining of CPAE cells with a LIVE/DEAD kit on day 5 after seeding on a control tissue culture polystyrene dish (A), nanostructured NCD films (B) and hierarchically organized micro- and nanostructured NCD (C). D: Number of MG 63 cells on substrates described in Figure 36. First column: total cell number, second column: number of viable cells. Mean ± S.E.M. (Standard Error of Mean) from 3 samples. ANOVA, Student-Newman-Keuls Method. Statistical significance: I, II, III, IV, V, VI: $p \leq 0.05$ in comparison with the sample of the same number [97].

The number of MG 63 and CPAE cells on days 3 and 5 after seeding on nanostructured and hierarchically micro-nanostructured O-terminated NCD was usually higher, or at least similar, in comparison with the values obtained on glass and polystyrene dishes (Figures 36, 37). Similarly as in the case of fullerene layers and nanotube-polymer composites, this favorable cell behavior was probably due to the surface nanostructure of both NCD layers. As mentioned above, the material nanostructure has been reported to be highly supportive for colonization with cells, preferentially osteoblasts. Nanostructured surfaces improve the adsorption of cell adhesion-mediating extracellular matrix molecules, e.g., fibronectin, vitronectin and collagen, from the serum of the culture medium or body fluids. These proteins are adsorbed in an appropriate amount and probably also in a favorable spatial conformation, enabling the accessibility of specific amino acid sequences by cell adhesion receptors, e.g., integrins. Preferential adhesion of osteoblasts over other cell types has been explained by preferential adsorption of vitronectin on nanostructured surfaces, due to its relatively small molecule, and its preferential recognition by osteoblasts [53,54,106-111,122].

In MG 63 cells, the microstructure of the NCD layers also seemed to have a positive effect on their adhesion. On hierarchically micro- and nanostructured surfaces, the MG 63 cells were better spread, i.e., they adhered over a significantly larger area than on the other tested substrates. This result is in good correlation with an earlier study reporting a supportive effect of the surface microroughness of dental titanium implants on the spreading of rat osteoblasts in primary cultures [181], as well as the development of a polygonal shape in osteogenic cells [182]. In addition, as mentioned above, the hierarchically organized micro- and nanostructure of a cell adhesion substrate is considered as a factor mimicking the architecture of natural tissues, and thus beneficial for cell colonization [116].

However, on micro- and nanostructured surfaces, the cell number tended to be lower than in purely nanostructured NCD, especially in endothelial CPAE cells. For example, on day 3 after seeding, the cell population density obtained on nanostructured NCD amounted to 44,000 ± 6,900 cells/cm^2, while on micro- and nanostructured NCD it was only 17,400 ± 2,400 cells/cm^2. A possible explanation is that on the latter surfaces, the spreading of CPAE cells, which are relatively large and have a flat polygonal morphology, and their subsequent growth, might be hampered by micro-sized surface irregularities [104,130-133]. It can be hypothesized that the idea of a hierarchically-organized micro- and nanoarchitecture of biomaterials, resembling the arrangement of natural tissues [116], may be useful mainly for constructing three-dimensional porous bioartificial implants rather than for planar two-dimensional surfaces. For example, nanodiamonds could be admixed into a polymeric matrix designed for the production of scaffolds for bone tissue engineering. The nanodiamonds would then decorate the walls of the pores, usually hundreds of micrometers in diameter, and the resulting nanostructure of the pore walls would facilitate the ingrowth of osteoblasts and bone tissue regeneration.

The cells on NCD films, especially those that are purely nanostructured, also displayed higher metabolic activity. As revealed by the XTT test, the absorbance of the formazan dye produced by MG 63 cells on nanostructured and hierarchically micro-nanostructured diamond films was significantly higher (by 260 ± 4 % and 102 ± 2 %, respectively) than on the control polystyrene dishes (absorbance normalized to 100 ± 8 %). The absorbance was not detectable on the silicon substrates, probably due to the very low number of viable cells on these materials. In addition, the color of the XTT reagent solution was altered, which may have been due to some non-specific reaction with the silicon substrates. Higher absorbance of formazane from cells cultivated on the two diamond layers indicates higher activity of mitochondrial enzymes in these cells than in those on the control cell culture surfaces (i.e., glass and polystyrene), and is in good correlation with the cell numbers found on both types of NCD films [96].

On boron-doped NCD films, the numbers of MG 63 cells were often lower than on undoped films modified only by oxidation and hydrogenation, though these cells were well spread and of normal morphology. Similarly as on the other NCD films, the cells on boron-doped NCD assembled dot- or streak-like focal adhesion plaques containing integrins alpha$_v$ and beta$_1$, i.e. receptors for vitronectin, fibronectin and collagen, as well as integrin-associated proteins talin and vinculin. These cells also developed a mesh-like beta actin cytoskeleton (Figure 38). However, the presence of boron might affect their proliferation ability, e.g., through an alteration to the calcium metabolism of these cells. Incubation of MG 63 cells with 2-aminoethoxydiphenylborate (2-APB) resulted in a blockage of store-operated Ca^{2+} channels, followed by lengthening of the S and G2/M phases of the cell cycle [187]. On the other hand, supplementation of rats with boron in drinking water improved the mechanical properties of the bone tissue [188]. In accordance with this finding, the MG 63 cells grown on boron-doped NCD surfaces were more brightly stained for osteocalcin, an important extracellular matrix glycoprotein involved in bone tissue mineralization (Figure 38).

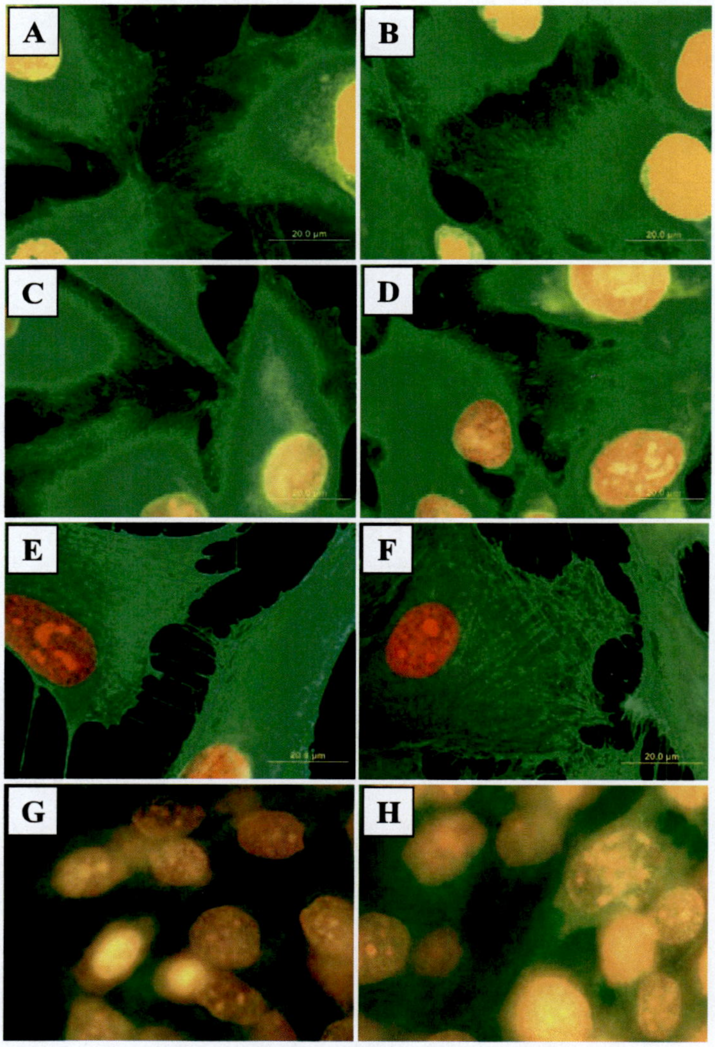

Figure 38. Immunofluorescence of $alpha_v$ integrins (A, B), talin (C, D), beta-actin (E, F) and osteocalcin (G, H) in human osteoblast-like MG 63 cells on day 1 (E, F), 3 (A-D) or 7 (G, H) after seeding on non-doped (A, C, E, G) and boron-doped (B, D, F, H) nanocrystalline diamond films deposited on microscopic glass slides. Olympus IX 50 microscope, DP 70 digital camera, obj. 100, bar = 20 μm.

In summary, nanocrystalline diamond films give good support for the adhesion and growth of osteogenic and endothelial cells and can be used in tissue engineering applications and *in vivo* experiments. This is a great advantage over other carbon nanoparticles, such as fullerenes and nanotubes, which, as mentioned above, can have a cytotoxic behavior, especially in a free form dispersed in cell culture media [10,11,82,47]. Virgin silicon substrates (typically used in the semiconductor industry) have been found to be cytotoxic and not suitable for biomedical applications, e.g., for biomaterial coatings or for *in situ* bio-sensor elements. Endothelial cells were usually more sensitive to surface roughness and their adhesion and growth was better on purely nanostructured than on hierarchically organized micro- and nanostructured NCD films. At the same time, MG 63 cells were more responsive to the chemical composition of the adhesion substrate, i.e., Si cytotoxicity [97].

5. Other Carbon-Based Layers for Potential Biomaterial Coating

These layers comprised films of amorphous hydrogenated carbon (a-C:H), deposited by by magnetron sputtering on carbon fiber-reinforced carbon composites (CFRC) to strengthen and smooth their surface in order to prevent a potential release of carbon particles from the material, and to enhance its colonization with cells. However, the adhesion and growth of MG 63 cells and primary vascular smooth muscle cells was not significantly increased in comparison with uncoated CFRC and reference cell culture polystyrene. This was most probably due to the relatively high hydrophobicity of these a-C:H films [189]. Similarly, a-C:H films have been used as bioinert coatings for blood-contacting implants and devices, as well as articular surfaces of joint replacements [103,105,122,190,191]. However, other a-C:H in our experiments, deposited on polyethyleneterephtalate (PET) by RF magnetron sputtering, enhanced adhesion and growth of mouse 3T3 fibroblasts [101]. Similarly, the a-C:H formed on polytetrafluoroethylene (PTFE) by chemical vapor deposition from acetylene induced by a UV-excimer lamp increased the colonization of the polymer with human umbilical vein endothelial cells (HUVEC, line EA.hy926) [102]. The explanation is that these a-C:H films contained additional oxygenic structures (Figure 39) and were moderately wettable (the static water drop contact angle decreased from 90° to approx. 65° with the deposition time, see Figure 40), which rendered them more attractive for cells. In addition, the material surface morphology was changed during the photo-deposition of carbon. The dependence of the surface morphology on the deposition time is shown in Figure 41 in AFM scans of 1×1 μm^2. In the first phase of deposition, the surface is covered by carbon and then the holes present in the material are filled. However, as is apparent from Figure 41, the surface nanoroughness of PTFE then increased up to 20 minutes of photo-deposition of carbon, after which it decreased. The increasing nanoscale surface roughness of the adhesion substrate has been shown to have a positive influence on cell functioning. Human osteoblast-like cells of the line SAOS-2 in cultures on nanocrystalline diamond layers showed increasing activity of cellular dehydrogenases with increasing surface nanoroughness (the root mean square roughness, rms, was in the range from 11 nm to 39 nm) [57]. Similarly, in our studies, the a-C:H layers photo-deposited on PTFE through a metallic mask for 20 minutes, were optimal for micropatterning the material surface with domains adhesive for cells, and thus for induction of regionally-selective cell adhesion and directed cell growth (Figure 42) [102].

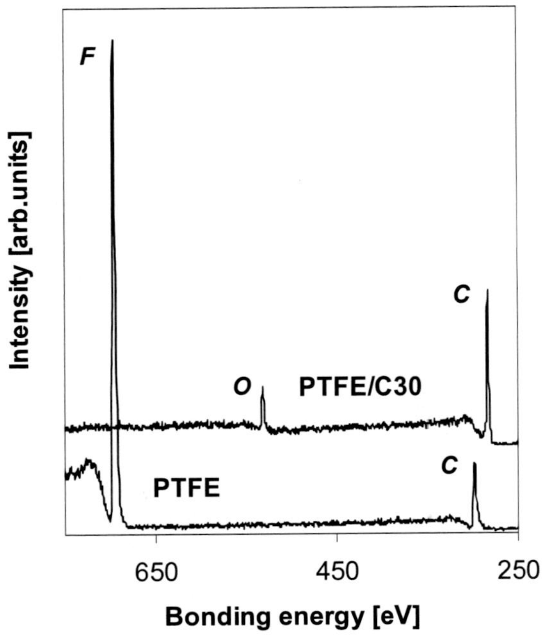

Figure 39. XPS spectra of the pristine polytetrafluoroethylene (PTFE) and PTFE with a carbon layer deposited for 30 min (PTFE/C30). A constant off-set is added to the latter spectrum [102].

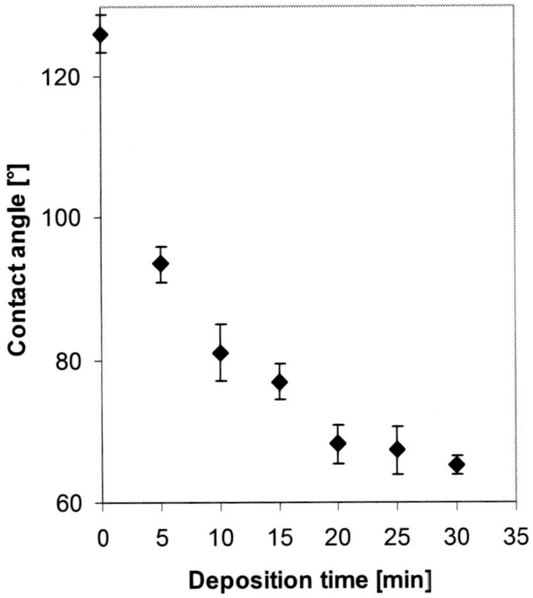

Figure 40. Dependence of water contact angle on CVD carbon deposition time on PTFE [102].

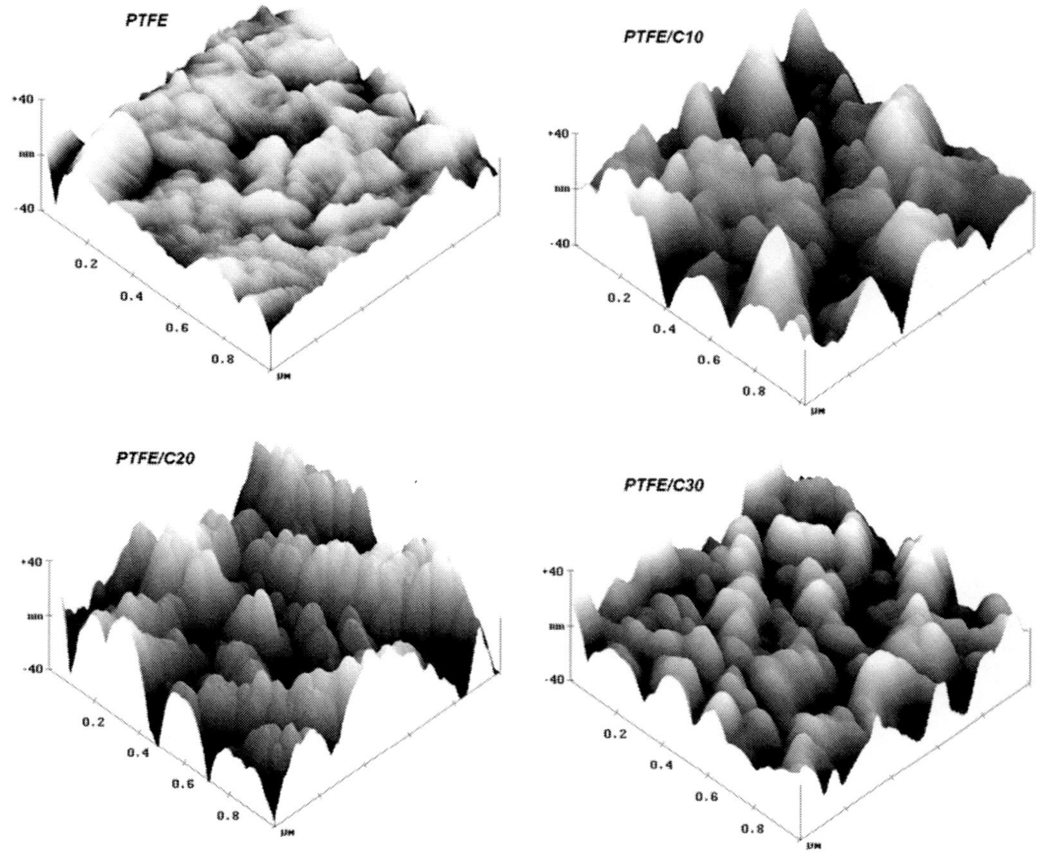

Figure 41. AFM images (scans of 1×1 μm^2) of the surface morphology of pristine polytetrafluoroethylene (PTFE) and PTFE with carbon layers deposited from 10 to 30 min (PTFE/C10 to PTFE/C30) [102].

The heterogeneous results obtained on a-C:H films indicate that the physical and chemical properties of these films, and their subsequent behavior in a biological environment can vary among different laboratories depending on the conditions of preparation, including the heterogeneity of the pristine substrates used for deposition. For example, due to the production process, the surfaces of pristine PTFE foils are typically not very homogeneous regarding surface roughnesses.

Enrichment of amorphous carbon films with Ti also stimulated cell adhesion and growth. Coating CFRC with a carbon-titanium layer deposited by the plasma-enhanced physical vapor deposition method (PECVD), using unbalanced d.c. planar magnetron sputtering, stimulated attachment, spreading, proliferation, protein synthesis and an increase in cell volume in MG 63 and vascular smooth muscle cells [104]. Similarly, nanocomposite hydrocarbon plasma polymer films with 20 at. % of Ti, deposited on microscopic glass slides and polystyrene dishes, also had a supportive effect on adhesion, growth and maturation of MG 63 cells and especially bovine vascular endothelial cells of the line CPAE [103]. On nanocomposite hydrocarbon plasma polymer films with 3 at. % of silver, focal adhesion plaques in CPAE cells were more brightly immunostained against vinculin than those in the cells on control Ag-free hydrocarbon films, glass or polystyrene dishes. The endothelial CPAE cells on films adhered, grew, formed a filamentous beta-actin cytoskeleton, developed Weibel-Palade bodies containing the von Willebrand factor, a marker of endothelial cell maturation, and created a confluent layer in an extent similar to that observed in cells growing on control surfaces. At the same time, films with 3at. % of Ag displayed bacteriostatic activity against *Escherichia coli* [105].

Pyrolytic carbon deposited on CFRC in the form of isotropic layers created in a tumbling bed reactor also supported adhesion and growth of rat vascular smooth muscle cells, human osteoblast-like MG 63 cells and human embryonal lung LEP fibroblasts [155,192,193]. However, these effects were more pronounced after coating CFRC with pyrolytic graphite prepared by decomposition of butane in a mixture with N_2 (pressure 4 Pa, temperature 1900 °C, exposure 300–400 min), especially if this treatment was combined with polishing CFRC with metallographic paper and diamond paste [132,133]. The MG 63 on these surfaces were better spread, grew more quickly, attained higher maximum population densities and contained higher concentrations of osteocalcin and osteopontin, i.e. indicators of osteogenic maturation. At the same time, the concentration of immunoglobulin molecules ICAM-1, markers of cell immune activation, in MG 63 cells remained similar as on reference culture substrates, represented by polystyrene dishes [131].

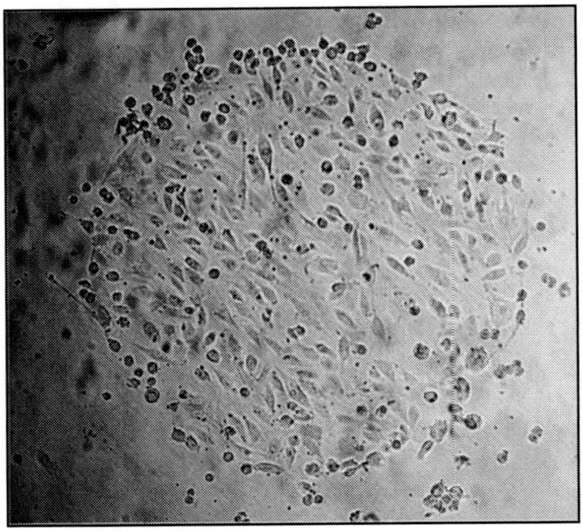

Figure 42. Phase-contrast micrographs of HUVEC (7 days after cell seeding) on a PTFE sample coated selectively with carbon for 20 min by photo-induced CVD through a contact mask (spot diameter 1.5 mm) [102].

6. Conclusion and Further Perspectives

Fullerenes C_{60} deposited as continuous films or layers micropatterned with grooves and bulges clearly supported the adhesion, growth, viability and maturation of human osteoblast-like MG 63 cells to a similar extent as standard tissue culture polystyrene dishes and microscopic glass coverslips. In addition, layers with fullerene prominences more than 1000 nm in height promoted regionally-selective adhesion and growth of MG 63 cells in the grooves located among the prominences, and thus they could be used as templates for directed cell adhesion and growth. Composites of C_{60} and metals, particularly Ti, are also promising as growth substrates for bone-derived cells.

The terpolymer of polytetrafluoroethylene, polyvinyldifluoride and polypropylene (PTFE/PVDF/PP), or polysulphone mixed with single- or multi-wall carbon nanotubes also provided good support for the adhesion and growth of MG 63 cells. In the case of PTFE/PVDF/PP, these parameters improved markedly in comparison with cells on the pure terpolymer. Carbon nanotubes could be added into polymeric matrices designed for the construction of porous 3D scaffolds for bone tissue engineering in order to stimulate cell ingrowth inside the material. However, the concentration of osteocalcin and osteopontin, indicators of osteogenic maturation, in cells on polymer-nanotube composites was often lower than on pure terpolymer or polystyrene dishes.

All nanocrystalline diamond films used in this study, showing different levels of surface organization, i.e., nanostructured (*rms* 8.2 nm or 18 nm) and hierarchically micro-nanostructured (*rms* 301 and 7.6 nm, respectively), as well as different chemical composition (i.e. hydrogenated, oxygen-terminated and boron-doped) provided good support for the adhesion, spreading, viability, growth and maturation of human osteoblast-like MG 63 cell in cultures on these materials. Thus, these NCD films could be used for the surface modification of bone implants (e.g., bone-anchoring parts of joint prostheses or bone replacements) in order to improve their integration with the surrounding bone tissue.

Other carbon allotropes and carbon-based materials, such as amorphous hydrogenated carbon, pyrolytic graphite, hydrocarbon plasma polymers, as well as their composites with metals can also serve for coating various materials in order to enhance their bioactivity.

The interactions of carbon nanoparticles with biological systems can be further intensified by the functionalization of these molecules with various atoms, chemical groups or molecules including ligands for cell adhesion receptors, such as RGD-containing oligopeptides or KRSR preferred by bone cells. However, at the same time, thorough studies on the potential cytotoxicity, immunogenicity, mutagenicity and carcinogenicity of carbon nanoparticles are needed for their potentional biomedical applications.

ACKNOWLEDGMENTS

This study was supported by the Grant Agency of the Czech Republic (Grants No. 204/06/0225 and 101/06/0226), by the Ministry of Education of the Czech Republic (Research Program No. LC 06041), and the Acad. Sci. CR (Grants No. KAN400480701, KAN200100801, KAN 101120701 and KAN 400100701). Mr. Robin Healey (Czech Technical University, Prague) and Mrs. Sherryl Ann Vacik, Pangrac & Associates, Aransas, Texas, U.S.A. are gratefully acknowledged for their language revision of the manuscript. We also thank Dr. Vera Lisa and Mrs. Ivana Zajanova (Inst. Physiol., Acad. Sci. CR) for their excellent assistance in cell culturing and immunocytochemical methods.

References

[1] Poreda, RJ; Becker L. Fullerenes and interplanetary dust at the Permian-Triassic boundary. *Astrobiology*, 2003, 3, 75-90.

[2] Komatsu, N; Ohe, T; Matsushige, K. A highly improved method for purification of fullerenes applicable to large-scale production. *Carbon*, 2004, 42, 163–167.

[3] Kroto, HW; Heath, JR; O'Brien, SC; Curl, RF; Smalley, RE. C_{60}: Buckminsterfullerene. *Nature*, 1985, 318, 162-163.

[4] Kroto, HW. Smaller carbon species in the laboratory and space. *Int. J. Mass Spectrometry and Ion Processes*, 1994, 138, 1-15.

[5] Foley, S; Crowley, C; Smaihi, M; Bonfils, C; Erlanger, BF; Seta, P; Larroque, C. Cellular localisation of a water-soluble fullerene derivative. *Biochem. Biophys. Res. Commun.*, 2002, 294, 116-119.

[6] Isobe, H; Nakanishi, W; Tomita, N; Jinno, S; Okayama, H; Nakamura, E. Gene delivery by aminofullerenes: structural requirements for efficient transfection. *Chem. Asian. J.*, 2006, 1, 167-175.

[7] Isobe, H; Nakanishi, W; Tomita, N; Jinno, S; Okayama, H; Nakamura, E. Nonviral gene delivery by tetraamino fullerene. *Mol. Pharm.*, 2006, 3, 124-134.

[8] Bakry, R; Vallant, RM; Najam-ul-Haq, M; Rainer, M; Szabo, Z; Huck, CW; Bonn, GK. Medicinal applications of fullerenes. *Int. J. Nanomedicine*., 2007, 2, 639-649.

[9] Guldi, DM; Prato, M. Excited-state properties of C60 fullerene derivatives. *Acc. Chem. Res.*, 2000, 33, 695-703.

[10] Sayes, CM; Gobin, AM; Ausman, KD; Mendez, J; West JL; Colvin,VL. Nano-C60 cytotoxicity is due to lipid peroxidation. *Biomaterials,* 2005, 26, 7587-7595.

[11] Dhawan, A; Taurozzi, JS; Pandey, AK; Shan, W; Miller, SM; Hashsham, SA; Tarabara, VV. Stable colloidal dispersions of C60 fullerenes in water: evidence for genotoxicity. *Environ. Sci. Technol.*, 2006, 40, 7394-7401.

[12] Isakovic, A; Markovic, Z; Todorovic-Markovic, B; Nikolic, N; Vranjes-Djuric, S; Mirkovic, M; Dramicanin, M; Harhaji, L; Raicevic, N; Nikolic, Z; Trajkovic, V. Distinct cytotoxic mechanisms of pristine versus hydroxylated fullerene. *Toxicol. Sci.*, 2006, 91, 173-183.

[13] Yamawaki, H; Iwai, N. Cytotoxicity of water-soluble fullerene in vascular endothelial cells. *Am. J. Physiol. Cell. Physiol.*, 2006, 290, C1495-C1502.

[14] Yamakoshi, Y; Umezawa, N; Ryu, A; Arakane, K; Miyata, N; Goda, Y; Masumizu, T; Nagano, T. Active oxygen species generated from photoexcited fullerene (C_{60}) as potential medicines: O_2^- versus $^1O_2$. *J. Am. Chem. Soc.*, 2003, 125, 12803-12809.

[15] Tang, YJ; Ashcroft, JM; Chen, D; Min, G; Kim, CH; Murkhejee, B; Larabell, C; Keasling, JD; Chen, FF. Charge-associated effects of fullerene derivatives on microbial structural integrity and central metabolism. *Nano Lett.* 2007, 7, 754-760.

[16] Satoh, M; Matsuo, K; Kiriya, H; Mashino, T; Hirobe, M; Takayanagi, I. Inhibitory effect of a fullerene derivative, monomalonic acid C60, on nitric oxide-dependent relaxation of aortic smooth muscle. *Gen. Pharmacol.*, 1997, 29, 345-351.

[17] Wilson, SR. *Fullerenes: Chemistry, Physics, and Technology*. Kadish KM and Ruoff RS (Eds.), John Wiley and Sons, 2000, 437-465.

[18] Lai, H-S; Chen, Y; Chen, W-J; Chang, K-J; Chiang, L-Y. Free radical scavenging activity of fullerenol on grafts after small bowel transplantation in dogs. *Transplant. Proc.*, 2000, 32, 1272–1274.
[19] Huang, SS; Tsai, SK; Chih, CL; Chiang, LY; Hsieh, HM; Teng, CM; Tsai, MC. Neuroprotective effect of hexasulfobutylated C_{60} on rats subjected to focal cerebral ischemia. *Free Radic. Biol. Med.*, 2001, 30, 643-649.
[20] Yang, DY; Wang, MF; Chen, IL; Chan, YC; Lee, MS; Cheng, FC. Systemic administration of a water-soluble hexasulfonated C(60) (FC(4)S) reduces cerebral ischemia-induced infarct volume in gerbils. *Neurosci. Lett.,* 2001, 311, 121-124.
[21] Nakamura, E; Isobe, H. Functionalized fullerenes in water. The first 10 years of their chemistry, biology, and nanoscience. *Acc. Chem. Res.* 2003, 36, 807-815.
[22] Chen, Y-W; Hwang, KC; Yen, C-C; Lai, YL. Fullerene derivatives protect against oxidative stress in RAW 264.7 cells and ischemia-reperfused lungs. *Am. J. Physiol. Regul. Integr. Comp. Physiol.,* 2004, 287, R21-R26.
[23] Kawase, T; Tanaka, K; Seirai, Y; Shiono N; Oda, M. Complexation of carbon nanorings with fullerenes: Supramolecular dynamics and structural tuning for a fullerene sensor. *Angew. Chem. Int. Ed. Engl.*, 2003, 42, 5597-5600.
[24] Bhattacharya, S; Tominaga, K; Kimura, T; Uno, H; Komatsu, N. A new metalloporphyrin dimer: Effective and selective molecular tweezers for fullerenes. *Chem. Phys. Lett.*, 2007, 433, 395-402.
[25] Simon, F. Studying single-wall carbon nanotubes through encapsulation: from optical methods till magnetic resonance. J. *Nanosci. Nanotechnol.*, 2007, 7, 1197-1220.
[26] Sitharaman, B; Tran, LA; Pham, QP; Bolskar, RD; Muthupillai, R; Flamm, SD; Mikos, AG; Wilson, LJ. Gadofullerenes as nanoscale magnetic labels for cellular MRI. *Contrast Media Mol. Imaging.*, 2007, 2, 139-146.
[27] Fumelli, C; Marconi, A; Salvioli, S; Straface, E; Malorni, W; Offidani, AM; Pellicciari, R; Schettini, G; Giannetti, A; Monti, D; Franceschi, C; Pincelli, C. Carboxyfullerenes protect keratinocytes from ultraviolet-B-induced apoptosis. *J. Invest. Dermatol.*, 2000, 115, 835-841.
[28] Isakovic, A; Markovic, Z; Todorovic-Markovic, B; Nikolic, N; Vranjes-Djuric, S; Mirkovic, M; Dramicanin, M; Harhaji, L; Raicevic, N; Nikolic, Z; Trajkovic, V. Distinct cytotoxic mechanisms of pristine versus hydroxylated fullerene. *Toxicol. Sci.*, 2006, 91, 173-183.
[29] Lai, YL; Murugan, P; Hwang, KC. Fullerene derivative attenuates ischemia-reperfusion-induced lung injury. *Life Sci.*, 2003, 72, 1271-1278.
[30] Wolff, DJ, Papoiu, AD; Mialkowski, K; Richardson, CF; Schuster, DI; Wilson, SR. Inhibition of nitric oxide synthase isoforms by tris-malonyl-C(60)-fullerene adducts. *Arch. Biochem. Biophys.*, 2000, 378, 216-223.
[31] Iijima, S. Helical microtubules of graphitic carbon. *Nature*, 1991, 354, 56-58.
[32] Iijima, S; Ichihashi, T. Cobalt-catalysed growth of carbon nanotubes with single-atomic-layer walls. *Nature*, 1993, 363, 603-605.
[33] Iijima, S. Carbon nanotubes: past, present, and future. *Physica B*, 2002, 323, 1–5.

[34] Fernandez-Alonso, F; Bermejo, FJ; Cabrillo, C; Loutfy, RO; Leon, V; Saboungi, ML. Nature of the bound states of molecular hydrogen in carbon nanohorns. *Phys. Rev. Lett.*, 2007, 98, 215503.

[35] Lee, H; Cho, J. Sn(78)Ge(22)@carbon core-shell nanowires as fast and high-capacity lithium storage media. *Nano Lett.*, 2007, 7, 2638-2641.

[36] Ando, Y; Zhao, X; Inoue, S; Iijima, S. Mass production of multiwalled carbon nanotubes by hydrogen arc discharge. *J. Crystal Growth*, 2002, 237–239, 1926–1930.

[37] Yudasaka, M; Ichihashi, T; Komatsu, T; Iijima, S. Single-wall carbon nanotubes formed by a single laser-beam pulse. *Chemical Physics Letters*, 1999, 299, 91–96.

[38] Okazaki, T; Saito, T; Matsuura, K; Ohshima, S; Yumura, M; Oyama, Y; Saito, R; Iijima, S. Photoluminescence and population analysis of single-walled carbon nanotubes produced by CVD and pulsed-laser vaporization methods. *Chem. Phys. Lett.*, 2006, 420, 286–290.

[39] Kagan, VE; Tyurina, YY; Tyurin, VA; Konduru, NV; Potapovich, AI; Osipov, AN; Kisin, ER; Schwegler-Berry, D; Mercer, R; Castranova, V; Shvedova, AA. Direct and indirect effects of single walled carbon nanotubes on RAW 264.7 macrophages: role of iron. *Toxicol. Lett.*, 2006, 165, 88-100.

[40] Pulskamp, K; Diabaté, S; Krug, HF. Carbon nanotubes show no sign of acute toxicity but induce intracellular reactive oxygen species in dependence on contaminants. *Toxicol. Lett.*, 2007, 168, 58-74.

[41] Pensabene, V; Vittorio, O; Raffa, V; Menciassi, A; Dario, P. Investigation of CNTs interaction with fibroblast cells. *Conf. Proc. IEEE Eng. Med. Biol. Soc.*, 2007, 1, 6620-6623.

[42] Wick, P; Manser, P; Limbach, LK; Dettlaff-Weglikowska, U; Krumeich, F; Roth, S; Stark, WJ; Bruinink, A. The degree and kind of agglomeration affect carbon nanotube cytotoxicity. *Toxicol. Lett.*, 2007, 168, 121–131.

[43] Islam, MF; Rojas, E; Bergey, DM; Johnson, AT; Yodh, AG. High weight fraction surfactant solubilization of single-wall carbon nanotubes in water. *Nano Letters*, 2003, 3, 269-273.

[44] Sayes, CM; Liang, F; Hudson, JL; Mendez, J; Guo, W; Beach, JM; Moore, VC; Doyle, CD; West, JL; Billups, WE; Ausman, KD; Colvin, VL. Functionalization density dependence of single-walled carbon nanotubes cytotoxicity in vitro. *Toxicol. Lett.*, 2006, 161, 135-142.

[45] Dillon, C; Gennett, T; Jones, KM; Alleman, JL; Parilla, PA; Heben, MJ. A simple and complete purification of single-walled carbon nanotube materials. *Adv. Mater.*, 1999, 11, 1354-1358.

[46] Chattopadhyay, D; Galeska, I; Papadimitrakopoulos, F. Complete elimination of metal catalysts from single wall carbon nanotubes. *Carbon*, 2002, 40, 985–988.

[47] Bottini, M; Bruckner, S; Nika, K; Bottini, N; Bellucci, S; Magrini, A; Bergamaschi, A; Mustelin, T. Multi-walled carbon nanotubes induce T lymphocyte apoptosis. *Toxicol. Lett.*, 2006, 160, 121-126.

[48] Magrez, A; Kasas, S; Salicio, V; Pasquier, N; Seo, JW; Celio, M; Catsicas, S; Schwaller, B; Forro, L. Cellular toxicity of carbon-based nanomaterials. *Nano Lett.*, 2006, 6, 1121-1125.

[49] Ebbesen, TW. Carbon nanotubes. *Phys. Today,* 1996, 49, 26-32.

[50] Yakobson, B; Smalley, RE. Fullerene Nanotubes: C[sub1,000,000) and Beyond. *Am. Sci.*, 1997, 85, 324-333.

[51] Shi, X; Sitharaman, B; Pham, QP; Liang, F; Wu, K; Edward Billups, W; Wilson, LJ; Mikos, AG. Fabrication of porous ultra-short single-walled carbon nanotube nanocomposite scaffolds for bone tissue engineering. *Biomaterials*, 2007, 28, 4078-4090.

[52] Abarrategi, A; Gutiérrez, MC; Moreno-Vicente, C; Hortigüela, MJ; Ramos, V; López-Lacomba, JL; Ferrer, ML; del Monte, F. Multiwall carbon nanotube scaffolds for tissue engineering purposes. *Biomaterials*, 2008, 29, 94-102.

[53] Price, RL; Waid, MC; Haberstroh, KM; Webster, TJ. Selective bone cell adhesion on formulations containing carbon nanofibers. *Biomaterials*, 2003, 24, 1877-1887.

[54] Price, RL; Ellison, K; Haberstroh, KM; Webster, TJ: Nanometer surface roughness increases select osteoblast adhesion on carbon nanofiber compacts.*J Biomed. Mater. Res. A*, 2004, 70, 129–138.

[55] Supronowicz, PR; Ajayan, PM; Ullmann, KR; Arulanandam, BP; Metzger, DW; Bizios, R. Novel current-conducting composite substrates for exposing osteoblasts to alternating current stimulation. *J. Biomed. Mater. Res.*, 2002, 59, 499-506.

[56] Zanello, LP; Zhao, B; Hu, H; Haddon, RC. Bone cell proliferation on carbon nanotubes. *Nano Lett.*, 2006, 6, 562-567.

[57] Kalbacova, M; Kalbac, M; Dunsch, L; Kromka, A; Vanecek, M; Rezek, B; Hempel, U; Kmoch, S: The effect of SWCNT and nano-diamond films on human osteoblast cells. *Phys. Stat. Sol.*, 2007, *(b)*, 1–4.

[58] Malarkey, EB; Parpura, V. Applications of carbon nanotubes in neurobiology. *Neurodegener. Dis.*, 2007, 4, 292-299, 2007.

[59] Mattson, MP; Haddon, RC; Rao, AM. Molecular functionalization of carbon nanotubes and use as substrates for neuronal growth. *J. Mol. Neurosci.*, 2000, 14, 175-182.

[60] http://www.cellnucleus.com/TEM.htm.

[61] Ohtsuki, T; Masumoto, K; Shikano, K; Sueki, K; Tanaka, T; Komatsu, K. Direct preparation of radioactive fullerenes as a tracer for applications. *Biol. Trace. Elem. Res.*, 1999, 71-72, 489-498.

[62] Bottini, M; Cerignoli, F; Dawson, MI; Magrini, A; Rosato, N; Mustelin, T. Full-length single-walled carbon nanotubes decorated with streptavidin-conjugated quantum dots as multivalent intracellular fluorescent nanoprobes. *Biomacromolecules*, 2006, 7, 2259-2263.

[63] Hoshino, A; Manabe, N; Fujioka, K; Suzuki, K; Yasuhara, M; Yamamoto, K. Use of fluorescent quantum dot bioconjugates for cellular imaging of immune cells, cell organelle labeling, and nanomedicine: surface modification regulates biological function, including cytotoxicity. *J. Artif. Organs.*, 2007, 10, 149-157.

[64] Hartman, KB; Wilson, LJ. Carbon nanostructures as a new high-performance platform for MR molecular imaging. *Adv. Exp. Med. Biol.*, 2007, 620, 74-84.

[65] Yinghuai, Z; Peng, AT; Carpenter, K; Maguire, JA; Hosmane, NS; Takagaki, M. Substituted carborane-appended water-soluble single-wall carbon nanotubes: new

approach to boron neutron capture therapy drug delivery. *J. Am. Chem. Soc.*, 2005, 127, 9875-9880.

[66] Escorcia, FE; McDevitt, MR; Villa, CH; Scheinberg, DA. Targeted nanomaterials for radiotherapy. *Nanomed.*, 2007, 2, 805-815.

[67] Singh, R; Pantarotto, D; McCarthy, D; Chaloin, O; Hoebeke, J; Partidos, CD; Briand, JP; Prato, M; Bianco, A; Kostarelos, K. Binding and condensation of plasmid DNA onto functionalized carbon nanotubes: toward the construction of nanotube-based gene delivery vectors. *J. Am. Chem. Soc.*, 2005, 127, 4388-4396.

[68] Yao, DS; Cao, H; Wen, S; Liu, DL; Bai, Y; Zheng, WJ. A novel biosensor for sterigmatocystin constructed by multi-walled carbon nanotubes (MWNT) modified with aflatoxin-detoxifizyme (ADTZ). *Bioelectrochemistry*, 2006, 68, 126-133.

[69] Du, D; Huang, X; Cai, J; Zhang, A; Ding, J; Chen, S. An amperometric acetylthiocholine sensor based on immobilization of acetylcholinesterase on a multiwall carbon nanotube-cross-linked chitosan composite. *Anal. Bioanal. Chem.*, 2007, 387, 1059-1065.

[70] Goyal, RN; Gupta, VK; Bachheti, N. Fullerene-C60-modified electrode as a sensitive voltammetric sensor for detection of nandrolone--an anabolic steroid used in doping. *Anal Chim Acta.*, 2007, 597, 82-89.

[71] Woolley, AT. Biofunctionalization of carbon nanotubes for atomic force microscopy imaging. *Methods Mol. Biol.*, 2004, 283, 305-319.

[72] Gibson, CT; Carnally, S; Roberts, CJ; Attachment of carbon nanotubes to atomic force microscope probes. *Ultramicroscopy*, 2007, 107, 1118-1122.

[73] Ding, F; Lin, Y; Krasnov, PO; Yakobson, BI. Nanotube-derived carbon foam for hydrogen sorption. *J. Chem. Phys.*, 2007, 127, 164703.

[74] Kowalczyk, P; Hołyst, R; Terrones, M; Terrones, H. Hydrogen storage in nanoporous carbon materials: myth and facts. *Phys. Chem. Chem. Phys.*, 2007, 9, 1786-1792.

[75] Nikitin, A; Li, X; Zhang, Z; Ogasawara, H; Dai, H; Nilsson, A. Hydrogen storage in carbon nanotubes through the formation of stable C-H bonds. *Nano Lett.*, 2008, 8, 162-167.

[76] Murata, K; Miyawaki, J; Yudasaka, M; Iijima, S; Kaneko, K. High-density of methane confined in internal nanospace of single-wall carbon nanohorns. *Carbon,* 2005, 43, 2817–2833.

[77] Kim, JW; Shashkov, EV; Galanzha, EI; Kotagiri, N; Zharov, VP. Photothermal antimicrobial nanotherapy and nanodiagnostics with self-assembling carbon nanotube clusters. *Lasers Surg. Med.*, 2007, 39, 622-34.

[78] Park, KH; Chhowalla, M; Iqbal, Z; Sesti, F. Single-walled carbon nanotubes are a new class of ion channel blockers. *J. Biol. Chem.*, 2003, 278, 50212-50216.

[79] Hinds, BJ; Chopra, N; Rantell, T; Andrews, R; Gavalas, V; Bachas, LG. Aligned multiwalled carbon nanotube membranes. *Science*, 2004, 303, 62-65.

[80] Majumder, M; Zhan, X; Andrews, R; Hinds, BJ. Voltage gated carbon nanotube membranes. *Langmuir*, 2007, 23, 8624-8631.

[81] van Hijkoop, VJ; Dammers, AJ; Malek, K; Coppens, MO. Water diffusion through a membrane protein channel: a first passage time approach. *J. Chem. Phys.*, 2007, 127, 085101.

[82] Cui, D; Tian, F; Ozkan, CS; Wang, M; Gao, H. Effect of single wall carbon nanotubes on human HEK293 cells. *Toxicol. Lett.*, 2005, 155, 73-85.

[83] Davoren, M; Herzog, E; Casey, A; Cottineau, B; Chambers, G; Byrne, HJ; Lyng, FM. In vitro toxicity evaluation of single walled carbon nanotubes on human A549 lung cells. *Toxicol. In Vitro*, 2007, 21, 438-448.

[84] Zhang, LW; Zeng, L; Barron, AR; Monteiro-Riviere, NA. Biological interactions of functionalized single-wall carbon nanotubes in human epidermal keratinocytes. *Int. J. Toxicol.*, 2007, 26, 103-113.

[85] Kisin, ER; Murray, AR; Keane, MJ; Shi, XC; Schwegler-Berry, D; Gorelik, O; Arepalli, S; Castranova, V; Wallace, WE; Kagan, VE; Shvedova, AA. Single-walled carbon nanotubes: geno- and cytotoxic effects in lung fibroblast V79 cells. *J. Toxicko. Environ. Health.*, 2007, *A* 70: 2071-2079.

[86] Iakoubovskii, K; Adriaenssens, GJ; Meykens, K; Nesladek, M; Vul, AY; Osipov, VY. Study of defects in CVD and ultradisperse diamond. *Diamond Relat. Mater.*, 1999, 8, 1476–1479.

[87] Qin, LC; Zhou, D; Krauss, AR; Gruen, DM. TEM Characterization of nanodiamond thin films. *NanoStructured Materials*, 1998, 10, 649-660.

[88] Bajaj, P; Akin, D; Gupta, A; Sherman, D; Shi, B; Auciello, O; Bashir, R. Ultrananocrystalline diamond film as an optimal cell interface for biomedical applications. *Biomed. Microdevices*, 2007, 9, 787–794.

[89] Hartl, A; Schmich, E; Garrido, JA; Hernando, J; Catharino, SC; Walter, S; Feulner, P; Kromka, A; Steinmüller, D; Stutzmann, M. Protein-modified nanocrystalline diamond thin films for biosensor applications. *Nat. Mater.*, 2004, 3, 736-742.

[90] Christiaens, P; Vermeeren, V; Wenmackers, S; Daenen, M; Haenen, K; Nesládek, M; vandeVen, M; Ameloot, M; Michiels, L; Wagner, P. EDC-mediated DNA attachment to nanocrystalline CVD diamond films. *Biosens. Bioelectron.*, 2006, 22, 170-177.

[91] Schrand, AM; Huang, H; Carlson, C; Schlager, JJ; Osawa, E; Hussain, SM; Dai, L. Are diamond nanoparticles cytotoxic? *J. Phys. Chem.*, 2007, 111, 2-7.

[92] Bakowicz, K; Mitura, S. Biocompatibility of NCD. *J. Wide Bandgap Mater.*, 2002, 9, 261.

[93] Fu, CC; Lee, HY; Chen, K; Lim, TS; Wu, HY; Lin, PK; Wei, PK; Tsao, PH; Chang, HC; Fann, W. Characterization and application of single fluorescent nanodiamonds as cellular biomarkers. *Proc. Natl. Acad. Sci. U. S. A.*, 2007, 104, 727-732.

[94] Papo, MJ; Catledge, SA; Vohra, YK; Machado, C. Mechanical wear behavior of nanocrystalline and multilayer diamond coatings on temporomandibular joint implants. *J. Mater. Sci. Mater. Med.*, 2004, 15, 773-777.

[95] Bacakova, L; Grausova, L; Vacik, J; Fraczek, A; Blazewicz, S; Kromka, A; Vanecek, M; Svorcik, V. Improved adhesion and growth of human osteoblast-like MG 63 cells on biomaterials modified with carbon nanoparticles. *Diamond Relat. Mater.*, 2007, 16, 2133-2140.

[96] Grausova, L; Bacakova, L; Kromka, A; Potocky, S; Vanecek, M; Nesladek, M; Lisa, V. Nanodiamond as a promising material for bone tissue engineering. *J. Nanosci. Nanotechnol.*, 2008, in press.

[97] Grausova, L; Kromka, A; Bacakova, L; Potocky, S; Vanecek, M; Lisa, V. Bone and vascular endothelial cells in cultures on nanocrystalline diamond films. *Diamond Relat. Mater.*, 2008, in press.

[98] Specht, CG; Williams, OA; Jackman, RB; Schoepfer, R. Ordered growth of neurons on diamond. *Biomaterials*, 2004, 25, 4073-4078.

[99] Rupprecht, S; Bloch, A; Rosiwal, S; Neukam, FW, Wiltfang, J. Examination of the bone-metal interface of titanium implants coated by the microwave plasma chemical vapor deposition method. *Int. J. Oral Maxillofac. Implants*, 2002, 17, 778-785.

[100] Xiao, X; Wang, J; Liu,C; Carlisle, JA; Mech, B; Greenberg, R; Guven, D; Freda, R; Humayun, MS; Weiland, J; Auciello, O. In vitro and in vivo evaluation of ultrananocrystalline diamond for coating of implantable retinal microchips. *J. Biomed. Mater. Res. B Appl. Biomater.*, 2006, 77, 273-281.

[101] Kubova, O; Bacakova, L; V. Svorcik, V. Biocompatibility of carbon layer on polymer. *Mater. Sci. Forum*, 2005, 482, 247-250.

[102] Kubova, O; Svorcik, V; Heitz, J; Moritz, S; Romanin, C; Matejka, P; Mackova, A. Characterization and cytocompatibility of carbon layers prepared by photo-induced chemical vapor deposition. *Thin Solid Films*, 2007, 515, 6765-6772.

[103] Grinevich, A; Bacakova, L; Choukourov, A; Boldyryeva, H; Pihosh, Y; Slavinska, D; Noskova, L; Skuciova, M; Lisa, V; Biederman, H. Nanocomposite Ti/hydrocarbon plasma polymer films from reactive magnetron sputtering as growth supports for osteoblast-like and endothelial cells. *J. Biomed. Mater. Res. Part A*, 2008, in press.

[104] Bacakova, L; Stary, V; Kofronova, O; Lisa, V. Polishing and coating carbon fiber-reinforced carbon composites with a carbon-titanium layer enhances adhesion and growth of osteoblast-like MG63 cells and vascular smooth muscle cells *in vitro*. *J. Biomed. Mater. Res.*, 2001, 54, 567-578.

[105] Bacakova, L; Koshelyev, H; Noskova, L; Choukourov, A; Benada, O; Mackova, A; Lisa, V; Biederman, H. Vascular endothelial cells in cultures on nanocomposite silver/hydrocarbon plasma polymer films with antimicrobial activity. *J. Optoelectronics and Advanced Materials*, 2008, in press.

[106] Webster, TJ; Ergun, C; Doremus, RH; Siegel, RW; Bizios, R. Enhanced functions of osteoblasts on nanophase ceramics. *Biomaterials*, 2000 , 21, 1803-1810.

[107] Webster, TJ; Ergun, C; Doremus, RH; Siegel, RW; Bizios, R. Specific proteins mediate enhanced osteoblast adhesion on nanophase ceramics. *J. Biomed. Mater. Res.*, 2000, 51, 475-483.

[108] Webster, TJ; Smith, TA. Increased osteoblast function on PLGA composites containing nanophase titania. *J. Biomed. Mater. Res. A*, 2005, 74, 677-686.

[109] Woo, KM; Chen,VJ; Ma, PX. Nano-fibrous scaffolding architecture selectively enhances protein adsorption contributing to cell attachment. *J. Biomed. Mater. Res. A*, 2003, 67, 531-537.

[110] Wei, G; Ma, PX. Structure and properties of nano-hydroxyapatite/polymer composite scaffolds for bone tissue engineering. *Biomaterials*, 2004, 25, 4749-4757.

[111] Wang, L; Zhao, G; Olivares-Navarrete, R; Bell, BF; Wieland, M; Cochran, DL; Schwartz, Z; Boyan, BD. Integrin beta$_1$ silencing in osteoblasts alters substrate-dependent responses to 1,25-dihydroxy vitamin D_3. *Biomaterials*, 2006, 27, 3716-3725.

[112] Dee, KC; Andersen, TT; Bizios, R. Design and function of novel osteoblast-adhesive peptides for chemical modification of biomaterials. *J. Biomed. Mater. Res.*, 1998, 40, 371-377.

[113] Gobin, S; West, JL. Val-ala-pro-gly, an elastin-derived non-integrin ligand: Smooth muscle cell adhesion and specificity. *J. Biomed. Mater. Res. A*, 2003, 67, 255–259.

[114] Shin, H; Jo, S; Mikos, AG. Biomimetic materials for tissue engineering. *Biomaterials*, 2003, 24, 4353–4364.

[115] Bacakova, L; Noskova, L; Koshelyev, H; Biederman, H; Vascular endothelial cells in cultures on metal/C:H composite films. *Engineering of Biomaterials*, 2004, 7 [37], 18-20.

[116] Tan, J; Saltzman, WM. Biomaterials with hierarchically defined micro- and nanoscale structure. *Biomaterials*, 2004, 25, 3593-3601.

[117] Bacakova, L; Grausova, L; Vacik, J; Svorcik, V. Human bone-derived cells in cultures on fullerene C_{60} layers of continuous and patterned morphology. *Conference NanoSMat* 2007, Algarve, Portugal, Abstract No. BBE20, p. 66.

[118] Grausova, L; Bilkova, P; Bacakova, L; Vacik, J; Svorcik, V.: Regionally-selective adhesion and growth of human osteoblast-like MG 63 cells on micropatterned fullerene C_{60} layers. 14th *International Conference on Plasma Physics and Applications (*14th *CPPA* 2007*)*, Brasov, Romania, p. 127.

[119] Levi, N; Hantgan, RR; Lively, MO; Carroll, DL; Prasad, GL. C_{60}-fullerenes: detection of intracellular photoluminescence and lack of cytotoxic effects. *J. Nanobiotechnol.*, 2006, 4, 14.

[120] Lin, JC; Wu, CH: Surface characterization and platelet adhesion studies on polyurethane surface immobilized with C_{60}. *Biomaterials*, 1999, 20, 1613-1620.

[121] Bacakova, L; Walachova, K; Svorcik, V; Hnatowitz, V. Adhesion and proliferation of rat vascular smooth muscle cells (VSMC) on polyethylene implanted with O(+) and C(+) ions. *J. Biomater. Sci. Polym. Ed.*, 2001, 12, 817-834.

[122] Bacakova, L; Filova, E; Rypacek, F; Svorcik, V; Stary, V. Cell adhesion on artificial materials for tissue engineering. *Physiol Res.*, 2004, 53 *Supp* , S35-45.

[123] Tsuchiya, T; Yamakoshi, YN; Miyata, N. A novel promoting action of fullerene C_{60} on the chondrogenesis in rat embryonic limb bud cell culture system. *Biochem. Biophys. Res. Commun.*, 1995, 206, 885-894.

[124] Straface, E; Natalini, B; Monti, D; Franceschi, C; Schettini, G; Bisaglia, M; Fumelli, C; Pincelli, C; Pellicciari, R; Malorni, W. C3-fullero-tris-methanodicarboxylic acid protects epithelial cells from radiation-induced anoikia by influencing cell adhesion ability. *FEBS Lett.*, 1999, 454, 335-340.

[125] Yamawaki, H; Iwai, N. Cytotoxicity of water-soluble fullerene in vascular endothelial cells. *Am. J. Physiol. Cell. Physiol.*, 2006, 290, C1495-C1502.

[126] Yamakoshi, Y; Umezawa, N; Ryu, A; Arakane, K; Miyata, N; Goda, Y; Masumizu, T; Nagano, T. Active oxygen species generated from photoexcited fullerene (C_{60}) as potential medicines: O_2^- versus $^1O_2$. *J. Am. Chem. Soc.*, 2003, 125, 12803-12809.

[127] Tang, YJ; Ashcroft, JM; Chen, D; Min, G; Kim, CH; Murkhejee, B; Larabell, C; Keasling, JD; Chen, FF. Charge-associated effects of fullerene derivatives on microbial structural integrity and central metabolism. *Nano Lett.*, 2007, 7, 754-760.

[128] Harhaji, L; Isakovic, A; Raicevic, N; Markovic, Z; Todorovic-Markovic, B; Nikolic, N; Vranjes-Djuric, S; Markovic, I; Trajkovic, V. Multiple mechanisms underlying the anticancer action of nanocrystalline fullerene. *Eur J Pharmacol.*, 2007, 568, 89-98.

[129] Harhaji, L; Isakovic, A; Vucicevic, L; Janjetovic, K; Misirkic, M; Markovic, Z; Todorovic-Markovic, B; Nikolic, N; Vranjes-Djuric, S; Nikolic, Z; Trajkovic, V. Modulation of tumor necrosis factor-mediated cell death by fullerenes. *Pharm. Res.*, 2007, in press.

[130] Bacakova, L; Stary, V; Hornik, J; Glogar, P; Lisa, V; Kofronova O. Osteoblast-like MG63 cells in cultures on carbon fibre-reinforced carbon composites. *Inzynieria Biomateriałów-Engineering of Biomaterials*, 2001, 4 [17-19], 11-12.

[131] Bacakova, L; Stary, V; Glogar, P; Lisa, V. Adhesion, differentiation and immune activation of human osteogenic cells in cultures on carbon-fibre reinforced carbon composites. *Inzynieria Biomateriałów-Engineering of Biomaterials*, 2003, 6, 8-9.

[132] Stary, V; Bacakova, L; Hornik, J; Chmelik, V. Bio-compatibility of the surface layer of pyrolytic graphite. *Thin Solid Films*, 2003, 433, 191-198.

[133] Stary, V; Glogar, P; Bacakova, L; Hnilica, F; Chmelik, V; Korinek, Z; Gregor, J; Mares, V; Lisa, V. A study of surface properties of composite materials and their influence on biocompatibility. *Acta Montana AB*, 2003, 11, 19-36.

[134] Haddow, DB; France, RM; Short, RD; MacNeil, S; Dawson, RA; Leggett, GJ; Cooper, E. Comparison of proliferation and growth of human keratinocytes on plasma copolymers of acrylic acid/1,7-octadiene and self-assembled monolayers. *J. Biomed. Mater. Res.*, 1999, 47, 379-387.

[135] Bacakova, L; Mares, V; Bottone, MG; Pellicciari, C; Lisa, V; Svorcik, V. Fluorine-ion-implanted polystyrene improves growth and viability of vascular smooth muscle cells in culture. *J. Biomed. Mater. Res.*, 2000, 49, 369-379.

[136] Bacakova, L; Mares, V; Lisa, V; Svorcik, V. Molecular mechanisms of improved adhesion and growth of an endothelial cell line cultured on polystyrene implanted with fluorine ions. *Biomaterials*, 2000, 21, 1173-1179.

[137] Mikulikova, R; Moritz, S; Gumpenberger, T; Olbrich, M; Romanin, Ch; Bacakova, L; Svorcik, V; Heitz, J. Cell microarrays on photochemically modified PTFE, *Biomaterials*, 2005, 26, 5572.

[138] Parizek, M; Bacakova, L; Lisa, V; Kubova, O; Svorcik, V; Heitz J. Vascular smooth muscle cells in cultures on synthetic polymers with adhesive microdomains. *Inzynieria Biomaterialow (Engineering of Biomaterials)*, 2006, *IX (*58-60*)*, 7-10.

[139] Parizek, M; Bacakova, L; Kasalkova, N; Svorcik, V; Kolarova, K. Improved adhesion and growth of vascular smooth muscle cells in cultures on Polyethylene modified by plasma discharge. *Inzynieria Biomaterialow (Engineering of Biomaterials)*, 2007, *X (*67-68*)*, 1-4.

[140] Kasalkova, N; Kolarova, K; Bacakova, L; Parizek, M; Svorcik, V. Cell adhesion and proliferation on modified polyethylene, *Mater. Sci. Forum*, 2007, 567-568, 269-272.

[141] Bacakova, L; Filova, E; Kubies, D; Machova, L; Proks, V; Malinova, V; Lisa, V; Rypacek, F. Adhesion and growth of vascular smooth muscle cells in cultures on bioactive RGD peptide-carrying polylactides. *J. Mater. Sci. Mater. Med.*, 2007, 18,1317-1323.

[142] Patrito, N; McCague, C; Norton, PR; Petersen, NO. Spatially controlled cell adhesion via micropatterned surface modification of poly(dimethylsiloxane). *Langmuir*, 2007, 23, 715-719.

[143] Hamilton, DW; Chehroudi, B; Brunette, DM. Comparative response of epithelial cells and osteoblasts to microfabricated tapered pit topographies in vitro and in vivo. *Biomaterials*, 2007, 28, 2281-2293.

[144] Heitz, J; Svorcik, V; Bacakova, L; Rockova, K; Ratajova, E; Gumpenberger, T; Bauerle, D; Dvorankova, B; Kahr, H; Graz, I; Romanin, C. Cell adhesion on polytetrafluoroethylene modified by UV-irradiation in an ammonia atmosphere. *J. Biomed. Mater. Res.*, 2003, 67*A*, 130-137.

[145] Charest, JL; Bryant, LE; Garcia, AJ; King, WP. Hot embossing for micropatterned cell substrates. *Biomaterials*, 2004, 25, 4767-4775.

[146] Kenar, H; Kose, GT; Hasirci, V. Tissue engineering of bone on micropatterned biodegradable polyester films. *Biomaterials*, 2006, 27, 885-895.

[147] Hamilton, DW; Brunette, DM. The effect of substratum topography on osteoblast adhesion mediated signal transduction and phosphorylation. *Biomaterials*, 2007, 28, 1806-1819.

[148] Papenburg, BJ; Vogelaar, L; Bolhuis-Versteeg, LA, Lammertink, RG; Stamatialis, D; Wessling, M. One-step fabrication of porous micropatterned scaffolds to control cell behavior. *Biomaterials*, 2007, 28,1998-2009.

[149] Guo, S; Fogarty, DP; Nagel, PM; Kandel, SA: Scanning tunneling microscopy of surface-adsorbed fullerenes: C_{60}, C_{70}, and C_{84}. *Surf. Sci.*, 2007, 601, 994-1000.

[150] Vacik, J; Naramoto, H; Narumi, K; Yamamoto, Y; Miyashita, K. Pattern formation induced by co-deposition of Ni and C60 on MgO.100. *Journal of Chemical Physics*, 2001, 114 [20], 9115 -9119.

[151] Vacik, J; Lavrentiev, V; Hnatowicz, V; Yamamoto, S; Vorlicek, V; Stadler, H. Spontaneous Partitioning of the Ni+C_{60} Thin Film Grown at RT. *Journal of Alloys and Compounds* 2008, in press.

[152] Engler, AJ; Sen, S; Sweeney, HL; Discher, DE. Matrix elasticity directs stem cell lineage specification. *Cell*, 2006, 126, 677-689.

[153] Hao, L; Lawrence, J; Chian, KS. Osteoblast cell adhesion on a laser modified zirconia based bioceramic. *J. Mater. Sci. Mater. Med.*, 2005, 16, 719-726.

[154] Zhao, G; Schwartz, Z; Wieland, M; Rupp, F; Geis-Gerstorfer, J; Cochran, DL; Boyan, BD. High surface energy enhances cell response to titanium substrate microstructure. *J. Biomed. Mater. Res. A*, 2005, 74, 49-58.

[155] Pesakova, V; Kubies, D; Hulejova, H; Himmlova, L. The influence of implant surface properties on cell adhesion and proliferation. *J. Mater. Sci. Mater. Med.*, 2007, 18, 465-473.

[156] Dos Santos, EA; Farina, M; Soares, GA; Anselme, K. Surface energy of hydroxyapatite and beta-tricalcium phosphate ceramics driving serum protein adsorption and osteoblast adhesion. *J. Mater. Sci. Mater. Med.*, 2008, in press.

[157] Grausova, L; Bacakova, L; Fraczek, A; Blazewicz, S; Blazewicz, M; Pepka, M. Biological effects of polymers modified with carbon nanotubes on human osteoblast-like MG 63 cells. *Engineering of Biomaterials*, 2006, 9[58-60], 14-16.

[158] Grausova, L; Bacakova, L; Fraczek, A; Filova, E; Blazewicz M. Adhesion, proliferation and viability of human osteoblast-like MG 63 cells on carbon nanotube-polysulfone composites. *Eight International Conference on the Science ad Application of Nanotubes, „Nanotubes* 2007*"*, Ouro Preto, Minas Gerais, Brasil, p. 181.

[159] Mailhot, JM; Sharawy, MM; Galal, M; Oldham, AM; Russell, CM. Porous polysulfone coated with platelet-derived growth factor-BB stimulates proliferation of human periodontal ligament fibroblasts. *J. Periodontol.*, 1996, 67, 981-985.

[160] Jozwiak, A; Karlik, W; Wiechetek, M; Werynski, A. Attachment and metabolic activity of hepatocytes cultivated on selected polymeric membranes. *Int. J. Artif. Organs*, 1998, 21, 460-466.

[161] Hoque, ME; Mao, HQ; Ramakrishna, S. Hybrid braided 3-D scaffold for bioartificial liver assist devices. *J. Biomater. Sci. Polym. Ed.*, 2007, 18, 45-58.

[162] Pamula, E; Bacakova, L; Filova, E; Buczynska, J; Dobrzynski, P; Noskova, L; Grausova, L: The influence of pore size on colonization of poly(L-lactide-glycolide) scaffolds with human osteoblast-like MG 63 cells in vitro. *J. Mater. Sci. Mater. Med.*, 2008, 19, 425-435.

[163] McDevitt, MR; Chattopadhyay, D; Jaggi, JS; Finn, RD; Zanzonico, PB; Villa, C; Rey, D; Mendenhall, J; Batt, CA; Njardarson, JT; Scheinberg, DA. PET imaging of soluble yttrium-86-labeled carbon nanotubes in mice. *PLoS ONE* 2, 2007, e907.

[164] Fraczek, A; Menaszek, E; Czajkowska, B; Bacakova, L; Blazewicz, M. In vitro and in vivo biocompatibility of single wall carbon nanotubes. *Engineering of Biomaterials*, 2007, 10*(*69-72*)*, 8-11.

[165] Tian, F; Cui, D; Schwarz, H; Estrada, GG; Kobayashi, H. Cytotoxicity of single-wall carbon nanotubes on human fibroblasts. *Toxicol. In Vitro*, 2006, 20, 1202-1212.

[166] Jia, G; Wang, H; Yan, L; Wang, X; Pei, R; Yan, T; Zhao, Y; Guo, X. Cytotoxicity of carbon nanomaterials: single-wall nanotube, multi-wall nanotube, and fullerene. *Environ. Sci. Technol.*, 2005, 39, 1378-1383.

[167] Radomski, A; Jurasz, P; Alonso-Escolano, D; Drews, M; Morandi, M; Malinski, T; Radomski, MW. Nanoparticle-induced platelet aggregation and vascular thrombosis. *Br. J. Pharmacol.*, 2005, 146, 882-893.

[168] Raja, PM; Connolley, J; Ganesan, GP; Ci, L; Ajayan, PM; Nalamasu, O; Thompson, DM. Impact of carbon nanotube exposure, dosage and aggregation on smooth muscle cells. *Toxicol. Lett.*, 2007, 169, 51-63.

[169] Dumortier, H; Lacotte, S; Pastorin, G; Marega, R; Wu, W; Bonifazi, D; Briand, JP; Prato, M; Muller, S; Bianco, A. Functionalized carbon nanotubes are non-cytotoxic and preserve the functionality of primary immune cells. *Nano Lett.*, 2006, 6, 1522-1528.

[170] Potocky, S; Kromka, A; Potmesil, J; Remes, Z; Polackova, Z; Vanecek, M. Growth of nanocrystalline diamond films deposited by microwave plasma CVD system at low substrate temperatures. *Phys. Stat. Sol. (a)*, 2006, 203, 3011-3015.

[171] Poruba, A; Fejfar, A; Remes, Z; Springer, J; Vanecek, M; Kocka, J; Meier, J; Torres, P; Shah, A. Optical absorption and light scattering in microcrystalline silicon thin films and solar cells. *J. Appl. Phys.*, 2000, 88, 148-160.

[172] Buhlman, S; Blank, E; Haubner, R; Lux, B: Characterization of ballas diamond depositions, *Diamond Rel. Mater.*, 1999, 8, 194-201.

[173] Zemek, J; Houdkova, J; Lesiak, B; Jablonski, A; Potmesil, J; Vanecek, M. Electron spectroscopy of nanocrystalline diamond surfaces. *J. Optoelectr. & Adv. Materials*, 2006, 8, 2133.

[174] Kim, HJ; Kim, SH; Kim, MS; Lee, EJ; Oh, HG; Oh, WM; Park, SW; Kim, WJ; Lee, GJ; Choi, NG; Koh, JT; Dinh, DB; Hardin, RR; Johnson, K; Sylvia, VL; Schmitz, JP; Dean, DD.Varying Ti-6Al-4V surface roughness induces different early morphologic and molecular responses in MG63 osteoblast-like cells. *J. Biomed. Mater. Res.* 2005, 74, 366-373.

[175] Zhao, G; Zinger, O; Schwartz, Z; Wieland, M; Landolt, D; Boyan, BD. Osteoblast-like cells are sensitive to submicron-scale surface structure. *Clin. Oral. Implants Res.*, 2006, 17, 258-264.

[176] Zhao, W; Xu, JJ; Qiu, QQ; Chen, HY. Nanocrystalline diamond modified gold electrode for glucose biosensing. *Biosens Bioelectron*, 2006, 22, 649-655.

[177] Boyan, BD; Bonewald, LF; Paschalis, EP; Lohmann, CH; Rosser, J; Cochran, DL; Dean, DD; Schwartz, Z; Boskey, AL. Osteoblast-mediated mineral deposition in culture is dependent on surface microtopography. *Calcif. Tissue Int.*, 2002, 71, 519-529.

[178] Lohmann, CH; Bonewald, LF; Sisk, MA; Sylvia, VL; Cochran, DL; Dean, DD; Boyan, BD; Schwartz, Z. Maturation state determines the response of osteogenic cells to surface roughness and 1,25-dihydroxyvitamin D3. *J. Bone Miner. Res.*, 2000, 15, 1169-1180.

[179] Schneider, GB; Zaharias, R; Seabold, D; Keller, J; Stanford, C. Differentiation of preosteoblasts is affected by implant surface microtopographies. *J. Biomed. Mater. Res. A*, 2004, 69, 462-468.

[180] Lossdorfer, S; Schwartz, Z; Wang, L; Lohmann, CH; Turner, JD; Wieland, M; Cochran, DL; Boyan, BD. Microrough implant surface topographies increase osteogenesis by reducing osteoclast formation and activity. *J. Biomed. Mater. Res. A*, 2004, 70, 361-369.

[181] Sammons, RL; Lumbikanonda, N; Gross, M; Cantzler, P. Comparison of osteoblast spreading on microstructured dental implant surfaces and cell behaviour in an explant model of osseointegration. A scanning electron microscopic study. *Clin. Oral. Implants. Res.*, 2005, 16, 657-666.

[182] Sader, MS; Balduino, A; Soares, G de A; Borojevic, R. Effect of three distinct treatments of titanium surface on osteoblast attachment, proliferation, and differentiation. *Clin. Oral. Implants Res.*, 2005, 16, 667-675.

[183] Mares, JJ; Hubik, P; Nesladek, M; Kristofik. Boron-doped diamond — Grained Mott's metal revealing superconductivity. *Diamond Relat. Mater.*, 2007, 16, 921-925.

[184] Chang, JS; Chang, KL; Hwang, DF; Kong, ZL. *In vitro* cytotoxicitiy of silica nanoparticles at high concentrations strongly depends on the metabolic activity type of the cell line. *Environ. Sci. Technol.*, 2007, 41, 2064-2068.

[185] Oner, G; Cirrik, S; Bakan, O. Effects of silica on mitochondrial functions of the proximal tubule cells in rats. *Kidney Blood Press Res.*, 2005, 28, 203-210.

[186] Xiao, X; Wang, J; Liu, C; Carlisle, JA; Mech, B; Greenberg, R; Guven, D; Freda, R; Humayun, MS; Weiland, J; Auciello, O. In vitro and in vivo evaluation of

ultrananocrystalline diamond for coating of implantable retinal microchips. *J. Biomed. Mater. Res. B Appl. Biomater.*, 2006, 77, 273-281.

[187] Labelle, D; Jumarie, C; Moreau, R. Capacitative calcium entry and proliferation of human osteoblast-like MG-63 cells. *Cell Prolif.*, 2007, 40, 866-884.

[188] Naghii, MR; Torkaman, G; Mofid, M. Effects of boron and calcium supplementation on mechanical properties of bone in rats. *Biofactors*, 2006, 28, 195-201.

[189] Bacakova, L; Stary, V., Glogar P.: Adhesion and growth of cells in culture on carbon-carbon composites with different surface properties. *Engineering of Biomaterials*, 1998, 1 [2], 3-5.

[190] Lappalainen, R; Selenius, M; Anttila, A; Konttinen, YT; Santavirta, SS. Reduction of wear in total hip replacement prostheses by amorphous diamond coatings. *J. Biomed. Mater. Res. B Appl. Biomater.*, 2003, 66, 410-413.

[191] Ma, WJ; Ruys, AJ; Mason, RS; Martin, PJ; Bendavid, A; Liu, Z; Ionescu, M; Zreiqat, H. DLC coatings: Effects of physical and chemical properties on biological response. *Biomaterials*, 2007, 28, 1620–1628.

[192] Bacakova, L; Balik, K; Zizka, S; Adhesion and growth of vascular smooth muscle cells in cultures on carbon fibre-reinforced carbon composites covered with pyrolytic carbon. *Engineering of Biomaterial*s, 1998, 1, [-], 19-22.

[193] Pesakova, V; Klezl, Z; Balik, K; Adam, A. Biomechanical and biological properties of the implant material carbon-carbon composite covered with pyrolytic carbon. *J. Mater. Sci.: Mater. Med.*, 2000, 11, 793-798.

In: New Cell Adhesion Research
Editors: Patrick Nott and Matthew Temple
ISBN 978-1-60692-378-8

Chapter VIII

Diabetes Increases Risk for Oral Carcinogenesis by Induction of Cell Proliferation and Reduction of Cell Adhesion: An Animal Model Study

Christos Yapijakis and Eleftherios Vairaktaris*
Department of Oral and Maxillofacial Surgery, University of Athens Medical School, Attikon Hospital, Athens, Greece.

Abstract

Oral cancer is the seventh most common malignancy, worldwide. It constitutes a major health problem because it is still associated with poor prognosis and about 50% five-year survival, despite advances in surgery, radiation and chemotherapy treatment over the past two decades. Oral squamous cell carcinoma (OSCC) accounts for more than 90% of oral cancer cases. In the development of OSCC have been implicated environmental factors such as tobacco smoking and heavy alcohol consumption, infectious agents such as some types of human papilloma virus, genetic alterations involving oncogenes and tumor suppressor genes, as well as epigenetic changes. Recently, some epidemiological studies have also incriminated diabetes mellitus as a risk factor for the development of oral premalignant lesions and OSCC.

In order to investigate the possible association of diabetes with increased risk for oral cancer, as well as the underlying molecular mechanism of this association an experimental model of OSCC induced in diabetic and normal Sprauge–Dawley rats was established. Diabetes type I was induced by a single intraperitoneal injection of streptozotocin in 20 rats and verified by measuring glucose levels in blood samples

* Correspondence concerning this article should be addressed to: Professor Eleftherios Vairaktaris, MD, DDS, PhD, Department of Oral and Maxillofacial Surgery, University of Athens Medical School, Kifissias and Perikleous Stavrou 9, Athens 11524, Greece. E-mail address: lvairakt@med.uoa.gr; Tel: +30 210 6443035; Fax: +30 210 6443803.

obtained by tail prick in a strip-operated sensor. In 14 of the diabetic rats, as well as in 12 rats without diabetes, chemical carcinogenesis was induced by a five-month application of the carcinogen 4-nitroquinoline N-oxide on the hard palate of the animals. Six diabetic rats as well as 6 non-diabetic rats were used as controls of normal oral tissue without carcinogenesis. The four groups of animals were used in order to study the putative effect of diabetes on signal transduction pathways involved in the sequential stages of oral carcinogenesis: normal mucosa, hyperplasia, dysplasia, early invasion, well differentiated OSCC and moderately differentiated OSCC.

The diabetic rat model was used to investigate whether the molecular basis for the putative association of diabetes with oral oncogenesis involves the insulin receptor and its substrate-1 (IR and IRS-1, respectively), as well as the influence of diabetes on signal transduction pathways involving oncogene proteins EGFR, erbB2, erbB3, FGFR-2, FGFR-3, FAK, c-myc, N-ras, ets-1, H-ras, c-fos and c-jun, tumor suppressor gene proteins p53 and p16, apoptosis-related markers Bax and Bcl-2, and cell proliferation-related marker Ki-67. The obtained findings suggest that the insulin and insulin-like growth factor-1/IR/IRS-1 signalling pathway is altered by diabetes, leading to reduced cell adhesion and possibly increased risk for oral cancer. In addition, diabetes seems to promote the activation of Ras/Raf/MAPK signal transduction pathway, mainly by induction of receptors erbB2 and erbB3, resulting in increased cell proliferation. Finally, diabetes does not seem to influence apoptotic mechanisms during oral carcinogenesis. An interesting extrapolation of these findings is that the type I diabetic patients who are not insulin-regulated may have a greater risk for developing oral cancer than the general population does. This hypothesis remains to be examined by large epidemiological studies.

Keywords: oral cancer, diabetes, animal model, oncogenes, tumor suppressor genes, cell proliferation, apoptosis, signal transduction pathways, stages of carcinogenesis.

1. Diabetes as a Risk Factor for Oral Cancer

As a common chronic and systemic disease, diabetes mellitus has been correlated with a variety of oral conditions as early as the 19th century. These include periodontal disease, which has been reported as the sixth most common complication of diabetes, dental caries, tooth loss, stomatitis, glossitis and sensory changes [1-3]. Oral manifestations of diabetes are so common that dentists are recognized as being among the first health care professionals to whom a previously undiagnosed diabetic patient may present [4,5]. The involvement of diabetes mellitus in the pathogenesis of oral cancer emerged only recently [6,7]. Several independent epidemiological studies have incriminated diabetes as a risk factor for the development of oral premalignant conditions, such as leukoplakia [8-11]. In addition, a Hungarian group has reported epidemiological findings suggesting that diabetes is a risk factor for the development of oral benign tumors and carcinomas [9,12]. Such an association of diabetes with risk for malignancy has already been reported. A number of studies have found an increased incidence of diabetes mellitus in patients with bladder carcinoma [13,14].

Oral cancer is the seventh most common malignancy worldwide and the eighth most common in the European Union [15-17]. It is characterized by severe morbidity and <50% five-year survival, despite advances in treatment with surgery, radiation and chemotherapy in the last two decades [17-19]. More than 90% of oral malignancies represent oral squamous cell carcinoma (OSCC), with 300,000 new cases being diagnosed each year worldwide [15-19]. Environmental factors such as tobacco smoking and heavy alcohol consumption, infectious agents such as some types of human papilloma virus, genetic alterations involving oncogenes and tumor suppressor genes, as well as epigenetic changes have been implicated in the development of OSCC [19,20]. In addition, diabetes recently emerged as a risk factor for OSCC, as already mentioned [21]. Oral oncogenesis is a multi-step process involving progression through several histological stages until the development of carcinomas [19,20]. The deciphering of the molecular mechanisms involved in the initiation and progression to malignancy (including the effect of diabetes) is expected to improve prophylactic measures as well as treatment procedures [19]. In addition, the screening identification of individuals at high risk for oral cancer is expected to offer the opportunity to change their behaviors and optimize the implementation of preventive and therapeutic measures [22].

2. Molecular Mechanisms of Oncogenesis

Unlimited cell proliferation and division may occur due to mutational activation, deregulation or amplification of oncogenes or mutational inactivation, deregulation or loss of tumor suppressor genes [23]. These events may influence key points of signal transduction cascades, which regulate cell proliferation, resulting in cancer progression [23]. The most common pathways leading to cell proliferation act through a cascade activation of cell membrane receptor tyrosine kinase (RTKs), intracellular second messengers and gene expression of transcription factors (Figure 1).

Extracellular mitogenic growth factors bind to RTKs that are activated and transduce a positive growth signal by phosphorylating intracellular second messengers [23,24]. The RTK substrates belong to a protein family with homologous domains, the most important members of which, in regard to cell signalling, are phospholipase-γ (PLCγ), phosphatidylinositol kinase (PI3K) and Grb2 [25-27]. PLC-γ generates the downstream messengers diacylglycerol (DAG) and inositol triphosphate (IP3) by hydrolyzing membrane phospholipids (44). PI3K catalyzes the formation of downstream messenger phosphatidylinositol triphosphate (PIP3) resulting in the ultimate activation of protein kinase. C [28]. Grb2 binds to the RTK and also interact with downstream messenger Son of Sevenless (SOS), which in turn activates Ras [29-31]. The localization of Ras proteins to the inner surface of the cytoplasmic membrane is critical for activation of several downstream effectors, most notably the kinase Raf, the first signalling element in the pathway of the mitogen activated protein kinases (MAPKs)[32]. When activated, Raf in turn activates the MAPKs by phosphorylation, initiating a cascade of activation of kinases that eventually results in the enhanced transcription of genes coding for proteins that regulate cell proliferation and apoptosis [32,33].

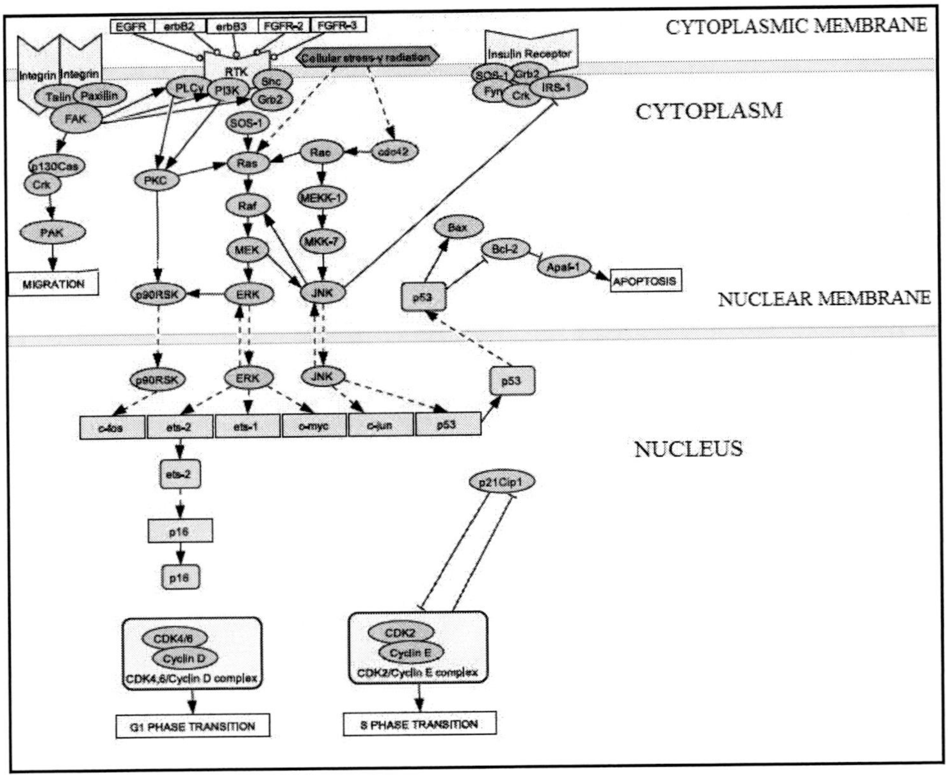

Figure 1. Signal transduction pathways in normal oral mucosa ([21] with permission). Broken or solid lines correspond to activation of specific protein molecules depending whether it occurs with or without passage via a membrane, respectively.

3. Animal Models of Oral Cancer

A number of animal models of induced oral oncogenesis have been generated, including hamster, rat and mouse models. These animal models allow reproducible pathological evaluation of all precancerous and cancer stages in comparison with normal oral mucosa [34,35].

The first demonstration of OSCC induction in an animal model was effected half a century ago in hamster cheek pouch with the aid of polycyclic aromatic carcinogens, such as 9,10-dimethyl-1,2-benzanthracene (DMBA), 20 methyl cholanthrene and 3,4-benzopyrene [36]. Thirty-five years ago, the rat model of oral oncogenesis was introduced using the water-soluble carcinogen 4-nitroquinoline-1-oxide (4NQO)[37-39]. A metabolic product of 4NQO binds predominantly at guanine residues of DNA producing a spectrum of precancerous and neoplastic lesions [40]. The rat model of 4NQO-induced carcinogenesis is considered as the preferred model for the study of OSCC, since it parallels oral oncogenesis in humans [41-47].

The use of carcinogens for induction of oral cancer in animal models is warranted, since chemical agents appear to be the dominant etiological factors in oral cavity. Animal models are a very useful tool towards understanding important mechanisms in oral oncogenesis.

4. The Model of Oral Oncogenesis in Diabetic Rats

An experimental animal model was used in order to study the effect of diabetes mellitus type I (also known as insulin-dependent diabetes) on the molecular pathways involved in the sequential stages of oral oncogenesis. Oral cancer was induced in diabetic type I and normal rats. The experimental procedures will be discussed briefly here, since they have been previously described in detail [8].

4.1. Animal Groups

Thirty-seven female Sprague-Dawley rats purchased from the Hellenic Pasteur Institute (Athens) at the age of six weeks and weighing approximately 135 g each were used in this study. The rats were handled in accordance with the Greek Legislation for the Care and Use of Laboratory Animals (P.D. 160/91 and Law 2015/92), conforming to Directive 86/609 of the European Union, as well as the Guide for the Care and Use of Laboratory Animals (U.S. National Institutes of Health Publication No. 85-23, revised 1996). The animals were randomly divided into four groups. A) Group D (n=6): Diabetic rats without carcinogenesis, B) Group Dc (n=13): Diabetic rats used for induced carcinogenesis, C) Group N (n=6): Normal rats without carcinogenesis, D) Group Nc (n=12): Normal rats used for induced carcinogenesis.

4.2. Induction of Diabetes

The induction of diabetes type I was performed in 19 previously overnight fasted animals by a single intraperitoneal injection of streptozotocin (STZ) dissolved in saline buffer at a dose of 70 mg/kg of weight (ZANOSAR, Pharmacia & Upjohn Co., USA) and confirmed by glucose levels in the blood after three weeks, as previously described [8].

4.3. Experimental Carcinogenesis

Oral cancer was induced in Dc and Nc animals by application of the carcinogen 4NQO at a concentration of 5% in propylene glycol 3 times per week for 5 months in the rats' hard palate (Fluca AG Chemische Fabrik, Switzerland), as described elsewhere [8]. Clinical signs of oral lesions putatively tumor-related were observed within 6 months after last application of carcinogen. After sacrifice of animals by ether treatment, the oral regions with cancer (mainly palate and tongue) of Dc and Nc rats and the respective regions of D and N rats were excised for immunohistochemical analysis [8].

4.4. Pathological and Immunohistochemical Evaluation

The histological status of the lesions was defined after examination of the complete section under light microscopy and the tissue profiles were classified into the following categories: normal mucosa, hyperplasia, dysplasia, early invasion, well- and moderately differentiated carcinoma. In every sample, a number of all possible representative histological lesions were analysed. The biopsies from the 37 animals were fixed in 10% neutralized formaldehyde solution and embedded in paraffin. Sections 4 μm-thick were prepared from each specimen and were mounted on Super Frost Plus-coated glass slides (Menzel and Co., Braunschweig, Germany). All sections were used for detection of cell membrane tyrosine kinase receptors (EGFR, erbB2, erbB3, FGFR-2, FGFR-3), a docking cell membrane protein (IRS-1), cytoplasmic proteins (FAK, H-ras, N-ras), apoptosis-related proteins (Bax, Bcl-2), a cell proliferation marker (Ki-67), nuclear transcriptional factors (p53, c-myc, c-fos, c-jun, ets-1), and a cell cycle protein (p16), using a standard immunohistochemical technique (DAB Immunoperoxidase Secondary Detection System) as described elsewhere [8,48-55].

4.5. Statistical Analysis

In order to evaluate the pattern of antibody expression in relation to the histological status, the various lesions were divided according to tumor progression into a) normal tissue, b) non-cancerous and precancerous conditions (hyperplasia, dysplasia), c) tumor (early invasion, well differentiated OSCC, moderately differentiated OSCC). The mean percentages of positively stained cells were calculated for all the different lesions present in each sample. These values were tabulated for diabetic animals (groups D and Dc) and normal ones (groups N and Nc). Using the two-tailed Student's *t*-test, the mean percentage of positively-stained cells from each non-cancerous and precancerous category was compared with those of the normal tissue, while the percentage of positively stained cells from each tumor category was compared with the average percentage of the non-cancerous and precancerous lesions. In all cases with no normal distribution, the results of both the Wilcoxon test and two-tailed Student's *t*-test provided the same level of significance.

5. Effect of Diabetes on Signal Transduction Pathways

Despite the recent epidemiological studies that associate diabetes and oral oncogenesis, no molecular basis of this association had been proposed in the literature. The model of oral oncogenesis in diabetic rats was employed to studying a possible mechanism linked to both diseases, as well as the expression of several proteins implicated in common signal transduction pathways that regulate cell proliferation and apoptosis. These studies allowed the deciphering of the involvement of 18 signal transduction factors in each sequential stage of oral oncogenesis both in the presence or absence of diabetes type I (Figure 2).

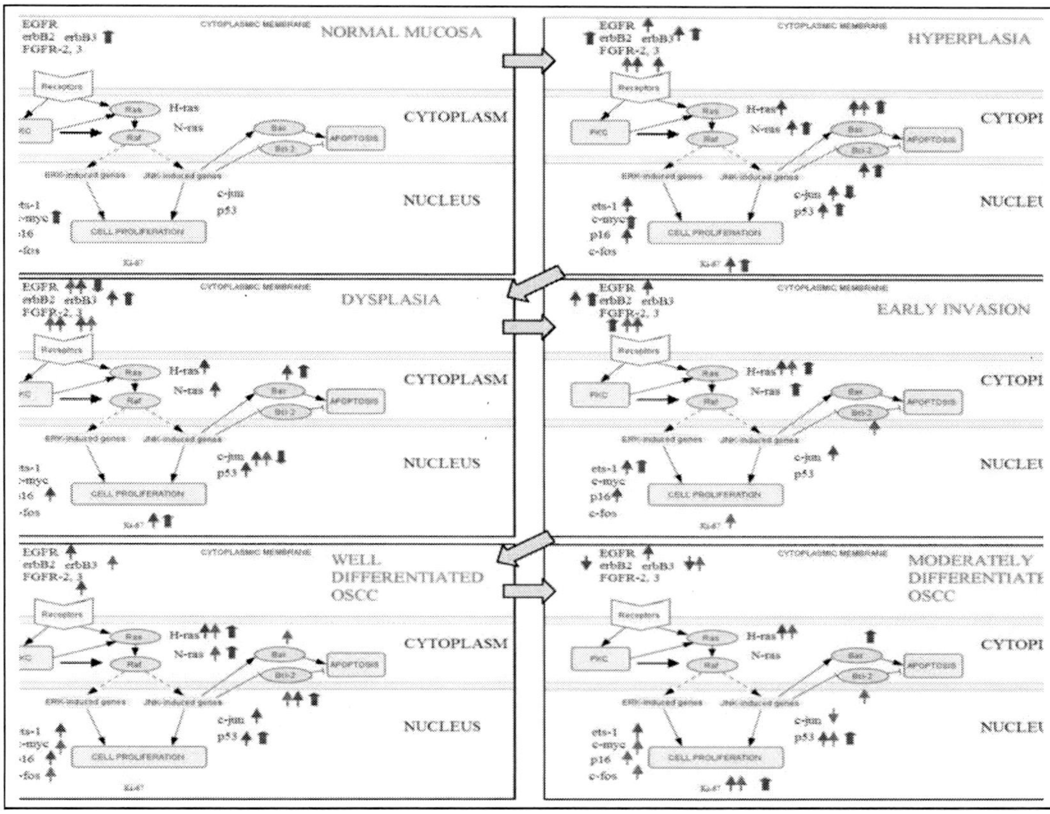

Figure 2. Signal transduction pathways during sequential stages of oral oncogenesis ([21] with permission). Broken or solid lines correspond to activation of specific proteins depending whether it occurs with or without passage via a membrane, respectively. See text (4.5) for statistical comparisons. The arrows by the studied proteins indicate the significant differences in their expression observed in the rat model. Red arrows indicate significant differences in diabetic rats, blue arrows indicate significant differences in normal rats, while green arrows indicate significant differences between normal and diabetic rats in the same stage. An arrow pointing upwards indicates a significant increase of expression in that stage, while one pointing downwards indicates a significant decrease.

5.1. Insulin Receptor and Its Substrate

A possible mechanism linked both to diabetes type I and cancer involves insulin receptor and its substrate (Figure 1). Binding of insulin to its cell membrane receptor stimulates tyrosine phosphorylation of insulin receptor substrate-1 (IRS-1), which in turn activates PI3K and starts a cascade resulting in cell growth [56]. IRS-1 is a docking cell membrane protein that plays an important role in the insulin signal transduction cascade and its expression is affected both in the insulin deficiency status (diabetes type I) and in the insulin resistance status (diabetes type II)[57,58]. One of the signalling molecules implicated in IRS-1 pathways is focal adhesion kinase pp125 (FAK), also known as cytosolic tyrosine kinase pp125 [59,60]. FAK interacts both with IRS-1 and with the transmembrane integrins, which are involved in cell migration and adhesion, apoptosis, wound repair and angiogenesis [61]. Following insulin stimulation, IRS-1 is detected co-immunoprecipitating with $\alpha_v{}^{3}{}_3$ integrin

[62]. Interaction of both FAK and integrins may induce activation of IRS-1 by tyrosine phosphorylation upon insulin stimulation, but at the same time the FAK/integrin complex may inhibit activation of IRS-1 in an insulin-independent phosphoinositolglycan-mediated fashion [59,63]. Over-expression of FAK and integrins is observed in OSCC and other malignancies, underlying their important role in oncogenesis [63-66]. FAK is a cytoskeletal-associated tyrosine kinase and its over-expression of leads to increased cell migration and survival under anchorage-independent conditions [64]. Interestingly, mutations in the insulin receptor alpha-subunit gene affecting its expression result in altered cell morphology associated with increased cell size, reduced cell adhesiveness and dephosphorylation of FAK in the absence of insulin [67].

In the rat model, expression of IRS-1 was significantly higher in diabetic (D) than normal animals, but it decreased in diabetic animals with tumor, especially in more advanced stages [8]. Furthermore, FAK expression was significantly higher in rats with cancer in comparison to the ones without it, regardless the diabetes status [8]. The expression of IRS-1 and the expression of FAK were not correlated during the sequential stages of oral oncogenesis (Table 1). Since the IRS-1 pathway plays an important role in the post-receptor signaling glucose transport in muscle and adipose tissues as well as in cell growth, the over-expression of IRS-1 in diabetic rats may be attributed to an attempt to cope in an environment of scarce insulin, since the abundance of IRS-1 might facilitate its activation by the relatively few receptors that bind to minimally available insulin, as previously suggested [68]. These data suggest that the role of IRS-1 as a signal-mediating molecule is diminished in an insulin-deprived malignant environment. This study showed that diabetes type I may indeed increase the risk for OSCC through alteration of the IRS-1/FAK pathway, which may possibly end up in cytoskeleton changes and reduced cell adhesion, an initial step towards neoplasia.

5.2. Receptor Tyrosine Kinases

Members of the EGFR family of RTKs include epidermal growth factor receptor (EGFR), erbB2 (also known as HER-2/neu) and erbB3 (also known as HER-3). Significant over-expression of EGFR has been reported in cases of head and neck cancer, correlated with tumor size and advanced stages of oncogenesis [20,69-71]. Over-expression of erbB2 has been significantly associated with shorter survival of patients with OSCC [71-73]. Accordingly, erbB3 over-expression has been correlated to the malignant progression of head and neck SCC [72-75]. Increased expression of both fibroblast growth factor receptors 2 and 3 (FGFR-2, FGFR-3) in oral dysplasia and OSCC has been reported [76].

In the diabetic rat model, there was a trend for reduced EGFR expression in diabetic compared to normal animals throughout oral oncogenesis (Figure 2)[53]. On the contrary, diabetes enhances erbB2 expression during nearly all stages of oral cancer development, as well as erbB3 expression in non-cancerous and precancerous stages (Figure 2)[51]. The expression pattern of FGFR-2 and FGFR-3 was not affected by diabetes during oral oncogenesis (Figure 2)[50].

Table 1. Expression of IRS-1 and of FAK during the sequential stages of oncogenesis in palate and tongue of normal and diabetic rats, as well as examination of their correlation (Spearman's correlation coefficient *r*). All possible histological regions were evaluated for every animal biopsy. Protein expression is given in mean percentage cells positive for immunochemical staining within all equivalent regions in each biopsy sample. Significant P values are shown in bold italics. A) Normal rats of groups N and Nc. B) Diabetic rats of groups D and Dc

A

Normal Rats – Palate

	Normal Sites		Precancerous Sites				Cancerous Sites			
			Hyperplasia		Dysplasia		Ca in Situ		Ca	
Pairs	**FAK**	**IRS-1**	**FAK**	**IRS-1**	**FAK**	**IRS-1**	**FAK**	**IRS-1**	**FAK**	**IRS-1**
1.	1.18	15.12	1.44	15.27	2.10	19.93	3.15	20.45	1.71	19.88
2.	2.34	21.44	2.91	19.19	3.40	16.82	2.53	21.49	5.06	22.16
3.	2.12	27.05	1.36	26.12	2.69	16.93	3.30	15.73	2.68	22.71
4.	2.10	12.36	2.89	20.19	1.49	15.58				
5.	0.96	18.84	2.00	17.14	1.40	15.26				
6.	1.41	17.07	3.40	18.19	1.25	14.66				
7.	3.41	14.53	1.03	10.13						
8.	0.91	9.15	1.33	16.12						
Mean	1.804	16.945	2.045	17.794	2.055	16.530	2.993	19.223	3.150	21.583
SD	0.854	5.566	0.901	4.559	0.850	1.888	0.408	3.070	1.724	1.501
r	0.205		0.299		0.458		-0.770		0.570	
P	0.626		0.472		0.361		0.441		0.430	
1.	23.66	26.85	10.10	10.33	20.19	21.16	19.31	18.13	23.61	22.03
2.	25.34	26.71	15.30	6.13	20.34	25.18	19.31	24.19	24.04	19.64
3.	20.52	25.65	14.91	11.37	24.17	19.12	20.90	26.18	21.09	21.10
4.	21.13	19.18	10.10	5.18	21.14	20.18			21.70	25.75
5.			9.76	10.11	21.92	16.70			22.99	22.14
6.			6.79	16.16	24.60	17.74			32.50	20.48
7.			20.14	21.19					23.04	28.87
8.			28.13	27.18					28.43	22.84
9.			20.71	9.16					19.80	20.74
10.			27.62	22.36					26.14	27.97
11.									26.17	22.30
12.									26.72	22.15
Mean	22.663	24.598	16.356	13.917	22.060	20.013	19.840	22.833	24.686	23.001
SD	2.244	3.651	7.567	7.443	1.909	2.999	0.918	4.193	3.522	2.954
r	0.568		0.692		-0.642		0.691		-0.098	
P	0.432		***0.027***		0.169		0.514		0.763	

Table 1. Continued

B

Diabetic Rats – Tongue										
	Normal Sites		Precancerous Sites				Cancerous Sites			
			Hyperplasia		Dysplasia		Ca in Situ		Ca	
Pairs	**FAK**	**IRS-1**	**FAK**	**IRS-1**	**FAK**	**IRS-1**	**FAK**	**IRS-1**	**FAK**	**IRS-1**
1.	4.27	25.30	3.14	29.14	4.24	23.73	5.06	25.69	3.24	26.44
2.	4.14	15.28	0.84	16.19	1.91	17.03	2.12	16.17	1.76	17.42
3.	2.32	21.73	1.41	28.16	2.61	27.15	4.24	18.40	2.25	17.32
4.	4.34	22.01	1.70	16.13	2.31	17.02	4.32	26.25	4.19	19.05
5.	3.29	17.59			4.17	14.17			5.73	18.55
6.	2.40	19.38								
7.	2.34	27.20								
8.	1.75	26.54								
9.	2.15	32.25								
Mean	3.000	23.031	1.773	22.405	3.048	19.820	3.935	21.628	3.434	19.756
SD	1.021	5.310	0.979	7.222	1.085	5.398	1.265	5.101	1.588	3.809
r	-0.480		0.630		0.015		0.789		0.091	
P	0.191		0.370		0.981		0.211		0.885	
1.	26.49	31.45	9.13	16.16	20.24	17.18	19.21	21.63	24.34	19.71
2.	27.39	21.64	10.33	13.19	18.12	24.38	20.69	20.77	30.17	17.09
3.	22.17	29.01	8.12	15.17	21.51	20.30	24.51	22.19	19.54	25.07
4.	22.79	29.25	10.13	9.86	25.12	21.31	19.14	19.17	24.19	24.17
5.	20.79	37.24	11.37	7.86	24.23	19.76	23.13	28.17	32.68	18.77
6.			17.35	12.13			20.14	24.73	25.32	17.36
7.			7.31	7.35			20.19	11.19	28.30	21.85
8.			12.27	6.18			-	23.28	22.38	24.84
9.			6.17	10.50			24.64	19.38	22.67	23.19
10.			24.14	5.16			20.13	28.66	26.59	20.02
11.			24.36	19.19						
12			19.15	20.18						
13.			24.73	31.18						
Mean	23.926	29.718	14.197	13.393	21.844	20.586	21.309	21.917	25.618	21.207
SD	2.863	5.603	6.863	7.166	2.871	2.612	2.184	4.999	3.924	3.029
r	-0.707		0.472		-0.272		0.129		-0.726	
P	0.182		0.103		0.658		0.742		*0.018*	

5.3. Cytoplasmic Proteins

Mutations in genes coding for Ras proteins seem to be quite rare in OSCC patients from Western countries and more frequent in patients from Southeast Asian countries, such as India and Taiwan [77]. Furthermore, involvement of N-ras in oral oncogenesis is an infrequent occurrence [78,79]. There are conflicting reports in the literature regarding the increased expression of Ras proteins at an early or a later stage in oral carcinogenesis [80,81].

In the rat model, there was a trend for increased H-ras expression in diabetic compared to normal animals during oncogenesis, which reached a statistical significant difference in early invasion and well-differentiated OSCC (Figure 2)[53]. N-ras expression increased with tumor advancement in diabetic rats, while in normal rats N-ras expression was not observed in premalignant stages and increased only in well-differentiated OSCC (Figure 2)[49]. The expression of both H-ras and N-ras seems to be positively affected by diabetes during oral oncogenesis.

5.4. Apoptosis and Cell Proliferation Markers

There are conflicting reports in the literature regarding the levels of the apoptosis-related proteins Bcl-2 and Bax in oral cancer, although in other tumors their levels are so well defined that they are used as prognostic markers [82-84]. Some studies of human OSCC biopsies reported increased levels of Bcl-2 as well as increased levels of Bcl-2/Bax ratio corresponding to reduced apoptosis in comparison to normal oral mucosa [85,86]. Other studies of human biopsies reported reduced Bcl-2 expression and Bcl-2/Bax ratio and increased apoptosis in the initial tumor stages, while Bax was not found to be significantly altered during the different stages of oral oncogenesis [87,88]. Tumor cells are known to manifest elevated levels of certain marker proteins that control cell cycle, such as Ki-67 [89,90]. The cell proliferation marker Ki-67 is highly expressed in initial stages of oral oncogenesis [91].

In the rat model, the Bcl-2/Bax ratio was relatively stable during the oncogenesis process in diabetic animals, contrary to normal rats which appear to have an increased Bcl-2/Bax ratio in the stage of moderately differentiated carcinoma, indicating reduced apoptosis levels (Figure 2)[52]. Furthermore, the expression of Ki-67 was higher in diabetic rats than in normal ones in almost all stages of oral oncogenesis (Figure 2)[52]. Diabetes seems to promote increased cell proliferation during oral oncogenesis, but this is accomplished without affecting the Bax/Bcl-2-mediated apoptotic pathways.

5.5. Nuclear Transcriptional Factors

Gene amplification leading to increased levels of the transcription factor c-myc occurs frequently in poorly differentiated OSCC [92,93]. One of the best studied tumor suppressor genes is *p53*, but its high frequency of alterations in human cancer has not fully elucidated since mutant p53 proteins constitute a complex family of several hundred proteins with heterogeneous properties [94-96]. Mutations that inactivate *p53* gene or impair function of its protein product have been detected in more than 60% of cases of OSCC [97]. The inactivated mutant p53 protein is more stable than the normal one and its accumulation increases gradually during oral oncogenesis [94,97-99]. Together c-fos and c-jun form the composite transcription factor activating protein-1 (AP-1), a mitogen-activated transactivator important for cell proliferation and differentiation [100]. In general, c-jun is considered to be an early marker of early stages of oncogenesis [101]. There are conflicting reports about c-fos, with some studies indicating elevated levels of c-fos in malignant oral lesions, others noting a stable expression at normal levels throughout oral oncogenesis, while still other studies reporting a gradual decrease of c-fos expression from normal oral mucosa towards advanced stages of oral cancer [102-105]. Up-regulation of *ets-1* gene expression has been observed in oral cancer, correlating with the grade of invasiveness and metastasis [106].

Based on the rat model studies, diabetes contributes to increased accumulation of mutant p53 resulting in increased tumor cell proliferation during oral oncogenesis (Figure 2)[48]. Furthemore, c-myc expression is enhanced by diabetes only in the initial stages of oral oncogenesis (Figure 2)[48]. Higher c-jun levels were observed only in non-cancerous and precancerous stages in normal rats compared to diabetic rats, indicating that diabetes does not affect in tumor stages the c-jun *N*-terminal kinase (JNK)/c-jun pathway (Figure 2)[54]. No statistical differences concerning c-fos expression were detected between diabetic and normal animals (Figure 2)[53]. The same pattern of increased ets-1 expression was observed both in diabetic and normal rats, but in cancerous stages this expression was higher in diabetic animals, indicating that diabetes may contribute to enhanced invasion and metastatic potential (Figure 2)[49].

5.6. Cell Cycle Regulators

The negative regulator of cell proliferation p16 is frequently abrogated in head and neck cancer. In OSCC point mutations of the *p16* gene are rare, but alternative mechanisms of p16 function loss have been reported, including homozygous deletions or methylation silencing of the promoter region of the *p16* gene [107-110]. In the experimental rat model of oral oncogenesis, the expression of p16 was not found to be affected by diabetes (Figure 2)[55].

6. Insights from the Diabetic Rat Model Studies

Activation of cell membrane RTKs (such as insulin receptor, EGFR, erbB2, erbB3, FGFR-2 and FGFR-3) leads to the recruitment of signalling molecules (such as IRS-1) to the cell membrane and the initiation of signalling cascades (such as the Ras/Raf/MAPK and PI3K cascades). The animal model studies described here indicate that diabetes mellitus type I seems to influence key points of signal transduction cascades that regulate important cell functions, thereby increasing risk for oral cancer. Specifically, the diabetic rat model provided us with the following insights:

Insulin deficiency seems to result in reduced amounts of insulin receptor and a counter-active increased expression of IRS-1. The latter may lead eventually to cytoskeleton changes and reduced cell adhesion, an initial step towards neoplasia.

Diabetes seems to promote the activation of Ras/Raf/MAPK signal transduction pathway by induction of receptors erbB2 and erbB3 mainly (and not EGFR) leading to increased cell proliferation.

Diabetes does not seem to influence levels of apoptosis related to the Bcl-2/Bax ratio.

An interesting extrapolation of these findings is that the type I diabetic patients who are not insulin-regulated may have a greater risk for developing oral cancer than the general population does. This hypothesis remains to be examined by large epidemiological studies.

ACKNOWLEDGMENTS

The authors wish to thank Professor Efstratios Patsouris,MD,PhD for reviewing this manuscript and for valuable suggestions. In addition, the authors thank all colleagues who participated in the studies presented here, especially Ph.D. students Lambros Goutzanis,MD,DDS for the effective development of the animal model, Sofia Spyridonidou,B.S. for excellent laboratory work and data interpretation contributions and Dimitris Avgoustidis,MD for competent assistance in statistical analysis. This work was co-funded by the European Social Fund and National Resources (EPEAEK II "Pythagoras" 70/3/7391) grant.

REFERENCES

[1] Taylor GW, Manz MC and Borgnakke WS: Diabetes, periodontal diseases, dental caries, and tooth loss: a review of the literature. *Compend Contin Educ Dent 25*(3):179-190, 2004.

[2] Kim J and Amar S: Periodontal disease and systemic conditions: a bidirectional relationship.*Odontology 94*(1):10-21, 2006.

[3] Löe H: Periodontal disease. The sixth complication of diabetes mellitus. *Diabetes Care 16*:329-334, 1993.

[4] Finney LS, Finney MO and Gonzalez-Campoy JM: What the mouth has to say about diabetes. Careful examinations can avert serious complications. *Postgrad Med 102*(6):117-126, 1997.

[5] Milosevic A: Eating disorders and the dentist. *Br Dent J 186*(3):109-113, 1999.

[6] Auluck A: Diabetes mellitus: an emerging risk factor for oral cancer? *J Can Dent Assoc 73*(6):501-503, 2007.

[7] Goutzanis L, Vairaktaris E, Yapijakis C, Kavantzas N, Nkenke E, Derka S, Vassiliou S, Acil Y, Kessler P, Stavrianeas N, Perrea D, Donta I, Skandalakis P and Patsouris E: Diabetes may increase risk for oral cancer through the insulin receptor substrate-1 and focal adhesion kinase pathway. *Oral Oncol 43*(2): 165-173, 2007.

[8] Albrecht M, Bánóczy J, Dinya E and Tamás G: Occurrence of oral leukoplakia and lichen planus in diabetes mellitus. *J Oral Pathol Med 21*:364-366, 1992.

[9] Ujpál M, Matos O, Bibok G, Somogyi A, Szabo G and Suba Z: Diabetes and oral tumors in Hungary: epidemiological correlations. *Diabetes Care 27*(3): 770-774, 2004.

[10] Dietrich T, Peter A. Reichart PA and Scheifele C. Clinical risk factors of oral leukoplakia in a representative sample of the US population. *Oral Oncol 40*:158-163, 2004.

[11] Dikshit RP, Ramdas K, Hashibe M, Thomas G, Somanathan T and Sankaranarayanan R: Association between diabetes mellitus and premalignant oral diseases: a cross sectional study in Kerala, India. *Int J Cancer 118*(2):453–457, 2006.

[12] Ujpál M, Barabás J, Kovalszky I, Szabó G, Németh Z, Gábris K and Suba Z: A preliminary comparative study of the prognostic implications of type 2 diabetes

mellitus for patients with primary gingival carcinoma treated with surgery and radiation therapy. *J Oral Maxillofac Surg. 65*(3):452-456, 2007.

[13] Kravchick S, Gal R, Cytron S, Peled R, Weissman Y, Mukamel E and Koren R. Increased incidence of diabetes mellitus in the patients with transitional cell carcinoma of urinary bladder. *Pathol Oncol Res 7*:56–59, 2001.

[14] Yeung NG, Husain J and Waterfall N. Diabetes mellitus and bladder cancer: an epidemiological relationship? *Pathol Oncol Res 9*:30–31, 2003.

[15] Parkin DM, Bray F, Ferlay J and Pisani P: Global cancer statistics, 2002. *CA Cancer J Clin 55*(2): 74-108, 2005.

[16] Boyle P and Ferlay J: Cancer incidence and mortality in Europe, 2004. *Annals of Oncol 16*: 481-488, 2005.

[17] Silverman S Jr: Demographics and occurrence of oral and pharyngeal cancers. The outcomes, the trends, the challenge. *J Am Dent Assoc 132*:7S-11S, 2001.

[18] Sudbo J: Novel management of oral cancer: a paradigm of predictive oncology. *Clin Med Res 2*(4): 233-242, 2004.

[19] Shah JP, Johnson NW and Batsakis JG (editors): *Oral cancer.* Martin Dunitz Publishers, London, UK.

[20] Mehrotra R and Yadav S: Oral squamous cell carcinoma: etiology, pathogenesis and prognostic value of genomic alterations. *Indian J Cancer 43*(2): 60-66, 2006.

[21] Vairaktaris E, Spyridonidou S, Goutzanis L, Vylliotis A, Lazaris A, Donta I, Perrea D, Yapijakis C and Patsouris E: Diabetes and oral oncogenesis. *Anticancer Res 27*: 4185-4194, 2007.

[22] Lavelle CL and Scully C: Criteria to rationalize population screening to control oral cancer. *Oral Oncol 41*(1):11-6, 2005.

[23] Bell HS and Ryan KM: Intracellular signalling and cancer: complex pathways lead to multiple targets. *Eur J Cancer 41*(2): 206-215, 2005.

[24] Cross M and Dexter TM: Growth factors in development, transformation, and tumorigenesis. *Cell 64*: 271-280, 1991.

[25] Margolis B: Proteins with SH2 domains: transducers in tyrosine kinase signaling pathway. *Cell Growth Differ 3*: 73-80, 1992.

[26] Karin M: Signal transduction from cell surface to nucleus in development and disease. *FASEB J 6*(8): 2581-290, 1992.

[27] Sandy JR: Signal transduction. *Br J Orthod 25*(4): 269-274, 1998.

[28] Fantl WJ, Escobedo JA, Martin GA, Turck CW, del Rosario M, McCormick F and Williams LT: Distinct phosphotyrosines on a growth factor receptor bind to specific molecules that mediate different signaling pathways. *Cell 69*: 413-423, 1992.

[29] Lowenstein EJ, Daly RJ, Batzer AG, Li W, Margolis B, Lammers R, Ullrich A, Skolnik EY, Bar-Sagi D and Schlessinger J: The SH2 and SH3 domain containing protein GRB2 links receptor tyrosine kinases to ras signaling. *Cell 70*: 431-442, 1992.

[30] Li N, Batzer A, Daly R, Yajnik V, Skolnik E, Chardin P, Bar-Sagi D, Margolis B and Schlessinger J: Guanine-nucleotide-releasing factor hsosl binds to Grb2 and links receptor tyrosine kinases to Ras signaling. *Nature 363*: 85-88, 1993.

[31] Porfiri E and McCormick F: Regulation of epidermal growth factor receptor signaling by phosphorylation of the ras exchange factor hSOS1. *J Biol Chem 271*: 5871 -5877, 1996.

[32] Beeram M, Patnaik A and Rowinsky EK: Raf: a strategic target for therapeutic development against cancer. *J Clin Oncol 23*(27): 6771-6790, 2005.

[33] Kamer AR, Krebs L, Hoghooghi SA and Liebow C: Proliferative and apoptotic responses in cancers with special reference to oral cancer. *Crit Rev Oral Biol Med 10*(1): 58-78, 1999.

[34] Tang XH, Knudsen B, Bemis D, Tickoo S and Gudas LJ: Oral cavity and esophageal carcinogenesis modeled in carcinogen-treated mice. *Clin Cancer Res 10(*1 Pt 1): 301-313, 2004.

[35] Vairaktaris E, Spyridonidou S, Papakosta V, Vylliotis A, Lazaris A, Perrea D, Yapijakis C and Patsouris E: The hamster model of sequential oral oncogenesis. *Oral Oncol* 2007 PMID 18061531 (in press).

[36] Salley JJ: Experimental carcinogenesis in the cheek pouch of the Syrian hamster. *J Dent Res 33*(2): 253-262, 1954.

[37] Boyd NM and Reade PC: Temporal alterations in cytokeratin expression during experimental oral mucosal carcinogenesis. *Carcinogenesis 12*(10): 1767-1771, 1991.

[38] Tanaka T, Kojima T, Okumura A, Yoshimi N and Mori H: Alterations of the nucleolar organizer regions during 4-nitroquinoline 1-oxide-induced tongue carcinogenesis. *Carcinogenesis 12*: 329-333, 1991.

[39] Vered M, Yarom N and Dayan D: 4NQO oral carcinogenesis: animal models, molecular markers and future expectations. *Oral Oncol 41*(4): 337-339, 2005.

[40] Kanojia D and Vaidya MM: 4-Nitroquinoline-1-oxide induced experimental oral carcinogenesis. *Oral Oncol 42*(7): 655-667, 2006.

[41] Itoh T, Yamamoto Y, Saka T, Inoue I and Takahashi H: Estimation of proliferative activity of experimental tongue carcinoma in rats. Immunohistochemical and DNA cytofluorometric analysis. *Acta Otolaryngol 113*: 568-574, 1993.

[42] Rich AM and Reade PC: Nuclear morphometry in experimental oral mucosal carcinogenesis. *Eur J Cancer B Oral Oncol 32B*: 169-175, 1996.

[43] Osugi Y: p53 Expression in various stages of 4-nitroquinoline 1-oxide induced carcinoma in rat tongue. *J Osaka Dent Univ 30*: 29-35, 1996.

[44] Dayan D, Hirshberg A, Kaplan I, Rotem N and Bodner L: Experimental tongue cancer in desalivated rats. *Oral Oncol 33*: 105-109, 1997.

[45] Nishimura A: Changes in Bcl-2 and Bax expression in rat tongue during 4-nitroquinoline 1-oxide-induced carcinogenesis. *J Dent Res 78*: 1264-1269, 1999.

[46] Kaplan I, Engelberg J and Dayan D: The effect of desalivation on 4-nitroquinoline-1-oxide-induced tongue carcinogenesis: a morphometric study of nucleolar organizer regions. *J Oral Pathol Med 30*: 48-52, 2001.

[47] Sakaki T, Tamura I, Katoda H and Kakudo K: Changing expression of E- and P-cadherin during rat tongue carcinogenesis induced by 4-nitroquinoline 1-oxide. *J Oral Pathol Med 32*: 530-537, 2003.

[48] Vairaktaris E, Kalokerinos G, Goutzanis L, Spyridonidou S, Vassiliou S, Derka S, Nkenke E, Yapijakis C, Vylliotis A, Lazaris A and Patsouris E: Diabetes alters

expression of p53 and c-myc in different stages of oral oncogenesis. *Anticancer Res 27*: 1465-1474, 2007.

[49] Vairaktaris E, Goutzanis L, Kalokerinos G, Yannopoulos A, Yapijakis C, Vassiliou S, Spyridonidou S, Vylliotis A, Nkenke E, Lazaris A, Tesseromatis C and Patsouris E: Diabetes increases both N-ras and ets-1 expression during rat oral oncogenesis resulting in enhanced cell proliferation and metastatic potential. *In Vivo 21*(4): 615-621, 2007.

[50] Vairaktaris E, Goutzanis L, Nkenke E, Spyridonidou S, Vassiliou S, Derka S, Vylliotis A, Yapijakis C, Lazaris A, Strantzias P, Kalimeras E and Patsouris E: Diabetes does not influence oral oncogenesis through fibroblast growth factor receptors. *In Vivo 21*(4): 623-628, 2007.

[51] Vairaktaris E, Goutzanis L, Vassiliou S, Spyridonidou S, Nkenke E, Papageorgiou G, Strantzias P, Lazaris A, Yapijakis C and Patsouris E: Enhancement of erbB2 and erbB3 expression during oral oncogenesis in diabetic rats. *J Cancer Res Clin Oncol* 2007 PMID: 17704947 (in press).

[52] Vairaktaris E, Kalokerinos G, Goutzanis L, Yapijakis C, Derka S, Vassiliou S, Spyridonidou S, Vylliotis A, Nkenke E, Lazaris A and Patsouris E: Diabetes enhances cell proliferation but not Bax/Bcl-2 mediated apoptosis during oral oncogenesis. *Int J Oral Maxillof Surg 37*(1): 60-65, 2008.

[53] Vairaktaris E, Goutzanis L, Yapijakis C, Vassiliou S, Spyridonidou S, Vylliotis A, Nkenke E, Lazaris A, Strantzias P and Patsouris E: Diabetes enhances the expression of H-ras and suppresses the expression of EGFR leading to increased cell proliferation. *In Vivo* 2007 (in press).

[54] Vairaktaris E, Goutzanis L, Kalokerinos G, Vassiliou S, Spyridonidou S, Avgoustidis D, Strantzias P, Lazaris A, Papageorgiou G, Ragos V, Yapijakis C and Patsouris E: Increased risk for oral cancer in diabetic animals is not associated with c-jun activation pathway. *J Craniomaxillofac Surg 35*(8): 382-387, 2007.

[55] Vairaktaris E, Goutzanis L, Nkenke E, Spyridonidou S, Vassiliou S, Derka S, Vylliotis A, Yapijakis C, Lazaris A, Strantzias P, Kalimeras E and Patsouris E: Expression of cell cycle regulator p16 is not affected by diabetes during oral oncogenesis. *In Vivo* 2007 *21*(5): 745-750, 2007.

[56] White MF and Kahn CR. The insulin signaling system. *J Biol Chem 269*:1-4, 1994.

[57] Folli F, Saad MJ, Backer JM and Kahn CR. Regulation of phosphatidylinositol 3-kinase activity in liver and muscle of animal models of insulin-resistant and insulin-deficient diabetes mellitus. *Clin Invest 92*:1787-1794, 1993.

[58] Kondo T and Kahn CR. Altered insulin signaling in retinal tissue in diabetic states. *J Biol Chem 279*:37997-38006, 2004.

[59] Lebrun P, Mothe-Satney I, Delahaye L, Van Obberghen E and Baron V. Insulin Receptor Substrate-1 as a Signaling Molecule for Focal Adhesion Kinase $pp125^{FAK}$ and $pp60^{src}$. *J Biol Chem 273*:32244-32253, 1998.

[60] Lebrun P, Baron V, Hauck CR, Schlaepfer DD and Van Obberghen E. Cell Adhesion and Focal Adhesion Kinase Regulate Insulin Receptor Substrate-1 Expression. *J Biol Chem 275*:38371-38377, 2000.

[61] Hanks SK, Calalb MB, Harper MC and Patel SK. Focal Adhesion Protein-Tyrosine Kinase Phosphorylated in Response to Cell Attachment to Fibronectin. *PNAS USA 89*:8487-8491, 1992.

[62] Vuori K and Ruoslahti E. Association of insulin receptor substrate-1 with integrins. *Science 266*:1576-1578, 1994.

[63] Muller G, Wied S and Frick W. Cross talk of pp125(FAK) and pp59(Lyn) non-receptor tyrosine to insulin mimetic signaling in adipocytes. Mol Cell Biol 20:4708-4723, 2000.

[64] Kornberg LJ. Focal adhesion kinase expression in oral cancers. *Head Neck 20*:634-639, 1998.

[65] Schneider GB Kurago Z, Zaharias R, Gruman LM, Schaller MD and Hendrix MJ. Elevated focal adhesion kinase expression facilitates oral tumor cell invasion. *Cancer 95*:2508-2515, 2002.

[66] Walker DM, Boey G and McDonald LA. The pathology of oral cancer. *Pathology 35*(5): 376-383, 2003.

[67] Konstantopoulos N and Clark S. Reduced cell attachment and phosphorylation of focal adhesion kinase associated with expression of a mutant insulin receptor. *J Biol Chem 27*:28960-28968, 1996.

[68] Friedman JE, Ishizuka T, Liu S, Farrell CJ, Bedol D, Koletsky RJ, Kaung HL and Ernsberger P. Reduced insulin receptor signaling in the obese spontaneously hypertensive Koletsky rat. *Am J Physiol 273*:1014-1023, 1997.

[69] Issing WJ, Wustrow TPU and Heppt W: Oncogenes related to head and neck cancer. *Anticancer Res 13*:2541-2552, 1993.

[70] Shin D, Ro JY, Hong WK and Hittelman WN: Dysregulation of epidermal growth factor receptor expression in premalignant lesions during head and neck tumorigenesis. *Cancer Res 54*: 3153-3159, 1994.

[71] Xia W, Lau YK, Zhang HZ, Liu AR, Li L, Kiyokawa N, Clayman GL, Katz RL and Hung MC: Strong correlation between c-erbB-2 overexpression and overall survival of patients with oral squamous cell carcinoma. *Clin Cancer Res 33*: 3-9, 1997.

[72] Xia W, Lau YK, Zhang HZ, Xiao FY, Johnston DA, Liu AR, Li L, Katz RL and Hung MC: Combination of EGFR, Her-2/neu, and Her-3 is a stronger predictor for the outcome of oral squamous cell carcinoma than any individual family members. *Clin Cancer Res 5*: 4164-4174, 1999.

[73] Charoenrat P, Rhys-Evans PH, Archer DJ and Eccles SA: C-erbB receptors in squamous cell carcinomas of the head and neck: clinical significance and correlation with matrix metalloproteinases and vascular endothelial growth factors. *Oral Oncol 38*:73-80, 2002.

[74] Funayama T, Nakanishi T, Takahashi K, Taniguchi S, Takigawa M and Matsumura T: Overexpression of c-erbB-3 in various stages of human squamous cell carcinomas. *Oncology 55*:161-167, 1998.

[75] Shintani S, Funayama T, Yoshihama Y, Alcalde RE and Matsumura T: Prognostic significance of ERBB3 overexpression in oral squamous cell carcinoma. *Cancer Lett 95*: 79-83, 1995.

[76] Wakulich C, Jackson-Boeters L, Daley TM and Wysocki GP: Immunohistochemical localization of growth factors fibroblast growth factor-1 and fibroblast growth factor-2

and receptors fibroblast growth factor receptor-2 and fibroblast growth factor receptor-3 in normal oral epithelium, epithelial dysplasias, and squamous cell carcinoma. *Oral Surg Oral Med Oral Pathol Oral Radiol Endod 93*: 573-579, 2002.

[77] Das N, Majumder J and DasGupta UB: ras Gene mutations in oral cancer in eastern India. *Oral Oncol 36*: 76-80, 2000.

[78] Matsuda H, Konishi N, Hiasa Y, Hayashi I, Tsuzuki T, Tao M, Kitahori Y, Yoshioka N, Kirita T and Sugimura M: Alterations of p16/CDKN2, p53 and ras genes in oral squamous cell carcinomas and premalignant lesions. *J Oral Pathol Med 25*: 232-238, 1996.

[79] Yeudall WA, Torrance LK, Elsegood KA, Speight P, Scully C and Prime SS: Ras gene point mutation is a rare event in premalignant tissues and malignant cells and tissues from oral mucosal lesions. *Eur J Cancer B Oral Oncol 29B*(1): 63-67, 1993.

[80] McDonald JS, Jones H, Pavelic ZP, Pavelic LJ, Stambrook PJ and Gluckman JL: Immunohistochemical detection of the H-ras, K-ras, and N-ras oncogenes in squamous cell carcinoma of the head and neck. *J Oral Pathol Med 23*(8): 342-346, 1994.

[81] Yarbrough WG, Shores C, Witsell DL, Weissler MC, Fidler ME and Gilmer TM: ras mutations and expression in head and neck squamous cell carcinomas. *Laryngoscope 104*(11): 1337-1347, 1994.

[82] Korkolopoulou P, Lazaris ACh, Konstantinidou AE, Kavantzas N, Patsouris E, Christodoulou P, Thomas-Tsagli E and Davaris P: Differential expression of bcl-2 family proteins in bladder carcinomas. Relationship with apoptotic rate and survival. *Eur Urol 41*(3):274-283, 2002.

[83] Konstantinidou AE, Korkolopoulou P and Patsouris E: Apoptotic markers for tumor recurrence: a minireview. *Apoptosis. 7*(5):461-470, 2002.

[84] Konstantinidou AE, Korkolopoulou P and Patsouris E. Apoptotic markers for primary brain tumor prognosis. *J Neurooncol 72*(2):151-156, 2005.

[85] Chen Y, Kayano T and Takagi M: Dysregulated expression of bcl-2 and bax in oral carcinomas: evidence of post-transcriptional control. *J Oral Pathol Med 29*: 63-69, 2000.

[86] Loro LL, Vintermyr OK, Liavaag PG, Jonsson R and Johannessen AC: Oral squamous cell carcinoma is associated with decreased bcl-2/bax expression ratio and increased apoptosis. *Hum Pathol 30*(9): 1097-1105, 1999.

[87] Tang XH, Knudsen B, Bemis D, Tickoo S and Gudas LJ: Oral cavity and esophageal carcinogenesis modelled in carcinogen-treated mice. *Clin Cancer Res 10*: 301-313, 2004.

[88] Singh BB, Chandler FW Jr, Whitaker SB and Forbes-Nelson AE: Immunohistochemical evaluation of bcl-2 oncoprotein in oral dysplasia and carcinoma. *Oral Surg Oral Med Oral Pathol Oral Radio Endod 85*: 692-698, 1998.

[89] Konstantinidou AE, Korkolopoulou P, Vassilopoulos I, Tsenga A, Thymara I, Agapitos E, Patsouris E and Davaris P: Reduced retinoblastoma gene protein to Ki-67 ratio is an adverse prognostic indicator for ovarian adenocarcinoma patients. *Gynecol Oncol 88*(3):369-378, 2003.

[90] Levidou G, Korkolopoulou P, Thymara I, Vassilopoulos I, Saetta AA, Gakiopoulou H, Konstantinidou A, Kairi-Vassilatou E, Pavlakis K and Patsouris E: Expression and

prognostic significance of cyclin D3 in ovarian adenocarcinomas. *Int J Gynecol Pathol 26*(4):410-417, 2007.

[91] Derka S, Vairaktaris E, Papakosta V, Vassiliou S, Acil Y,Vylliotis A, Spyridonidou S, Lazaris AC, Mourouzis C, Kokkori A, Moulavasili P, Perrea D, Donta I, Yapijakis C and Patsouris E: Cell proliferation and apoptosis culminate in early stages of oral oncogenesis. *Oral Oncol 42*(5): 540-550, 2006.

[92] Spandidos DA, Lamothe A and Field JK: Multiple transcriptional activation of cellular oncogenes in human head and neck solid tumors. *Anticancer Res 5*: 221-224, 1985.

[93] Schantz SP: Basic science advances in head and neck oncology: the past decade. *Sem Surg Oncol 11*: 272-279, 1995.

[94] Soussi T: p53 alterations in human cancer: more questions than answers. *Oncogene 26*(15):2145-2156, 2007.

[95] Goudopoulou A, Saetta A, Korkolopoulou P, Patsouris E, Fanourakis G, Miaouli M, Thomas-Tsagli E and Davaris PS: p53 mutations detection in urinary bladder cancer in the Greek population: application of the NIRCA assay. *J Exp Clin Cancer Res 22*(1):99-105, 2003.

[96] Saetta AA, Aroni K, Stamatelli A, Lazaris AC and Patsouris E: Expression of mismatch repair enzymes, hMLH1 and hMSH2 is not associated with microsatellite instability and P53 protein accumulation in basal cell carcinoma. *Arch Dermatol Res 297*(3):99-107, 2005.

[97] Sakai E, Rikimaru K, Ueda M, Matsumoto Y, Ishii N, Enomoto S, Yamamoto H and Tsuchida N: The p53 tumor-suppressor gene and ras oncogene mutations in oral squamous cell carcinoma. *Int J Cancer 52*: 867-872, 1992.

[98] Tanaka N, Sugihara K, Odajima T, Mimura M, Kimijima Y and Ichinose S: Oral squamous cell carcinoma: electron microscopic and immunohistochemical characteristics. *Med Electron Microse 35*: 127-138, 2002.

[99] Carlos de Vicente J, Junquera Gutierrez LM, Zapatero AH, Fresno Forcelledo MF, Hernandez-Vallejo G and Lopez Arranz JS: Prognostic significance of p53 expression in oral squamous cell carcinoma without neck node metastasis. *Head Neck 26*: 22-30, 2004.

[100] Shaulian E and Karin M: AP-1 in cell proliferation and survival. *Oncogene 20*: 2390-2400, 2001.

[101] Szabo E, Riffe ME, Steinberg SM, Birrer MJ and Linnoila RI: Altered c-JUN expression: an early event in human lung carcinogenesis. *Cancer Res 56*: 305-315, 1996.

[102] De Sousa SO, Mesquita RA, Pinto DS JR and Gutkind S: Immunolocalization of c-Fos and c-Jun in human oral mucosa and in oral squamous cell carcinoma. *J Oral Pathol Med 31*: 78-81, 2002.

[103] Mangone FR, Brentani MM, Nonogaki S, Begnami MD, Campos AH, Walder F, Carvalho MB, Soares FA, Torloni H, Kowalski LP and Federico MH: Overexpression of Fos-related antigen-1 in head and neck squamous cell carcinoma. *Int J Exp Pathol 86*: 205-212, 2005.

[104] Ohyama M, Hirayama Y, Tanuma JI, Hirano M, Semba I, Shisa H, Hiai H, Sugihara K and Kitano M: Expressions of junB and c-fos are enhanced in 4-nitroquinoline 1-oxide-induced rat tongue cancers. *Pathol Int 54*: 35-40, 2004.

[105] Turatti E, Da Costa Neves A, De Magalhaes MH and De Sousa SO: Assessment of c-Jun, c-Fos and cyclin D1 in premalignant and malignant oral lesions. *J Oral Sci 47*: 71-76, 2005.

[106] Oikawa T and Yamada T: Molecular biology of the ETS family of transcription factors. *Gene 303*: 11-34, 2003.

[107] Cairns P, Mao L, Merlo A, Lee DJ, Schwab D, Eby Y, Tokino K, van der Riet P, Blaugrund JE and Sidransky D: Rates of p16 (MTS1) mutations in primary tumors with 9p loss. *Science 265*(5170): 415-417, 1994.

[108] Merlo A, Herman JG, Mao L, Lee DJ, Gabrielson E, Burger PC, Baylin SB and Sidransky D: 5' CpG island methylation is associated with transcriptional silencing of the tumor suppressor p16/CDKN2/MTS1 in human cancers. *Nat Med 1*(7): 686-692, 1995.

[109] Reed AL, Califano J, Cairns P, Westra WH, Jones RM, Koch W, Ahrendt S, Eby Y, Sewell D, Nawroz H, Bartek J and Sidransky D: High frequency of p16 (CDKN2/MTS-1/INK4A) inactivation in head and neck squamous cell carcinoma. *Cancer Res 56*(16): 3630-3633, 1996.

[110] Prime SS, Eveson JW, Guest PG, Parkinson EK and Paterson IC: Early genetic and functional events in the pathogenesis of oral cancer. *Radiat Oncol Investig 5*(3): 93-96, 1997.

In: New Cell Adhesion Research
Editors: Patrick Nott and Matthew Temple
ISBN 978-1-60692-378-8

Chapter IX

EXPRESSION OF MOLECULES WITH A POTENTIAL FOR MODULATING INTERACTION WITH EXTRACELLULAR MATRICES ON HEPATIC STELLATE CELLS: NEURAL CELL ADHESION MOLECULES AND OSTEONECTIN

Kazuki Nakatani[1,*], Kazuo Ikeda[1], Yuji Nakajima[1] and Shuichi Seki[2]

Department of [1]Anatomy and [2]Hepatology, Graduate School of Medicine, Osaka City University, Japan.

ABSTRACT

Liver fibrosis is a pathological feature induced by the progression of chronic liver injuries and is characterized by the accumulation of an excess amount of extracellular matrices in the damaged lesions, as an abnormal wound healing response. Activated hepatic stellate cells have been postulated as the cells responsible for the advancement of liver fibrogenesis. Therefore, the molecular function of activated hepatic stellate cells has been investigated vigorously in considering the cells potential therapeutic targets for preventing and curing liver fibrosis. Hepatic stellate cells have been shown to express various molecules, such as cytokines, cytokine receptors, extracellular matrices and adhesion molecules. Here, we review two molecules in the liver with the potential of cell-cell and/or cell-matrix interaction: neural cell adhesion molecules (NCAM) and osteonectin. NCAM is a glycoprotein mediating cell-cell and cell-matrix interaction in a

* Correspondence concerning this article should be addressed to: Kazuki Nakatani, M.D., Department of Anatomy, Graduate School of Medicine, Osaka City University, 1-4-3 Asahimachi, Abeno-ku, Osaka 545-8585, Japan. Tel: ++81-6-6645-3706, Fax: ++81-6-6646-3603, E-mail: kazuki@med.osaka-cu.ac.jp.

homophilic and heterophilic manner, respectively. In the human liver, NCAM expression was found in the stellate cells of undamaged or slightly damaged livers in addition to nerve fibers and natural killer cells. In the rat liver, NCAM was also detected in the activated stellate cells, but not in the quiescent stellate cells. However, the quiescent stellate cells came to express NCAM following treatment with several cytokines, such as transforming growth factor beta-1, platelet-derived growth factor, epidermal growth factor and insulin-like growth factor. The second molecule described here, osteonectin, is a calcium-binding matricellular glycoprotein and has varied functional potential for modulating cell-matrix interaction, for example its counteradhesive effects. Osteonectin has been demonstrated to be expressed in the various interstitial cells of injured and fibrotic tissues. It has also been found in the myofibroblastic cells located in chronically injured liver tissues, however only scantly in intact stellate cells. In addition, the number of osteonectin-positive human stellate cells in livers with chronic hepatitis increased in comparison with the number found in undamaged or slightly damaged livers and in livers with cirrhosis. Osteonectin has also been found in the myofibroblastic cells of the tumor capsule and along the capillaries within hepatocellular carcinoma. Further investigation of the function of these molecules in stellate cells is imperative to elucidate the molecular mechanisms of the communication between stellate cells and matrices.

Introduction

Acute viral infection or temporal treatment with hepato-toxic agents, such as carbon tetrachloride and thioacetamide, result in partial necrosis of liver parenchyma, but recovery is relatively quick. Repeated liver damage, however, would bring about fibrosis as a wound healing response, and then lead ultimately to cirrhosis in the form of nodule formation and organ contraction [1]. Liver fibrosis is characterized by abnormal accumulation of an excess amount of extracellular matrices (ECMs) in the damaged lesion [2,3]. In the liver parenchyma, hepatic stellate cells (HSCs, also referred to as Ito cells, fat-storing cells or lipocytes) naturally produce and maintain many kinds of ECMs, such as collagens and fibronectin [4]. HSCs are activated due to chronic liver damage and deposit an excess amount of ECMs in the injured liver tissues; liver fibrosis progresses as a result [2,5]. HSCs have therefore been considered possible therapeutic targets for preventing liver fibrosis [6]. Indeed, we have reported that preventing the activation and/or proliferation of HSCs results in dramatic attenuation of liver fibrosis [7-9]. Various approaches to elucidate the molecular mechanism of HSCs activation have been advanced, including investigation of the molecules expressed in HSCs.

Until the early 1980's, HSCs were identified by location, morphological feature and detection of vitamin A [10-13]. HSCs are located in the space of Disse, a narrow space between the sinusoidal wall and hepatocytes, and have unique stellate shapes that allow them to be identified morphologically by transmission electron microscopy, but not by HE staining for light microscopy. HSCs have rich fat-droplets that include vitamin A, so HSCs can be found by methods used for the detection of vitamin A such as Golgi's silver impregnation method or vitamin A autofluorescence. HSCs without vitamin A predominantly located in the centrilobular area (zone III) cannot, however, be detected by these methods. The presence of desmin, a kind of intermediate cytoskeletal filament, in HSCs of normal rat liver was first described by Yokoi et al. [14]. Further, the number or size of desmin-positive HSCs has been reported to increase in acute or chronic liver injury [15,16]. However, it has also been indicated that there are many desmin-negative HSCs in the liver parenchyma [17]. Subsequently, Gard et al. were first to report that HSCs of normal rat liver possess glial fibrillary acidic protein (GFAP), another kind of intermediate filament [18]. Although the expression of GFAP in HSCs attracted little attention until the mid 1990's, it was then reappraised by several researchers [17,19-21]. GFAP has been reported to be the molecule detected in quiescent HSCs [18], and its expression was found to decrease due to HSCs activation [19,21]. On the other hand, the unique molecules expressed in activated rat HSCs have too been the subject of investigation. Ramadori et al. found that activated rat HSCs have unique expression of alpha-smooth muscle actin (αSMA), while quiescent HSCs do not [22]. Other cytoskeletal proteins, vimentin [23] and nestin [24], have been reported in rat HSCs.

We have shown HSCs to express several molecules, such as heat shock protein 47 [25] and c-myb [26]. We have also attempted to elucidate any unique molecules expressed in activated HSCs by comparison of mRNA in quiescent HSCs and activated HSCs [27], as well as proteins expressed in both [28]. From these comparisons, we found cellular prion protein (Prp) expressed in activated HSCs by using the suppression subtractive hybridization method [27]. Moreover, the new hemoprotein cytoglobin/ stellate cells-associated protein (Cygb/STAP) was discovered using the proteomics approach [29]. Activated HSCs possess a large amount of Cygb/STAP, while quiescent HSCs contain a comparatively small amount of this new protein [29,30]. In addition, cellular Prp and Cygb/STAP, as well as αSMA, were confirmed in human HSCs, unlike desmin and GFAP which have not been detected in human HSCs [31-33].

From another perspective, HSCs have been shown to express nerve-related proteins such as GFAP, cellular Prp, nestin and Cygb/STAP. GFAP is detected in the astrocytes [34]. Cellular Prp is predominantly localized in neurons and glias in the central nervous system [32,35]. Nestin was originally defined as a protein in neural stem cells [24,36]. Gene expression of Cygb/STAP was shown in the brain [37,38], and its protein was detected in brain nerve cells [39]. Furthermore we found, for the first time, that human HSCs of intact or mildly damaged liver have the expression of neural cell adhesion molecules (NCAM; CD56/Leu19) [40,41]. Expression of this molecule was also confirmed in activated rat HSCs, but not in quiescent ones [42,43]. In addition we reported that, during chronic liver inflammation, activated human HSCs have an increased expression of osteonectin (ON) [44,45], which is also expressed in the nerve fibers [44] and in the brain [46]. These two molecules have also been described to potentially bind with collagens [43,44]. In this review, we describe the molecular structure and characteristics of NCAM and ON, and give an overview of their distribution and potential function in the liver.

Molecular Structure and Characteristics of NCAM

NCAM is the first molecule mediating cell-cell adhesion to be identified on the basis of functional criteria [47]. NCAM is classified into the immunoglobulin (Ig) superfamily, members of which are characterized by possessing Ig-like domains, and are implicated in many developmental processes of neuronal tissues, including growth, guidance and migration of neural crest cells and nerve cells, axon bundling and formation of neuromuscular junctions [48-52]. NCAM is expressed as three principal isoforms generated by alternative splicing of a primary transcript from a single gene [53-55]. Two isoforms are transmembrane forms with either a long or short cytoplasmic domain, respectively named NCAM180 and NCAM140. Another form lacking a cytoplasmic domain, NCAM120, is anchored to the plasma membrane by glycophosphatidylinositol (GPI) linkage. These isoforms were named after their molecular weight. In addition, there have been reports of several other isoforms, some of which are tissue and/or developmental stage-specific, generated by alternative splicing in the extracellular domain [53,54,56-59]. The molecular weight of NCAM also varies following post-translational glycosylation [60]. All of the three major NCAM isoforms, NCAM180, NCAM140 and NCAM120, have five C2-type Ig-like domains and two fibronectin type III (FNIII) domains in the extracellular region [54,55] (Figure 1). A famous alternative splicing variant of NCAM has the variable alternatively spliced exon (VASE) in the fourth Ig-like domain, and has been shown to arise postnatally and to serve downmodulating axon growth [59,61]. NCAM can also be post-translationally modified by the addition of polysialic acid (PSA) to the fifth Ig-like domain [62]. PSA is a homopolymer composed of alpha-2, 8-linked sialic acid residues [63]. Long chains of PSA are extensively located in the embryonic nervous system, whereas PSA expression is more restricted during neonatal development and in the adult brain [64,65]. Long chains of PSA attached to NCAM are suggested to inhibit NCAM-mediated cell adhesion because of their big molecular size and/or the negative charge of PSA. In general, polysialylation of NCAM has been implicated in outgrowth, elongation and guidance of neurite, migration of neuronal cells and synaptic formation [65-67].

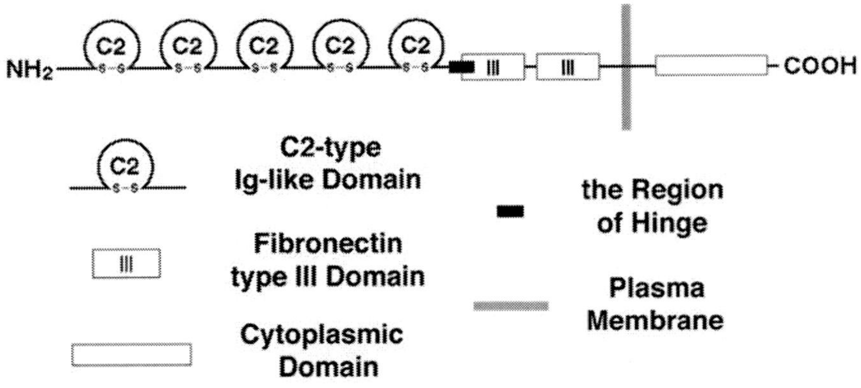

Figure 1. Molecular structure of neural cell adhesion molecule (NCAM). NCAM has five C2-type Ig-like domains and two fibronectin type III domains in the extracellular region. Of the three major NCAM isoforms, NCAM180 differs from NCAM140 by the inclusion of exon 18, producing a longer cytoplasmic domain. NCAM120 does not have a cytoplasmic domain, but is anchored to the plasma membrane by glycophosphatidylinositol linkage.

NCAM mediates cell-cell and cell-matrix interaction in a homophilic or heterophilic manner [68,69]. Moreover, this molecule not only interacts with the molecules on other cell surfaces (trans interaction), but also binds to molecules within the same cell membrane (cis interaction, which has been suggested to implicate intracellular signal transduction) [70]. NCAM binds heterophilically to heparan sulfate proteoglycans (HSPGs), including agrin, through heparin-binding sites in the first and second Ig-like domains, and to several chondroitin sulfate proteoglycans (CSPGs) including neurocan and phosphacan [71-75]. NCAM also interacts with collagen type I, II, III, IV and IX [52,71]. In addition, NCAM is capable of lateral binding to neural cell adhesion molecule L1 with the fourth Ig-like domain of NCAM, which facilitates homophilic binding between L1 molecules apposed in trans [76-78]. NCAM has also been shown to interact with fibroblast growth factor (FGF) receptors (FGFR) in cis, and to stimulate neurite outgrowth via activation of phospholipase C-gamma (PLCγ) and diacylglycerol (DAG) lipase [79-81]. It has been proposed that a short sequence of about 20 amino acids in the FGFR, termed the CAM-homology domain (CHD), could bind directly with a related short sequence within the VASE-encoded region of NCAM, which shares its homology with CHD [79,82]. On the other hand, the intracellular domain of NCAM has been described to interact with some cytoskeletal and cytoskeleton-associated molecules, such as spectrins, alpha and beta-tubulin and alpha-actinin [83-86]. NCAM180 binds to the microtubulin-associated protein 1A, beta-actin and tropomyosin, but NCAM140 does not [86]. In addition, only NCAM180 interacts with the Rho A-binding kinase-alpha [86], and only NCAM140 interacts with nonreceptor tyrosine kinase Fyn and focal adhesion kinase (FAC) [87,88].

NCAM has been shown to contribute toward many neuro-developmental processes, including migration of nerve cells, outgrowth, elongation and guidance of neurites and formation of synapses [50]. Nevertheless, NCAM null mice are viable, fertile and present only minor defects in brain development and in behaviour: e.g., a reduction in the size of the olfactory bulb by about two-thirds, leading to defects in olfactory neuron migration, and loss of space learning ability [62]. An equivalent defect is observed following elimination of only NCAM180 [89]. In addition, targeted mutation of NCAM, which produces only a secreted form of this molecule consisting of its entire extracellular region, has a lethal effect in the embryo [90].

NCAM Signaling Mechanisms

Interaction between NCAM and FGFR Activates a Signal Transduction Cascade Involving Calcium

Intracellular signaling pathways of NCAM (Figure 2) have gradually been elucidated by vigorously studying NCAM-mediated neuronal tissue developmental events and NCAM-induced bearing of cultured neuronal cells, especially neurite outgrowth [69,91]. The first signaling molecule described to be associated with the NCAM-mediated signal transduction pathway was calcium ion (Ca^{2+}) [92]. Ca^{2+} was known as a key signaling molecule in the regulation of growth cone guidance [93]. The addition of antibodies against NCAM, as well as cell-cell adhesion, was shown to lead to an increase in cytoplasmic Ca^{2+} concentration, while both L-type and N-type Ca^{2+}–channel blockers completely inhibited NCAM-induced neurite outgrowth in PC12 cells [92,94]. Furthermore, a recent study suggested that mobilization of Ca^{2+} from intracellular stores is required for NCAM-mediated neurite outgrowth and probably acts in concert with a Ca^{2+} influx through a voltage-dependent Ca^{2+} channel on the plasma membrane [91,95]. In addition, the lipid metabolite arachidonic acid (AA) was shown to induce NCAM-mediated neurite outgrowth and to work as an upstream activator of voltage-dependent Ca^{2+} channels in NCAM-mediated signaling [96-99]. AA is liberated from monoacylglycerol (MAG) by MAG lipase [100]. MAG is produced from DAG by DAG lipase, and DAG is produced from phosphatidyl inositol bisphosphate (PIP_2) by PLCγ [82,100]. PLCγ was considered to act upstream of the release of AA in NCAM-mediated signaling, because NCAM-mediated neurite outgrowth was disturbed by PLC inhibitor but was able to be rescued by treatment with AA [97,101]. On the other hand, AA has also been described to be generated from phospholipids due to the direct cleavage activity of phospholipase A_2 [102], but phospholipase A_2 was not found to be involved in the NCAM-mediated neurite outgrowth response [96]. PLCγ is also known to mobilize Ca^{2+} from intracellular stores through liberation of inositol triphosphate (IP_3) from PIP_2 [91,92,103].

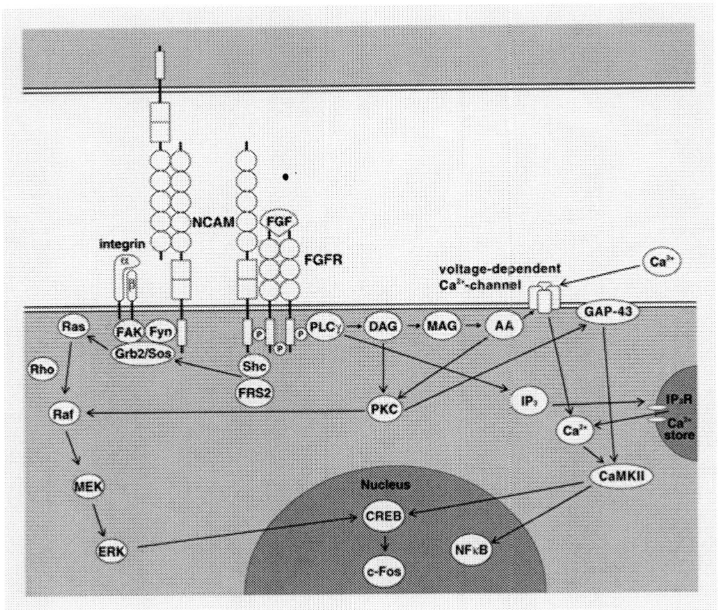

Figure 2. Presumed intracellular signaling pathway of NCAM. *AA*, arachidonic acid; *Ca^{2+}*, calcium ion; *CaMKII*, Ca^{2+}/ calmodulin-dependent kinase II; *CREB*, cyclic adenosine monophosphate response element binding protein; *DAG*, diacylglycerol; *ERK*, extracellular signal regulated kinase; *FAK*, focal adhesion kinase; *FGF*, fibroblast growth factor; *FGFR*, fibroblast growth factor receptor; *FRS2*, fibroblast growth factor receptor substrate 2; *GAP-43*, growth associated protein-43; *Grb2*, growth factor receptor bound protein 2; *IP_3*, inositol triphosphate; *IP_3R*, inositol triphosphate receptor; *MAG*, monoacylglycerol; *MEK*, mitogen-activated protein kinase/ extracellular signal regulated kinase kinase; *NFκB*, nuclear factor kappa B; *PKC*, protein kinase C; *PLCγ*, phospholipase C gamma; *Shc*, Src homology and collagen; *Sos*, son of sevenless. (Ref. [69,91,100]).

What activates PLCγ in the NCAM-mediated signaling pathway? The major NCAM-associated signaling molecule acting upstream from Ca^{2+} channel activation is thought to be FGFR [91]. It is widely expressed in the developing nervous system. Activation of FGFR stimulates neurite outgrowth from various primary cultured nerve cells and PC12 cells, and this effect is similar to the NCAM-activated cascade via Ca^{2+} signaling [79,81]. Numerous investigations of inhibition of NCAM-mediated neurite outgrowth following treatment with blocking antibodies against FGFR or peptides consisting of a part of FGFR, which are presumed to be involved in binding to NCAM, and transfection of dominant negative FGFR, have shown that FGFR acts upstream of PLCγ, AA, and activation of the Ca^{2+} channels in the NCAM signaling pathway [79,91,101]. In addition, FGFR-induced neurite outgrowth in cerebellar neurons was inhibited by a DAG lipase inhibitor or combination of L- and N-type Ca^{2+} channel antagonists [81,104]. NCAM is considered to bind directly to FGFR through interaction between VASE of NCAM and CHD of FGFR, as mentioned above, and to be able to activate FGFR [79,82]. One of the downstream molecules in the NCAM-FGFR-PLCγ-DAG signaling pathway is protein kinase C (PKC) [91]. PKC has been shown to be activated via DAG and/or AA, to activate the Raf-mitogen-activated protein (MAP) kinase pathway, and to phosphorylate growth associated protein (GAP)-43 in NCAM-mediated signal transduction [97,105,106].

Several molecules have been suggested as the downstream effectors of the NCAM-mediated Ca^{2+}–influx, i.e., Ca^{2+}/calmodulin-dependent kinase (CaMK) II and GAP-43 [91]. CaMKII is one of the most abundant protein kinases in the mammalian brain, and has been implicated in a variety of cellular events that utilize Ca^{2+} as a second messenger, such as neurotransmitter synthesis and cytoskeletal function [107-110]. CaMKII has also been described to stimulate neurite outgrowth when overexpressed in neuroblastoma cell lines [110,111]. It has been shown that neurite outgrowth induced by NCAM as well as by AA is prevented after treatment with CaMKII inhibitor, indicating that CaMKII may act downstream of the Ca^{2+}–influx in the signaling pathway initiated by the NCAM-FGFR interaction [112]. CaMKII is known to phosphorylate cyclic adenosine monophosphate (cAMP) response element binding protein (CREB) [113] and to be implicated in activation of nuclear factor kappa B (NFκB) [114,115]. Moreover, CaMKII may have direct effects on the cytoskeletal dynamics of the neuronal cell [91,116].

GAP-43 is a membrane-associated protein which has been implicated in axon growth and growth cone guidance, and in synaptic plasticity [117-119]. Unphosphorylated GAP-43 is bound to calmodulin, whereas GAP-43 phosphorylated by PKC, which becomes sufficiently activated with increasing intracellular levels of Ca^{2+}, AA or DAG, is activated and simultaneously releases calmodulin, which becomes available to activate other target proteins, such as CaMKII [117]. Phosphorylated GAP-43 can directly associate with actin filaments and modulate its stabilization [106]. Both FGF and NCAM failed to stimulate neurite outgrowth in GAP-43 knockout mice [120]. Furthermore, stimulation of cerebellar neurons from wild-type mice with FGF or NCAM has resulted in increased phosphorylation of GAP-43, and this effect could be blocked by a DAG lipase inhibitor, indicating that GAP-43 is located downstream of the NCAM-FGFR-DAG pathway.

NCAM Signaling via the MAP Kinase-Pathway: The Role of Nonreceptor Tyrosine Kinases

It has been demonstrated that phosphorylated MAP kinases, extracellular signal regulated kinase (ERK) 1 and ERK2, are increased following NCAM-NCAM homophilic interaction or clustering of NCAM on the cell surface by means of antibodies against NCAM or NCAM-Fc fusion protein in several neuronal cells [97,121]. Furthermore, it has been shown that NCAM-induced neurite outgrowth is prevented following treatment with pharmacological inhibitors of MAP kinase/ERK kinase (MEK), which is the upstream activator of the ERKs, or following transfection of dominant negative forms of Ras and Raf, which are the upstream activators of MEK, in PC12 cells [97]. Additionally, NCAM-mediated neurite outgrowth in PC12 cells has also been shown to be stimulated after transfection of constitutively active forms of Ras, Raf, and MEK as well as neurite outgrowth without NCAM. Therefore, NCAM is suggested to initiate activation of a Ras-Raf-MEK-ERK pathway, and activation of this pathway is involved in the neuritogenic response to NCAM. In addition to Ras, the small G protein Rho also participates in NCAM140-mediated activation of the MAP kinase pathway, although where it acts within the pathway has not yet been clarified [121].

FGFR is known to activate the MAP kinase pathway through binding FGF, and this activation is involved in the neurite outgrowth response to FGF [122]. FGFR lacks the consensus binding sequence for growth factor receptor bound protein (Grb) 2, which makes the complex with son of sevenless (Sos) and activates Ras, whereas activated FGFR can interact with the adapter proteins Src homology and collagen (Shc) and FGFR substrate 2 (FRS2), which subsequently are phosphorylated by the activated FGFR tyrosine kinase, thereby creating docking sites for the Src homology 2 (SH2) domain in Grb2 [123-125]. In addition, NCAM-FGFR interaction may activate Raf in a Ras-dependent way via PKC [97,126].

NCAM-induced activation of the MAP kinase pathway may not only be led by FGFR activation. It has been suggested that this pathway is activated by a signal involving Fyn, which is a member of the Src family of nonreceptor tyrosine kinase [69,91]. NCAM140-mediated neurite outgrowth was demonstrated to depend critically on the Fyn as shown in cultures of cerebellar and sensory neurons from Fyn null mice [87]. Fyn constitutively associates with NCAM140 in the plasma membrane, whereas it does not associate with NCAM180 or NCAM120 [88]. Clustering of NCAM140 on the cell surface by means of antibodies against NCAM, as mimics of NCAM homophilic binding, induces autophosphorylation of the associated Fyn molecules, leading to recruitment and subsequent phosphorylation of FAK, the nonreceptor tyrosine kinase which is a critical molecule in the integrin signal transduction and a constituent of focal adhesion contacts [88]. FAK phosphorylated on a specific tyrosine by Src kinases, such as Fyn, can constitute a binding site for the Grb2/Sos adapter complex, and this adapter complex activates Ras [127]. Therefore, it was suggested that NCAM-induced Ras-MAP kinase cascade is activated via the pathway involving Fyn/FAK. However, the cytoplasmic part of NCAM140 does not contain any known phosphorylatable tyrosine domains which could act as binding sites for SH2 domain of Fyn, or any polyproline motifs for the binding of an SH3 domain [128]. NCAM140 has been described to co-immunoprecipitate with Fyn and to be able to interact with the N-terminal domain of Fyn [88], whereas the mechanism of activation of Fyn via NCAM140 remains unclear.

NCAM-induced MAP kinase pathway results in phosphorylation of the transcription factor CREB on serine residue 133 [121]. Activation of CREB can result in stimulation of transcription of c-Fos. A study using the dominant negative form of c-Fos demonstrated that c-Fos is involved with NCAM-mediated neurite outgrowth [98]. Additionally, Fyn was described to associate with NCAM-induced NFκB activation in studies using an inhibitor of protein tyrosine kinase of the Src family or knock out mice deficient in either Fyn or Src [114].

A new perspective has recently been proposed with regard to the mechanism for reconciling the divergent pathway of NCAM signaling [69]. Niethammer et al. described that distinct MAP kinase pathways occur based on compartmentalization of different isoforms of NCAM in or out of membrane lipid microdomains (MLMs) within the plasma membrane [129]. MLMs are known to have high concentrations of cholesterol and glycosphingolipids, and thereby a higher resistance toward low concentrations of nonionic detergents, such as TritonX-100, in comparison to other membrane compartments [129,130]. Thus, MLMs have been called by various names, such as detergent-resistant microdomains, cholesterol-enriched membrane microdomains, and glycosphingolipid-enriched membrane microdomains. MLMs were also named "lipid rafts" by Simons and Ikonen, who considered that MLMs may drift within the plasma membrane because their lipid construction impedes their mixing with other membrane compartments [131]. Recent studies have shown that various molecules involved in signal transduction, such as Src, Fyn [132], Grb2, Ras, Raf [133], Rho, FAK [134], Sos [135], and GAP-43 [136], are located in MLMs [137]. NCAM120, an isoform anchored to the plasma membrane by GPI linkage, was known to be located in MLMs, and coexist with Fyn [138,139]. However, because NCAM120 has no intracellular domain, NCAM120 cannot directly activate Fyn. On the other hand, NCAM140, which is able to interact directly with Fyn, was also detected in MLMs [119,129]. Additionally, NCAM140 in MLMs was shown to transduce signals through Fyn and FAK leading to activation of the MAP kinase pathway, whereas NCAM140 not located in MLMs requires FGFR for activation of MAP kinase signal transduction [129]. In addition, NCAM140 in MLMs was demonstrated to coexist with GAP-43 [119], and NCAM-mediated signaling can phosphorylate GAP-43 in MLMs.

NCAM Expression in Extra-neuronal Tissues

NCAM has been studied with great interest concerning the development of various neuronal tissues and constitution of neural networks [140], since first found by Thiery et al. through research using chick retinal cells [47], and it has also been investigated in extra-neuronal tissues, including muscular and endocrinal tissues [141,142]. NCAM has also been known as CD56/Leu19, a marker of human natural killer (NK) cells [143,144]. Rats or mice NK cells, however, have no expression of NCAM/CD56/Leu19 [144].

NCAM expression in muscular tissues was found in embryonic skeletal muscle [145]. There, NCAM is uniformly present on the surface of the myotubes, although downregulated as the myotubes mature [146,147]. Furthermore, NCAM expression concentrates in the neuromuscular junctions, after birth [146]. However, after denervation or during disease-associated muscle degeneration, NCAM expression is again upregulated [148,149]. Additionally, NCAM expression is upregulated during aging in normal rat skeletal muscle [150]. The embryonic myocardium also has NCAM expression [151-153]. NCAM expression rapidly decreases and is restricted in the neuromuscular junctions, after birth [151], whereas this expression is again upregulated by hypertrophic change or degeneration of myocardium during heart disease [154,155]. In addition, NCAM expression is also upregulated after denervation in the transplanted human heart [153]. On the other hand, NCAM is expressed in vascular smooth muscle cells during the development of the vascular system, whereas NCAM expression rapidly decreases after birth [156,157]. NCAM expression is also detectable in embryonic intestinal smooth muscle cells, while this expression rapidly decreases and disappears postnatally [142,156,158,159]. However, Filiz et al. reported that NCAM expression is detectable in adult rat bronchial smooth muscle cells [142].

NCAM is expressed in several endocrine tissues, such as chromaffin cells of the adrenal medulla [141,160], pituitary endocrine cells [161] and endocrine pancreatic cells in islets of Langerhans [162-164]. Additionally, Leydig cells of the testis express NCAM [165]. On the other hand, NCAM expression of fibroblasts has been poorly investigated [141]. NCAM expression has been shown to be undetectable in fibroblasts of normal adult peripheral nerve in endo-, peri- and epi-neural situations, and also at all developmental ages studies. Gatchalian et al. described that NCAM is expressed by perisynaptic fibroblasts in denervated skeletal muscle [166]. NCAM has also been demonstrated to be detected in activated fibroblasts during the skin injury wound healing process [167], and in activated fibroblasts after treatment with lipopolysaccharide in the intestine [168]. As mentioned above, NCAM expression in fibroblasts has only been detected in pathological states, whereas we were able to demonstrate for the first time that NCAM is expressed in the normal human HSCs, which belong to liver-specific fibroblast lineages [40,41] (Figures 3, 4).

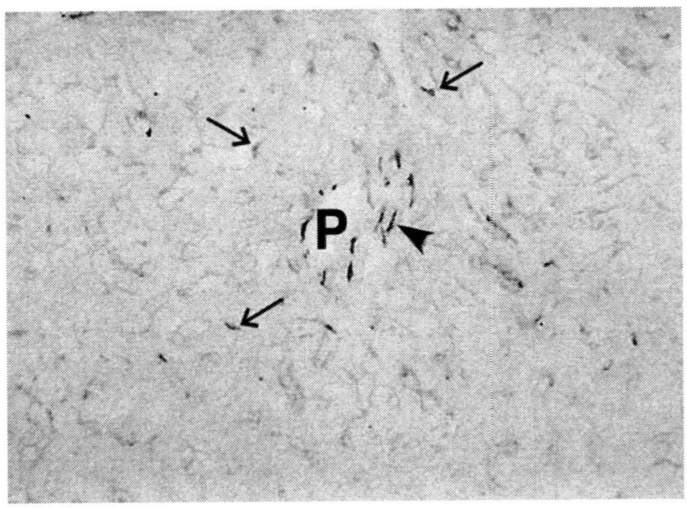

Figure 3. Immunohistochemistry for NCAM in a human liver obtained from a patient with Gilbert's disease. NCAM is detected in the sinusoidal lining cells (*arrows*). Nerve fibers (*arrowhead*) are also positive for NCAM. *P*, portal vein. x250.

NCAM Expression in the Liver

NCAM Expression in Diseased Bile Ductules

It is well known that nerve fibers and NK cells, called pit cells, are distributed in liver tissue [144,169]. Whiteside et al. were first to demonstrate that in the adult liver, NCAM/CD56/Leu19 is expressed in NK cells as a human tumor-infiltrating lymphocyte [170]. Roskams et al. also described that nerve fibers in the portal tracts and along the sinusoids are positive for NCAM [171]. Moreover, they found that NCAM is expressed in the reactive bile ductules in cholestatic liver disease such as primary biliary cirrhosis (PBC), primary sclerosing cholangitis (PSC) and extrahepatic biliary obstruction. These NCAM-positive reactive bile ductules were also seen in the regenerating human liver tissues after submassive liver necrosis [172] and in the liver tissues of chronic hepatitis [173] and of congenital extrahepatic biliary atresia [174]. According to further investigation, reactive bile ductules with atypical morphology, suggested to derive from ductular metaplasia of hepatocytes and/or from activated facultative bipotential stem cells, were demonstrated to coexpress NCAM and B-cell leukemia lymphoma-2 protein (Bcl-2), but to be negative for the proliferation marker Ki67 [175]. On the other hand, typical reactive ductules, derived from the proliferating well-differentiated biliary epithelial cells, were negative for either NCAM or Bcl-2. These NCAM-positive reactive bile ductules have been suggested to possess neuroendocrine features [171,172], but the functional significance of NCAM in these cells remains unknown.

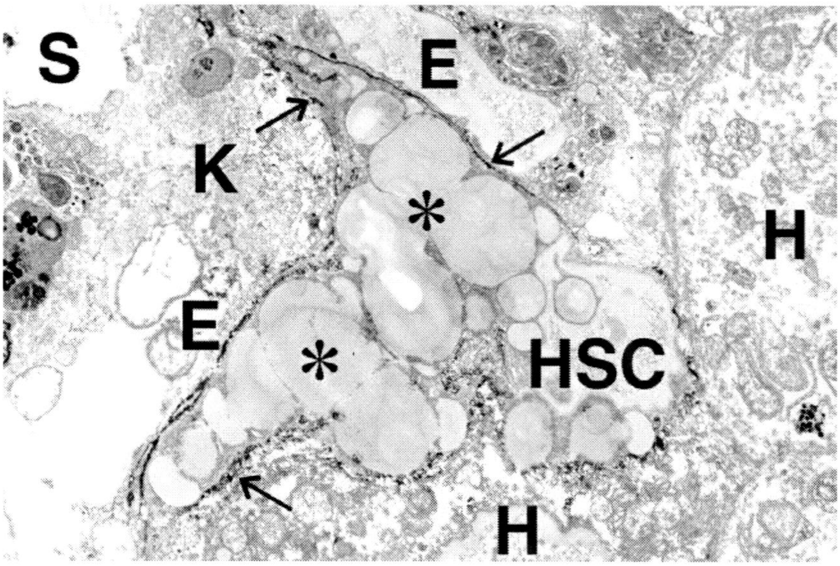

Figure 4. Immunoelectron microscopic observation of the NCAM-positive human hepatic stellate cell (*HSC*) in the non-cancerous portion of a metastatic liver tumor. NCAM-positive immunoprecipitates (*arrows*) are found on the surface of stellate cells with numerous lipid droplets (*asterisks*). *E*, sinusoidal endothelial cells; *H*, hepatocytes; *K*, a Kupffer cell; *S*, sinusoid. x5,400.

NCAM Expression in HSCs

The human liver parenchyma has been known to be abundantly innervated, while, in the rat and mouse liver, nerve fibers are only distributed in the portal tracts [169,176]. In the liver lobules, nerve fibers run along the sinusoidal wall in space of Disse and make contacts with hepatocytes, sinusoidal endothelial cells, Kupffer cells and HSCs [12,177]. The nerve fibers distributed in liver parenchyma gradually decrease during chronic liver damage; nerve fibers diminish more in the cirrhotic liver than in the liver of chronic hepatitis [178-180]. The function of innervation to HSCs is not yet well known, although it has been hypothesized that these nerves may be involved with sinusoidal microcirculation through managing HSC contractility [181,182]. We demonstrated for the first time that human HSCs possess NCAM, and suggested that HSCs may interact with nerve terminals via NCAM [40,41] (Figures 3, 4). NCAM-positive HSCs are heterogeneously located in the liver lobules, i.e., they are predominant in the periportal and intermediate area (zone I and II) compared to in the zone III. This heterogeneity of HSCs in liver lobules has been well known; for example, the number of lipid droplets in HSCs were greater in zone I than in zone III, while the size ratio was opposite [11,12]. Although we have suggested the mediating interaction between HSCs and nerve terminals as a functional significance of NCAM in HSCs, we have also demonstrated that hepatocytes, sinusoidal endothelial cells and Kupffer cells, reported previously to contact with nerves, are negative for NCAM. Thus it is difficult to explain the function of NCAM in HSCs via only innervation. NCAM is also expressed in NK cells residing in the human liver sinusoid. We demonstrated that rat HSCs may interact with lymphocytes in the fibrous septa during liver fibrogenesis [183]. NK cells in human liver might interact with HSCs, but evidence of this has not been shown. We undertook our study using the rat model of liver fibrosis to investigate further the function of NCAM expressed in HSCs [43].

Unexpectedly, NCAM expression was only seen as thin fibrous structures restricted within the portal tracts in the intact rat liver [43]. On the contrary, in the fibrotic liver produced after repeated administration of carbon tetrachloride or thioacetamide, NCAM-positive interstitial cells had accumulated in the fibrous area (Figure 5). Immunoelectron microscopy confirmed that these cells contained small lipid droplets, suggesting they were activated HSCs. Additionally, myofibroblasts in the expanded portal area of a rat model with bile duct ligation were also positive for NCAM (unpublished data). Furthermore, Knittel et al. ascertained that NCAM is expressed in cultured activated rat HSCs, and is upregulated in quiescent HSCs following treatment with transforming growth factor-beta (TGFβ)-1, insulin-like growth factor (IGF)-1, platelet-derived growth factor (PDGF) and epidermal growth factor (EGF), while upregulation of NCAM by these growth factors is not detected in activated HSCs caused by long time culture [42,184]. In addition, hepatocyte growth factor (HGF) can upregulate NCAM expression in activated HSCs. Conversely, NCAM expression in the activated HSCs was downregulated after treatment with tumor necrosis factor-alpha (TNFα) and interferon-gamma (IFNγ) [184]. The question of the nature of NCAM function in rat HSCs has thus been raised but it remains unanswered. Meanwhile, NCAM has been demonstrated to interact with ECMs, such as HSPGs, CSPGs and collagens [71-75]. HSCs in the fibrotic liver are activated and are simultaneously transformed into myofibroblastic phenotypes, and then they begin to produce large amounts of ECMs, especially collagens [2,5]. Thus, NCAM in the activated rat HSCs may interact with ECMs, and might be involved with regulation of HSC function. In addition, NCAM140-transfected glioma cell lines were demonstrated to decrease production of matrix metalloproteinases (MMPs) [185]. It was also demonstrated that NCAM is expressed in fibroblasts during wound healing due to skin injury, and may be involved with the healing process of skin tissues [167]. Liver fibrogenesis can be considered a kind of wound healing process, thus NCAM in HSCs may bear the same function as skin fibroblasts. On the other hand, although nerve fibers are rarely distributed in intact liver lobules in rat, as mentioned above [176], Akiyoshi and Terada demonstrated that in fibrotic liver, numerous acetylcholinesterase-containing nerve fibers are found in the fibrous septa forming a network, and building close contacts with mast cells and myofibroblasts, i.e., activated HSCs and portal fibroblasts [186]. Moreover, Akiyoshi and Terada observed that nerve terminals are also situated in close contact with myofibroblasts in the periseptal sinusoidal wall of cirrhotic nodules. These findings led to the hypothesis that NCAM can mediate between these myofibroblasts and nerve fibers. This perspective warrants further investigation. On the other hand, NCAM signal transduction in HSCs also remains unknown. HSCs have been demonstrated to express FGFR [187], and MAP kinase cascades in HSCs have been investigated vigorously [188]. However, the signaling pathway associated with Fyn has not yet been examined.

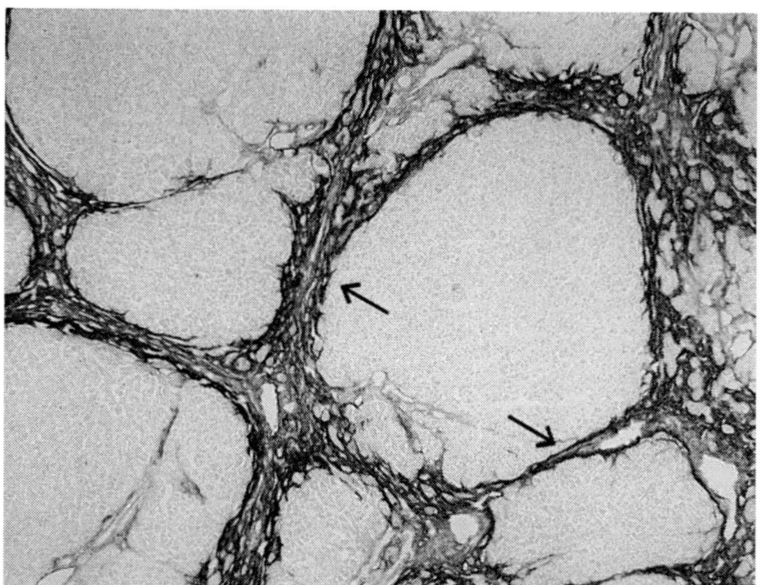

Figure 5. Immunohistochemistry for NCAM in a rat fibrotic liver, which was produced following repeated treatment with thioacetamide. NCAM-positive interstitial cells are found to accumulate in the fibrous septa (*arrows*). x100.

HSCs have been considered to be a member of the splanchnic fibroblast lineages, which Yamada and Hirosawa proposed to be classified as a "vitamin A-storing cell system" [189]. One of characteristics of HSCs is the storage of a large amount of vitamin A, as mentioned above [11], whereas various fibroblasts in splanchnic organs such as pancreas, kidney, digestive tracts, spleen, lung and heart, have a very small amount of vitamin A [11,190,191]. In mammals, vitamin A is accumulated in the liver, especially HSCs, while in lower animals such as the lampreys, vitamin A is decentralized and stored in splanchnic organs [192,193]. Thus, HSCs are considered to belong to the same lineage of fibroblasts in other splanchnic organs, known as extrahepatic stellate cells, under the supposition that these extrahepatic stellate cells and HSCs have common molecules. Indeed, we demonstrated that extrahepatic stellate cells possess Cygb/STAP expression, which is also detected in HSCs [30]. So, to elucidate whether extrahepatic stellate cells have NCAM, we preliminarily investigated pancreatic stellate cells (PSCs). PSCs have recently attracted attention, and it has been demonstrated that they have several marker molecules common to HSCs, such as αSMA and PDGF receptor-beta (PDGFRβ) [194-196]. In rats with spontaneously progressive chronic pancreatitis, Wistar Bonn/Kobori (WBN/Kob) rats, interstitial cells in the necrotic and fibrous lesion were shown to be positive for NCAM and to have the same distribution of αSMA-positive interstitial cells, suggested to be activated PSCs (unpublished data). However, we have not yet confirmed that these interstitial cells are identical to PSCs.

Molecular Structure and Characteristics of Osteonectin (ON)

For the first time, the calcium binding matricellular glycoprotein ON has been found as a noncollagenous protein of bone matrix [197]; it has also been termed "secreted protein, acidic and rich in cysteine (SPARC)" as a glycoprotein produced by mouse embryo parietal endoderm, specially acting in the synthesis of a rapidly expanding basement membrane [198], or named BM-40 as a component of the matrix of a basement membrane tumor [199]. Matricellular proteins comprise a nonhomologous group of extracellular regulatory macromolecules that mediate cell-matrix interactions but may not contribute significantly to extracellular matrix structure [200,201]. They have been known to include ON, thrombospondins 1 and 2, tenascins C and X and osteopontin, and to perform related functions, for example their counteradhesion effects lead to cell rounding and changes in cell shape resulting in the disruption of cell-matrix interactions, although structurally they are mostly unrelated. Thus, they are often required at events characterized by changes in cell shape and motility such as tissue renewal, tissue remodeling and embryonic development [201]. They can be also distinguished from traditional ECMs, such as fibronectin, laminin, fibrillar collagens and vitronectin, all of which are adhesive proteins and contribute to the structural stability of other ECMs.

ON is encoded by a single gene with a high degree of evolutionary conservation [202]. A secreted form of ON is a 43-kDa [203]. The human ON consists of 286 amino acid residues divided into three distinct domains (Figure 6): domain I is an acidic region rich in aspartic acids and glutamic acids on the N-terminal side, domain II is a cysteine-rich follistatin-like domain, and domain III is a largely alpha-helical region [201]. Domain I binds several calcium ions with low affinity and interacts with hydroxyapatite [204,205]. Domain II has two copper binding sites, one of which stimulates cell proliferation and angiogenesis [206-208]. Domain III is called the extracellular calcium-binding (EC) domain and contains a canonical pair of EF-hands with high-affinity calcium-binding sites [209]. Domain III also has binding sites for collagen type I and IV [210,211].

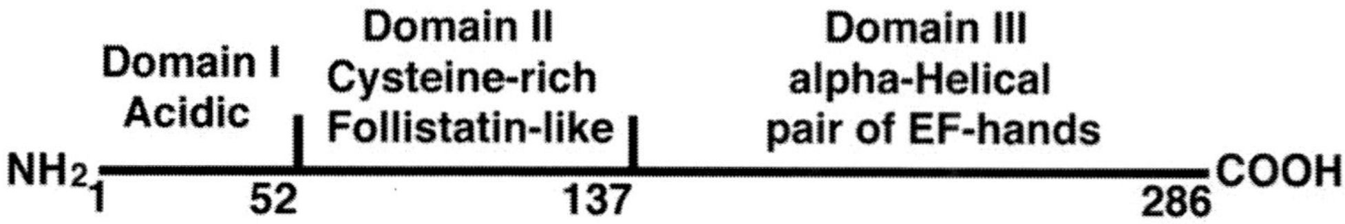

Figure 6. Molecular structure of osteonectin. Osteonectin consists of three domains: domain I (amino acid residues 1-52) is an acidic region, domain II (residues 53-137) is a cysteine-rich follistatin-like domain, and domain III (residues 138-286) is an alpha-helical region with a pair of EF-hands.

ON is considered to be a protein coded by a late-acting gene in differentiated cells that is responsive *in vitro* to retinoic acid, dexamethasone, and heat shock [202,211-214]. Many cytokines and growth factors have been described as potential regulators of ON mRNA production. It has been indeed demonstrated that TGFβ [215], bone morphogenetic protein (BMP)-2 [216] and IGF-1 [217] upregulate ON mRNA expression, while some other molecules downregulate ON expression, including basic FGF (bFGF) [218], interleukin (IL)-1, TNFα [216] and EGF [219]. Opinions concerning the effect of PDGF for expression of ON are diametrically opposed. Chandrasekhar et al. described that PDGF stimulates ON synthesis in the cultured rabbit articular chondrocytes [219], whereas Shiba et al. showed that PDGF decreases ON expression in human dental pulp cells in culture [217]. The reasons for this discrepancy have not yet been elucidated.

Vertebrate ON has capacity binding to a number of different ECM components, such as thrombospondin I [220], vitronectin [221], fibrillar collagens (type I, II, III, and V) [197,222], and collagen type IV [210], and could possess the potential to contribute to the organization of matrices in connective tissues as well as basement membranes [223]. ON has also been known to be able to bind to some growth factors. ON is able to bind PDGF dimers AB and BB, but not AA [214]. ON has been shown to interact specifically with the B-chain of PDGF, and to prevent binding of PDGF to its receptor on the cells. ON has also been shown to inhibit PDGF-mediated proliferation of mesangial cells, which secrete ON during experimental mesangial proliferative nephritis [225]. ON secreted by the cells, in which PDGF upregulates ON expression, may act as the molecule for negative feedback of PDGF [44,225]. Vascular endothelial growth factor (VEGF) is known to possess low but significant sequence homology with PDGF [226]. Kupprion et al. demonstrated that ON binds to VEGF and interferes with the binding of VEGF to human microvascular endothelial cells via a weakening of the association of VEGF with its cell surface receptor Flt-1 [227]. In addition, ON inhibits endothelial cell proliferation induced by VEGF. On the other hand, ON has been shown to diminish the response of bovine aortic endothelial cells to bFGF in regard to migration and proliferation, although physical interaction of ON with this growth factor remains to be determined [228]. Additionally, ON has been described to be able to modulate the activity of TGFβ in cultured mouse mesangial cells [229]. ON-null mice mesangial cells showed significantly decreased synthesis of TGFβ-1 mRNA, and addition of ON to these cells in culture restored the expression of TGFβ-1 to levels typical of wild-type cells. Whether ON directly interacts with TGFβ also remains to be elucidated.

ON has been considered to be a secreted glycoprotein, as a component of connective tissues and basement membranes, whereas recent studies have indicated that it may also work intracellularly [230]. Gooden et al. showed that ON accumulates within the nuclear matrix in embryonic chick cells and is translocated to the nucleus in a cell-cycle-dependent manner [231]. ON has been found to translocate to the nuclear matrix in active, proliferating embryonic cells, but not in mature or postmitotic cells, indicating that ON may play a role in the regulation of mitosis during development [201]. We observed by immunoelectron microscopy that ON-positive immunoreaction is detected on the nuclear membrane in activated human HSCs [44] (Figure 7).

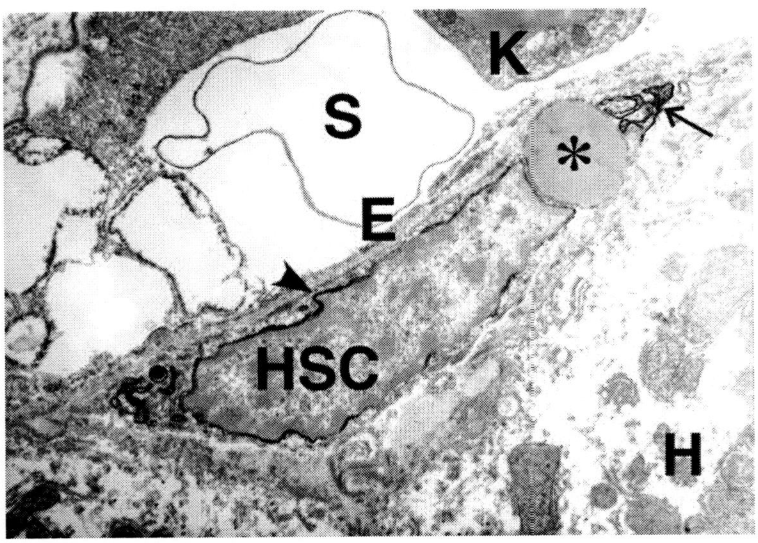

Figure 7. Immunoelectron microscopy for osteonectin in the liver parenchyma of a patient with chronic hepatitis. Osteonectin-positive immunoprecipitates are found in dilated endoplasmic reticulum (*arrow*) and nuclear membrane (*arrowhead*) of the hepatic stellate cell (*HSC*) located in the space of Disse. *Asterisks*, a lipid droplet; *E*, sinusoidal endothelial cells; *H*, a hepatocyte; *K*, a Kupffer cell; *S*, sinusoid. x7,500.

FUNCTION OF ON

Anti-Proliferation

Early studies indicated that ON is a potent cell cycle inhibitor that arrests cells in mid-G_1 phase, independently of discernible changes in cell shape [232,233]. ON also inhibits the proliferation of endothelial, smooth muscle, mesangial and fibroblastic cells, stimulated with VEGF, PDGF and bFGF, as mentioned above [225,227,228]. This anti-proliferative property of ON has been confirmed by investigation using the cells obtained from ON-null mice. Fibroblasts, mesangial cells and smooth muscle cells isolated from ON-null mice exhibited a higher rate of proliferation than their wild-type counterparts, and were more sensitive to the inhibition of cell cycle progression induced by recombinant ON [201,229,234]. This anti-proliferative function of ON has been suggested to be partially dependent on a signaling pathway through a G-protein-coupled receptor [235].

Counteradhesion

ON is well known for its counteradhesive function, which is probably related to the dissolution of focal adhesion complexes and reorganization of actin stress fibers [236]. Cell rounding was induced in various cultured cells after treatment with ON. Studies using tyrosine kinase inhibitors have demonstrated that cell rounding activity via ON in endothelial cells is mediated through a tyrosine phosphorylation-dependent pathway [235]. Cell adhesion and subsequent spreading on a compatible substrate are essential requirements for proper growth and survival of cells. In addition, disruption of the ECMs could lead to cell rounding and apoptosis [201]. In contrast, changes in cell shape are necessary temporarily for cells undergoing migration and proliferation. Counteradhesive proteins might regulate cell survival by their control of cell shape. Additionally, ON has been shown to induce shape changes in nerve cells, and this counteradhesive activity may facilitate neuronal rearrangement, including synaptic plasticity with its attendant physiological and behavioral implications [223].

ON and Dermal Wound Healing

Dermal wound healing involves many processes, such as ECM turnover and reassembly, cell migration and proliferation, and recruitment and/or differentiation of interstitial cells, and these processes could be affected by ON [223]. Basu et al. recently demonstrated that repair of excisional dermal wounds in ON-null mice is impaired [237]. They produced 25mm oblong full thickness excisional wounds, including the striated muscle layer, on the dorsal skin of wild-type and ON-null mice. This model of large excisional wounds indicated that healing of dermal wounds in ON-null mice is delayed compared to in wild-type mice. Moreover, granulation tissue formation and production of ECMs, such as collagen type I and fibronectin, were delayed in small wounds, 6mm full thickness circular excisional wounds, in ON-null mice compared to in wild-type mice, whereas lesion restoration in ON-null mice did not differ from that in wild-type mice. Additionally, the experiments using *in vitro* wound-healing assays showed that wound closure of ON-null primary dermal fibroblasts is deferred in comparison to wild-type cells, but purified ON treatment restored the wound closure to the same time course as in wild-type cells. These effects were confirmed to result from the prevention of cell migration, but not cell proliferation. Based on these results, Basu et al. suggested that ON plays a role in wound healing by promoting fibroblast migration and thus granulation tissue formation.

On the other hand, Bradshaw et al. reported that closure of excisional dermal wounds in ON-null mice is accelerated compared to in wild-type mice [238]. They made 5mm circular excisional wounds in wild-type and ON-null mice, and measured the size of the skin lesions over time. The size of the skin defects in ON-null mice reduced more rapidly than in wild-type mice. Bradshaw et al. also investigated the amount of deposits of ECMs, such as collagen type I and IV and fibronectin, at the wound site, but did not find gross differences between the ON-null and wild-type mice. However, the collagen content of unwounded skin appeared to be substantially lower in the ON-null mice. Additionally, it was confirmed by observation under conventional electron microscopes that in the unwounded skin and the capsular lesion produced by implanting biomaterials to skin, the collagen fibrils in dermis of ON-null mice were smaller and more uniform in size than in wild-type mice [239,240]. Bradshaw et al. also showed that primary cultured ON-null dermal fibroblasts displayed accelerated migration, relative to wild-type cells, in *in vitro* wound-healing assays, while substantial differences in the rate of proliferating cells were not apparent [238].

The reasons for the discrepancy mentioned above remain to be elucidated. Basu et al. described that size-dependent differences of speed of wound closure may cause this inconsistence in the results of Bas et al. and Bradshaw et al. [237]. However, this hypothesis cannot explain the contradiction of the results of *in vitro* wound-healing assays. Careful further examination is considered to be a necessity.

ON and Kidney: Functional Significance under Normal and Pathological Conditions

ON expression and function in the kidney during renal diseases, especially glomerular injuries, has been investigated diligently [241]. In the normal kidney, ON has been detected clearly in the glomeruli [242], and ON-positive glomerular cells have been identified as glomerular epithelial cells, named podocytes [243,244]. It was also reported that ON was expressed in the distal cortical tubules and medullary tubules [245]. Furthermore, ON production in podocytes has been reported to elevate during glomerular injury and correlated with the collagen type I deposition in the glomeruli [246]. Apart from podocytes, ON was expressed in mesangial cells with experimental mesangioproliferative glomerulonephritis [225]. ON mRNA levels in cultured mesangial cells increased following treatment with PDGF and bFGF, as mentioned above. Moreover, recombinant ON inhibited PDGF-induced proliferation of the mesangial cells.

Francki et al. demonstrated that the primary cultured mesangial cells obtained from ON-null mice display diminished expression of collagen type I and TGFβ1, whereas treatment with recombinant ON restored the expression of collagen type I mRNA to 70% and TGFβ1 mRNA to 100% of wild-type levels [229]. Bassuk et al. also described that administration of recombinant ON to rats with acute glomerulonephritis increases expression of TGFβ1, collagen type I and fibronectin [247]. Addition of recombinant ON in primary cultured rat mesangial cells also increased in TGFβ1 mRNA and in total and secreted TGFβ1 protein. A recent study using ON-null mice revealed that absence of ON induces an increase in the amounts of IGF-1 and –2, as well as IGF-1 receptor synthesized by mesangial cells, which can respond to IGF-1 [248]. Addition of recombinant ON to ON-null cells prevented IGF-1-stimulated MAP kinase activation and DNA synthesis.

Walpen et al. revealed that ON expression in cultured rat mesangial cells is diminished by stimulation for induction of inducible nitric oxide (NO) synthase (iNOS), such as IL-1β and TNFα treatment [249]. It is well known that mesangial cells are stimulated by IL-1β and TNFα to produce high levels of NO through expression of iNOS. These effects were annulled after treatment with an NOS inhibitor, and thus ON was hypothesized to be an NO-regulated gene product.

ON Expression in the Liver

Frizell et al. first conducted detailed analysis of ON expression in the mouse, rat and human liver [250]. ON mRNA expression could be confirmed in the normal liver of every species, while levels of ON mRNA were significantly increased in diseased or fibrotic livers. Moreover, the examination using isolated rat liver constituents revealed that an HSCs-rich fraction predominantly contained ON mRNA. Extracts from a Kupffer-endothelial cell fraction included a little amount of ON mRNA, but expression of ON mRNA was absent in extracts from freshly isolated hepatocytes. Therefore, Frizell et al. suggested that the main source of ON in the liver is HSC. These observations have been confirmed in supplementary examinations by several researchers [251,252]. In addition to in fibrotic livers, ON deposition has been investigated in the regenerating liver [253]. ON deposition at the sinusoidal surface of normal liver was found be highly enriched relative to whole liver lysates, whereas its deposition at the sinusoidal surface was reduced by 50% at 3 days after partial hepatectomy.

Investigations have been undertaken using human liver specimens obtained from patients with various liver diseases, such as alcoholic liver diseases, chronic viral hepatitis, liver cirrhosis, PBC and biliary atresia [44,45,250,252,254,255]. Blazejewski et al. showed that a part of αSMA-positive myofibroblastic cells located at the edge of the fibrous septa and at the expanded areas of piecemeal necrosis exhibited increased expression of ON mRNA, as well as scattered cells in the sinusoidal walls [252]. They also confirmed that cultured human liver myofibroblasts display ON mRNA expression. Lamireau et al. also revealed that markedly increased signals recognized as ON mRNA expression were found in fibrous septa, especially at its edge, and in bile duct proliferation areas of the liver with biliary atresia [254]. ON mRNA was also seen along the sinusoidal wall, predominantly at the periphery of the lobules. Grenard et al. showed for the first time that ON protein expression was detected in the fibrous septa and along the sinusoids [255].

We further undertook immunohistochemical investigation of localization of ON protein in the liver with chronic hepatitis [44,45]. As described by other investigators, ON expression was detected in the sparse sinusoidal lining cells in the undamaged or slightly damaged human livers, such as livers with Gilbert's disease and non-cancerous portions of metastatic liver tumors. Additionally, we found that in cirrhotic livers the small number of sinusoidal lining cells, still greater than in undamaged livers, and interstitial cells in fibrous septa are positive for ON, in contrast with other researchers' observations. However, ON expression in the sinusoidal lining cells of the liver with chronic hepatitis was increased to a striking degree (Figure 8). These sinusoidal lining cells were confirmed to be HSCs by immunoelectron microscopy (Figure 7). Immunoprecipitates produced by reaction with anti-ON antibodies were found in the nuclear membrane and rough endoplasmic reticulum (rER). ON-positive rERs were often dilated. In this study, we could not detect extracellular deposits of ON. The number of ON-positive sinusoidal lining cells in liver with chronic hepatitis has no significant differences among etiology, grades of activity, stages of fibrosis, and the histological activity index (HAI). We also compared ON-positive cells with αSMA-positive or PDGFRβ-positive cells. ON-positive sinusoidal lining cells appeared in a distribution similar to PDGFRβ–positive ones. Additionally, a significant correlation was found between the number of ON-positive sinusoidal lining cells and αSMA-positive ones in late stages of fibrosis (moderate and sever fibrosis stages), but not in early stages (no and mild fibrosis stages). However, the number of ON-positive interstitial cells in the fibrous septa was very small as compared with PDGFRβ–positive and αSMA-positive ones, unlike the previous reports of other investigators.

On the other hand, ON expression was explored in the human hepatocellular carcinoma (HCC) [256,257]. Various neoplastic cells, especially metastatic tumor cells, were shown to express ON [223]. Indeed, ON was proposed as a diagnostic marker of invasive meningiomas [258]. Additionally, the inhibition of ON expression in human melanoma cells diminished both their adhesive and invasive capacities *in vitro* and *in vivo* [259]. Le Bail et al. demonstrated that ON expression is detected in the myofibroblastic cells, as confirmed by positive staining for αSMA, in the tumor capsule, in the fibrous bands, and along capillaries within HCC [256]. Furthermore, Goldenberg et al. recently reported that ON mRNA expression is detected in HCCs by analysis using cDNA micro-arrays techniques [257].

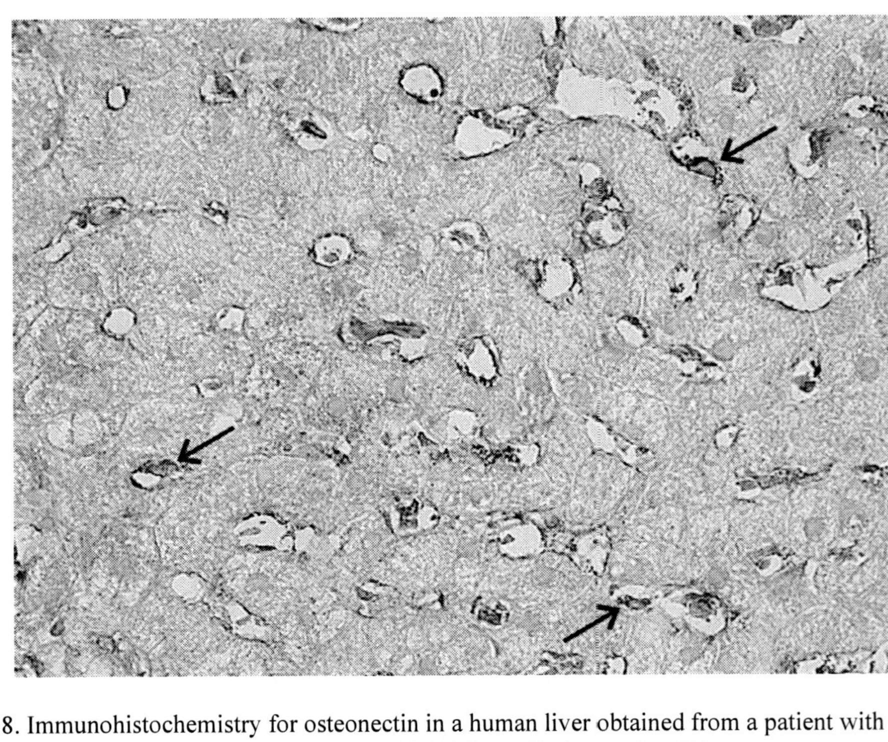

Figure 8. Immunohistochemistry for osteonectin in a human liver obtained from a patient with chronic hepatitis. Osteonectin is detected in the sinusoidal lining cells (*arrows*). x600.

Concluding Remarks

Many investigations have been performed with vigor to elucidate and to control HSCs function, because HSCs have been recognized as the cells responsible for the progression of liver fibrogenesis [6,260]. In analyzing HSCs function, exploring the molecules mediating cell-cell and/or cell-matrix interaction, such as the aforementioned NCAM and ON, is considered to be important. HSCs require interaction with a scaffold to activate their function. NCAM on HSCs might play a role interacting with ECMs, especially collagens. ON might interact with HSCs in a counteradhesive manner. At least, further investigations should be performed to elucidate the function of NCAM and ON in the liver. Other molecules with the ability of modulating cell-cell and cell-matrix interaction have been shown in HSCs. They include intercellular adhesion molecules (ICAM) 1, vascular cell adhesion molecules (VCAM) 1 [184] and integrins [261], which belong to the category "adhesion molecules". In addition to these adhesion molecules, the expression of a receptor tyrosine kinase with the ability to associate with collagens, named discoidin domain receptor (DDR) 2, in HSCs has recently gained attention [262,263]. DDR2 has been considered a collagen receptor because of its properties mediating signal transduction through binding to collagens [264]. It has been shown that DDR2 mRNA and protein in HSCs are indicated by activation of HSCs via primary culture and *in,vivo* during liver injury, whereas their expression is downregulated in milieus keeping or indicating cellular quiescence [263]. DDR2 was also shown to increase expression of MMP-2 in activated HSCs via the signaling pathway related to Src kinase and Shc [263,265]. In addition to these experimental models, investigation of DDR2 expression in the diseased human liver has begun [266]. As mentioned above, studies to elucidate molecular mechanisms of cell-cell and/or cell-matrix interaction associated with HSCs, including intracellular signaling pathways, are considered to be important, and should be pursued as a strategy for curing liver fibrosis.

Acknowledgments

This work was supported in part by Grant-in-Aid for Scientific Research from the Ministry of Education, Culture, Sports, Science and Technology of Japan (No. 10770009 and No. 15790110). We thank Dr. Kenji Kaneda of Houai Hospital (Osaka, Japan), Dr. Norifumi Kawada of Osaka City University and Dr. Nobuyoshi Kawakita of Kawakita Clinic (Osaka, Japan) for their valuable comments on this work. We also thank Miss Shoko Uoya of Osaka City University for her valuable technical support.

References

[1] Safadi R, Friedman SL. Hepatic fibrosis--role of hepatic stellate cell activation. *Med Gen Med 4*: 27, 2002.

[2] Reeves HL, Friedman SL. Activation of hepatic stellate cells--a key issue in liver fibrosis. *Front Biosci 7*: d808-826, 2002.

[3] Bedossa P, Paradis V. Liver extracellular matrix in health and disease. *J Pathol 200*: 504-515, 2003.

[4] Friedman SL, Roll FJ, Boyles J, Bissell DM. Hepatic lipocytes: the principal collagen-producing cells of normal rat liver. *Proc Natl Acad Sci U S A 82*: 8681-8685, 1985.

[5] Kawada N. The hepatic perisinusoidal stellate cell. *Histol Histopathol 12*: 1069-1080, 1997.

[6] Kawada N. Molecular approach for the hepatic fibrogenesis. *Recent Res Devel Biophys Biochem 3*: 471-479, 2003.

[7] Wang YQ, Ikeda K, Ikebe T, Hirakawa K, Sowa M, Nakatani K, Kawada N, Kaneda K. Inhibition of hepatic stellate cell proliferation and activation by the semisynthetic analogue of fumagillin TNP-470 in rats. *Hepatology 32*: 980-989, 2000.

[8] Okuyama H, Shimahara Y, Kawada N, Seki S, Kristensen DB, Yoshizato K, Uyama N, Yamaoka Y. Regulation of cell growth by redox-mediated extracellular proteolysis of platelet-derived growth factor receptor beta. *J Biol Chem 276*: 28274-28280, 2001.

[9] Matsui H, Ikeda K, Nakajima Y, Horikawa S, Kawada N. Sulfur-containing amino acids attenuate the development of liver fibrosis in rats through down-regulating stellate cell activation. *J Hepatol*, in press.

[10] Wake K. "Sternzellen" in the liver: perisinusoidal cells with special reference to storage of vitamin A. *Am J Anat 132*: 429-62, 1971.

[11] Wake K. Perisinusoidal stellate cells (fat-storing cells, interstitial cells, lipocytes), their related structure in and around the liver sinusoids, and vitamin A-storing cells in extrahepatic organs. *Int Rev Cytol 66*: 303-353, 1980.

[12] Wake K. *Sinusoidal structure and dynamics. In: Vidal-vanaclocha F (eds) Functional heterogeneity of liver tissue: from cell lineage diversity to sublobular compartment-specific pathogenesis*, Chapman and Hall, NY, pp57-67, 1997.

[13] Wake K. Hepatic *stellate cells. In: Tanikawa K, Ueno T (eds) Liver diseases and hepatic sinusoidal cells,* Springer-Verlag Tokyo, Tokyo, Japan, pp56-65, 1999.

[14] Yokoi Y, Namihisa T, Kuroda H, Komatsu I, Miyazaki A, Watanabe S, Usui K. Immunocytochemical detection of desmin in fat-storing cells (Ito cells). *Hepatology 4*: 709-714, 1984.

[15] Burt AD, Robertson JL, Heir J, MacSween RN. Desmin-containing stellate cells in rat liver; distribution in normal animals and response to experimental acute liver injury. *J Pathol 150*: 29-35, 1986.

[16] Ballardini G, Fallani M, Biagini G, Bianchi FB, Pisi E. Desmin and actin in the identification of Ito cells and in monitoring their evolution to myofibroblasts in experimental liver fibrosis. *Virchows Arch B 56*: 45-49, 1988.

[17] Niki T, De Bleser PJ, Xu G, Van Den Berg K, Wisse E, Geerts A. Comparison of glial fibrillary acidic protein and desmin staining in normal and CCl4-induced fibrotic rat livers. *Hepatology 23*: 1538-1545, 1996.

[18] Gard AL, White FP, Dutton GR. Extra-neural glial fibrillary acidic protein (GFAP) immunoreactivity in perisinusoidal stellate cells of rat liver. *J Neuroimmunol 8*: 359-375, 1985.

[19] Neubauer K, Knittel T, Aurisch S, Fellmer P, Ramadori G. Glial fibrillary acidic protein--a cell type specific marker for Ito cells in vivo and in vitro. *J Hepatol 24*: 719-730, 1996.

[20] Buniatian G, Gebhardt R, Schrenk D, Hamprecht B. Colocalization of three types of intermediate filament proteins in perisinusoidal stellate cells: glial fibrillary acidic protein as a new cellular marker. *Eur J Cell Biol 70*: 23-32, 1996.

[21] Buniatian G, Hamprecht B, Gebhardt R. Glial fibrillary acidic protein as a marker of perisinusoidal stellate cells that can distinguish between the normal and myofibroblast-like phenotypes. *Biol Cell 87*: 65-73, 1996.

[22] Ramadori G, Veit T, Schwogler S, Dienes HP, Knittel T, Rieder H, Meyer zum Buschenfelde KH. Expression of the gene of the alpha-smooth muscle-actin isoform in rat liver and in rat fat-storing (ITO) cells. *Virchows Arch B 59*: 349-357, 1990.

[23] Takase S, Leo MA, Nouchi T, Lieber CS. Desmin distinguishes cultured fat-storing cells from myofibroblasts, smooth muscle cells and fibroblasts in the rat. *J Hepatol 6*: 267-276, 1988.

[24] Niki T, Pekny M, Hellemans K, Bleser PD, Berg KV, Vaeyens F, Quartier E, Schuit F, Geerts A. Class VI intermediate filament protein nestin is induced during activation of rat hepatic stellate cells. *Hepatology 29*: 520-527, 1999.

[25] Kawada N, Kuroki T, Kobayashi K, Inoue M, Nakatani K, Kaneda K, Nagata K. Expression of heat-shock protein 47 in mouse liver. *Cell Tissue Res 284*: 341-346, 1996.

[26] Kitada T, Seki S, Nakatani K, Kawada N, Kuroki T, Monna T. Hepatic expression of c-Myb in chronic human liver disease. *Hepatology 26*: 1506-1512, 1997.

[27] Ikeda K, Kawada N, Wang YQ, Kadoya H, Nakatani K, Sato M, Kaneda K. Expression of cellular prion protein in activated hepatic stellate cells. *Am J Pathol 153*: 1695-1700, 1998.

[28] Kristensen DB, Kawada N, Imamura K, Miyamoto Y, Tateno C, Seki S, Kuroki T, and Yoshizato K. Proteome analysis of rat hepatic stellate cells. *Hepatology 32*: 268-277, 2000.

[29] Kawada N, Kristensen DB, Asahina K, Nakatani K, Minamiyama Y, Seki S, Yoshizato K. Characterization of a stellate cell activation-associated protein (STAP) with peroxidase activity found in rat hepatic stellate cells. *J Biol Chem 276*: 25318-25323, 2001.

[30] Nakatani K, Okuyama H, Shimahara Y, Saeki S, Kim DH, Nakajima Y, Seki S, Kawada N, Yoshizato K. Cytoglobin/STAP, its unique localization in splanchnic fibroblast-like cells and function in organ fibrogenesis. *Lab Invest 84*: 91-101, 2004.

[31] Enzan H, Himeno H, Iwamura S, Saibara T, Onishi S, Yamamoto Y, Hara H. Immunohistochemical identification of Ito cells and their myofibroblastic transformation in adult human liver. *Virchows Arch 424*: 249-256, 1994.

[32] Kitada T, Seki S, Ikeda K, Nakatani K, Sakaguchi H, Kawada N, Kadoya H, Kaneda K. Clinicopathlogical characterization of prion: a novel marker of activated human hepatic stellate cells. *J Hepatol 33*: 751-757, 2000.

[33] Asahina K, Kawada N, Kristensen DB, Nakatani K, Seki S, Shiokawa M, Tateno C, Obara M, Yoshizato K. Characterization of Human stellate cell activation-associated

protein (STAP) and its expression in human liver. *Biochim Biophys Acta 1577*: 471-475, 2002.

[34] Buniatian GH. Further similarities between astrocytes and perisinusoidal stellate cells of liver (Ito cells): colocalization of desmin and glial fibrillary acidic protein in astroglial primary cultures. *Biol Cell 89*: 169-177, 1997.

[35] Wechselberger C, Wurm S, Pfarr W, Hoglinger O. The physiological functions of prion protein. *Exp Cell Res 281*: 1-8, 2002.

[36] Lendahl U, Zimmerman LB, McKay RD. CNS stem cells express a new class of intermediate filament protein. *Cell 60*: 585-595, 1990.

[37] Burmester T, Ebner B, Weich B, and Hankeln T. Cytoglobin: a novel globin type ubiquitously expressed in vertebrate tissues. *Mol Biol Evol 19*: 416-421, 2002.

[38] Trent JT 3rd, and Hargrove MS. A ubiquitously expressed human hexacoordinate hemoglobin. *J Biol Chem 277*: 19538-19545, 2002.

[39] Schmidt M, Gerlach F, Avivi A, Laufs T, Wystub S, Simpson JC, Nevo E, Saaler-Reinhardt S, Reuss S, Hankeln T, Burmester T. Cytoglobin is a respiratory protein in connective tissue and neurons that is up-regulated by hypoxia. *J Biol Chem*, in press.

[40] Nakatani K, Seki S, Kawada N, Sakaguchi H, Kawakita N, Yanai A, Kitada T, Sakai Y, Yamada T, Masuichi H, Kuroki T, Kobayashi K, Kaneda K. *Expression of neural cell adhesion molecule (N-CAM) on the stellate cells of human liver*. In: Wisse E, Knook DL, Wake K (eds) Cells of the hepatic sinusoid, vol. 5, Kupffer Cell Foundation, Leiden, the Netherlands, pp267-269, 1995.

[41] Nakatani K, Seki S, Kawada N, Kobayashi K, Kaneda K. Expression of neural cell adhesion molecule (N-CAM) in perisinusoidal stellate cells of the human liver. *Cell Tissue Res 283*: 159-165, 1996.

[42] Knittel T, Aurisch S, Neubauer K, Eichhorst S, Ramadori G. Cell-type-specific expression of neural cell adhesion molecule (N-CAM) in Ito cells of rat liver. Up-regulation during in vitro activation and in hepatic tissue repair. *Am J Pathol 149*: 449-62, 1996.

[43] Nakatani K, Kawada N, Kadoya H, Masuichi H, Kitada T, Sakai Y, Yamada T, Kawakita N, Sakaguchi H, Seki S, Kuroki T, Kobayashi K, Morikawa H, Kaneda K. *Neural cell adhesion molecule (N-CAM) as a new marker of activated stellate cells of the rat*. In: Wisse E, Knook DL, Balabaud C (eds) Cells of the hepatic sinusoid, vol. 6, Kupffer Cell Foundation, Leiden, the Netherlands, pp57-58, 1997.

[44] Nakatani K, Seki S, Kawada N, Kitada T, Yamada T, Sakaguchi H, Kadoya H, Ikeda K, Kaneda K. Expression of SPARC by activated hepatic stellate cells and its correlation with the stages of fibrogenesis in human chronic hepatitis. *Virchows Arch 441*: 466-474, 2002.

[45] Nakatani K, Seki S, Kawada K, Fujii H, Kobayashi S, Kadoya H, Ikeda K, Kaneda K. *Expression of osteonectin in human hepatic stellate cells during chronic liver inflammation*. In: Wisse E, Knook DL, de Zanger R, Arthur MJP (eds) Cells of the hepatic sinusoid, vol. 8, Kupffer Cell Foundation, Leiden, the Netherlands, pp 231-232, 2001.

[46] Porter PL, Sage EH, Lane TF, Funk SE, Gown AM. Distribution of SPARC in normal and neoplastic human tissue. *J Histochem Cytochem 43*: 791-800, 1995.

[47] Thiery JP, Brackenbury R, Rutishauser U, Edelman GM. Adhesion among neural cells of the chick embryo. II. Purification and characterization of a cell adhesion molecule from neural retina. *J Biol Chem 252*: 6841-6845, 1977.

[48] Rutishauser U. Developmental biology of a neural cell adhesion molecule. *Nature 310*: 549-554, 1984.

[49] Rieger F, Grumet M, Edelman GM. N-CAM at the vertebrate neuromuscular junction. *J Cell Biol 101*: 285-293, 1985.

[50] Walsh FS, Doherty P. Neural cell adhesion molecules of the immunoglobulin superfamily: role in axon growth and guidance. *Annu Rev Cell Dev Biol 13*: 425-456, 1997.

[51] Isacke CM, Horton MA. NCAM. In: *The adhesion molecule facts book*, 2nd ed., Academic Press, London, UK, pp133-135, 2000.

[52] Cavallaro U, Niedermeyer J, Fuxa M, Christofori G. N-CAM modulates tumour-cell adhesion to matrix by inducing FGF-receptor signalling. *Nat Cell Biol 3*: 650-657, 2001.

[53] Owens GC, Edelman GM, Cunningham BA. Organization of the neural cell adhesion molecule (N-CAM) gene: alternative exon usage as the basis for different membrane-associated domains. *Proc Natl Acad Sci USA 84*: 294-298, 1987.

[54] Cunningham BA, Hemperly JJ, Murray BA, Prediger EA, Brackenbury R, Edelman GM. Neural cell adhesion molecule: structure, immunoglobulin-like domains, cell surface modulation, and alternative RNA splicing. *Science. 236*: 799-806, 1987.

[55] Goridis C, Brunet JF. NCAM: structural diversity, function and regulation of expression. *Semin Cell Biol 3*: 189-197, 1992.

[56] Santoni MJ, Barthels D, Barbas JA, Hirsch MR, Steinmetz M, Goridis C, Wille W. Analysis of cDNA clones that code for the transmembrane forms of the mouse neural cell adhesion molecule (NCAM) and are generated by alternative RNA splicing. *Nucleic Acids Res 15*: 8621-8641, 1987.

[57] Santoni MJ, Barthels D, Vopper G, Boned A, Goridis C, Wille W. Differential exon usage involving an unusual splicing mechanism generates at least eight types of NCAM cDNA in mouse brain. *EMBO J 8*: 385-392, 1989.

[58] Small SJ, Haines SL, Akeson RA. Polypeptide variation in an N-CAM extracellular immunoglobulin-like fold is developmentally regulated through alternative splicing. *Neuron 1*: 1007-1017, 1988.

[59] Small SJ, Akeson R. Expression of the unique NCAM VASE exon is independently regulated in distinct tissues during development. *J Cell Biol 111*: 2089-2096, 1990.

[60] Hoffman S, Sorkin BC, White PC, Brackenbury R, Mailhammer R, Rutishauser U, Cunningham BA, Edelman GM. Chemical characterization of a neural cell adhesion molecule purified from embryonic brain membranes. *J Biol Chem 257*: 7720-7729, 1982.

[61] Doherty P, Moolenaar CE, Ashton SV, Michalides RJ, Walsh FS. The VASE exon downregulates the neurite growth-promoting activity of NCAM 140. *Nature 356*: 791-793, 1992.

[62] Cremer H, Lange R, Christoph A, Plomann M, Vopper G, Roes J, Brown R, Baldwin S, Kraemer P, Scheff S, Barthels D, Rajewsky K, Wille W. Inactivation of the N-CAM

gene in mice results in size reduction of the olfactory bulb and deficits in spatial learning. *Nature 367*: 455-459, 1994.

[63] Finne J, Finne U, Deagostini-Bazin H, Goridis C. Occurrence of alpha 2-8 linked polysialosyl units in a neural cell adhesion molecule. *Biochem Biophys Res Commun. 112*: 482-487,1983.

[64] Seki T, Arai Y. Distribution and possible roles of the highly polysialylated neural cell adhesion molecule (NCAM-H) in the developing and adult central nervous system. *Neurosci Res 17*: 265-290, 1993.

[65] Rutishauser U, Landmesser L. Polysialic acid in the vertebrate nervous system: a promoter of plasticity in cell-cell interactions. *Trends Neurosci 19*: 422-427, 1996.

[66] Kiss JZ, Rougon G. Cell biology of polysialic acid. *Curr Opin Neurobiol 7*: 640-646, 1997.

[67] Durbec P, Cremer H. Revisiting the function of PSA-NCAM in the nervous system. *Mol Neurobiol 24*: 53-64, 2001.

[68] Rutishauser U, Acheson A, Hall AK, Mann DM, Sunshine J. The neural cell adhesion molecule (NCAM) as a regulator of cell-cell interactions. *Science 240* : 53-57, 1988.

[69] Panicker AK, Buhusi M, Thelen K, Maness PF. Cellular signalling mechanisms of neural cell adhesion molecules. *Front Biosci 8:* d900-911, 2003.

[70] Crossin KL, Krushel LA. Cellular signaling by neural cell adhesion molecules of the immunoglobulin superfamily. *Dev Dyn 218*: 260-79, 2000.

[71] Probstmeier R, Kuhn K, Schachner M. Binding properties of the neural cell adhesion molecule to different components of the extracellular matrix. *J Neurochem 53*: 1794-1801, 1989.

[72] Reyes AA, Akeson R, Brezina L, Cole GJ. Structural requirements for neural cell adhesion molecule-heparin interaction. *Cell Regul 1*: 567-576, 1990.

[73] Margolis RK, Rauch U, Maurel P, Margolis RU. Neurocan and phosphacan: two major nervous tissue-specific chondroitin sulfate proteoglycans. *Perspect Dev Neurobiol 3*: 273-90, 1996.

[74] Storms SD, Rutishauser U. A role for polysialic acid in neural cell adhesion molecule heterophilic binding to proteoglycans. *J Biol Chem 273*: 27124-27129, 1998.

[75] Prag S, Lepekhin EA, Kolkova K, Hartmann-Petersen R, Kawa A, Walmod PS, Belman V, Gallagher HC, Berezin V, Bock E, Pedersen N. NCAM regulates cell motility. *J Cell Sci 115*: 283-292, 2002.

[76] Kadmon G, Kowitz A, Altevogt P, Schachner M. The neural cell adhesion molecule N-CAM enhances L1-dependent cell-cell interactions. *J Cell Biol 110*: 193-208, 1990.

[77] Kadmon G, Kowitz A, Altevogt P, Schachner M. Functional cooperation between the neural adhesion molecules L1 and N-CAM is carbohydrate dependent. *J Cell Biol 110*: 209-218, 1990.

[78] Horstkorte R, Schachner M, Magyar JP, Vorherr T, Schmitz B. The fourth immunoglobulin-like domain of NCAM contains a carbohydrate recognition domain for oligomannosidic glycans implicated in association with L1 and neurite outgrowth. *J Cell Biol 121*: 1409-1421, 1993.

[79] Williams EJ, Furness J, Walsh FS, Doherty P. Activation of the FGF receptor underlies neurite outgrowth stimulated by L1, N-CAM, and N-cadherin. *Neuron 13:* 583-594, 1994.

[80] Doherty P, Furness J, Williams EJ, Walsh FS. Neurite outgrowth stimulated by the tyrosine kinase inhibitor herbimycin A requires activation of tyrosine kinases and protein kinase C. *J Neurochem 62*: 2124-2131, 1994.

[81] Williams EJ, Furness J, Walsh FS, Doherty P. Characterisation of the second messenger pathway underlying neurite outgrowth stimulated by FGF. *Development 120*: 1685-1693, 1994.

[82] Doherty P, Walsh FS. CAM-FGF receptor interactions: a model for axonal growth. *Mol Cell Neurosci 8*: 99-111, 1996.

[83] Pollerberg GE, Schachner M, Davoust J. Differentiation state-dependent surface mobilities of two forms of the neural cell adhesion molecule. *Nature 324*: 462-465, 1986.

[84] Pollerberg GE, Burridge K, Krebs KE, Goodman SR, Schachner M. The 180-kD component of the neural cell adhesion molecule N-CAM is involved in a cell-cell contacts and cytoskeleton-membrane interactions. *Cell Tissue Res 250*: 227-236, 1987.

[85] Leshchyns'ka I, Sytnyk V, Morrow JS, Schachner M. Neural cell adhesion molecule (NCAM) association with PKCbeta2 via betaI spectrin is implicated in NCAM-mediated neurite outgrowth. *J Cell Biol 161*: 625-639, 2003.

[86] Büttner B, Kannicht C, Reutter W, Horstkorte R. The neural cell adhesion molecule is associated with major components of the cytoskeleton. *Biochem Biophys Res Commun 310*: 967-971, 2003.

[87] Beggs HE, Soriano P, Maness PF. NCAM-dependent neurite outgrowth is inhibited in neurons from Fyn-minus mice. *J Cell Biol 127*: 825-833, 1994.

[88] Beggs HE, Baragona SC, Hemperly JJ, Maness PF. NCAM140 interacts with the focal adhesion kinase p125(fak) and the SRC-related tyrosine kinase p59(fyn). *J Biol Chem 272:* 8310-8319, 1997.

[89] Tomasiewicz H, Ono K, Yee D, Thompson C, Goridis C, Rutishauser U, Magnuson T. Genetic deletion of a neural cell adhesion molecule variant (N-CAM-180) produces distinct defects in the central nervous system. *Neuron 11:* 1163-1174, 1993.

[90] Rabinowitz JE, Rutishauser U, Magnuson T. Targeted mutation of NCAM to produce a secreted molecule results in a dominant embryonic lethality. *Proc Natl Acad Sci USA 93:* 6421-6424, 1996.

[91] Povlsen GK, Ditlevsen DK, Berezin V, Bock E. Intracellular signaling by the neural cell adhesion molecule. *Neurochem Res 28*: 127-141, 2003.

[92] Schuch U, Lohse MJ, Schachner M. Neural cell adhesion molecules influence second messenger systems. *Neuron 3*: 13-20, 1989.

[93] Zheng JQ. Turning of nerve growth cones induced by localized increases in intracellular calcium ions. *Nature 403*: 89-93, 2000.

[94] Doherty P, Ashton SV, Moore SE, Walsh FS. Morphoregulatory activities of NCAM and N-cadherin can be accounted for by G protein-dependent activation of L- and N-type neuronal Ca2+ channels. *Cell 67*: 21-33, 1991.

[95] Rønn LC, Dissing S, Holm A, Berezin V, Bock E. Increased intracellular calcium is required for neurite outgrowth induced by a synthetic peptide ligand of NCAM. *FEBS Lett 518*: 60-66, 2002.

[96] Williams EJ, Walsh FS, Doherty P. The production of arachidonic acid can account for calcium channel activation in the second messenger pathway underlying neurite outgrowth stimulated by NCAM, N-cadherin, and L1. *J Neurochem 62*: 1231-1234, 1994.

[97] Kolkova K, Novitskaya V, Pedersen N, Berezin V, Bock E. Neural cell adhesion molecule-stimulated neurite outgrowth depends on activation of protein kinase C and the Ras-mitogen-activated protein kinase pathway. *J Neurosci 20*: 2238-2246, 2000.

[98] Jessen U, Novitskaya V, Pedersen N, Serup P, Berezin V, Bock E. The transcription factors CREB and c-Fos play key roles in NCAM-mediated neuritogenesis in PC12-E2 cells. *J Neurochem 79*: 1149-1160, 2001.

[99] Liu L, Barrett CF, Rittenhouse AR. Arachidonic acid both inhibits and enhances whole cell calcium currents in rat sympathetic neurons. *Am J Physiol Cell Physiol 280*: C1293-1305, 2001.

[100] Dunican DJ, Doherty P. The generation of localized calcium rises mediated by cell adhesion molecules and their role in neuronal growth cone motility. *Mol Cell Biol Res Commun 3*: 255-263, 2000.

[101] Saffell JL, Williams EJ, Mason IJ, Walsh FS, Doherty P. Expression of a dominant negative FGF receptor inhibits axonal growth and FGF receptor phosphorylation stimulated by CAMs. *Neuron 18*: 231-242, 1997.

[102] Piomelli D. Arachidonic acid in cell signaling. *Curr Opin Cell Biol 5*: 274-280, 1993.

[103] Lodish HF, Berk A, Matsudaira P, Kaiser CA, Krieger M, Scott MP, Zipursky SL, Darnell J. *Signaling pathways that control gene activity*. In: Molecular Cell Biology, 5th ed, W. H. Freeman and Co, NY, pp571-610, 2003.

[104] Archer FR, Doherty P, Collins D, Bolsover SR. CAMs and FGF cause a local submembrane calcium signal promoting axon outgrowth without a rise in bulk calcium concentration. *Eur J Neurosci 11*: 3565-3573, 1999.

[105] Nishizuka Y. Protein kinase C and lipid signaling for sustained cellular responses. *FASEB J 9*: 484-496, 1995.

[106] He Q, Dent EW, Meiri KF. Modulation of actin filament behavior by GAP-43 (neuromodulin) is dependent on the phosphorylation status of serine 41, the protein kinase C site. *J Neurosci 17*: 3515-3524, 1997.

[107] Yamauchi T, Fujisawa H. Evidence for three distinct forms of calmodulin-dependent protein kinases from rat brain. *FEBS Lett 116*: 141-144, 1980.

[108] Yamauchi T, Fujisawa H. Purification and characterization of the brain calmodulin-dependent protein kinase (kinase II), which is involved in the activation of tryptophan 5-monooxygenase. *Eur J Biochem 132*: 15-21, 1983.

[109] Bennett MK, Erondu NE, Kennedy MB. Purification and characterization of a calmodulin-dependent protein kinase that is highly concentrated in brain. *J Biol Chem 258*: 12735-12744, 1983.

[110] Goshima Y, Ohsako S, Yamauchi T. Overexpression of Ca2+/calmodulin-dependent protein kinase II in Neuro2a and NG108-15 neuroblastoma cell lines promotes neurite outgrowth and growth cone motility. *J Neurosci 13*: 559-567, 1993.

[111] Yamauchi T, Yoshimura Y, Nomura T, Fujii M, Sugiura H. Neurite outgrowth of neuroblastoma cells overexpressing alpha and beta isoforms of Ca2+/calmodulin-dependent protein kinase II-effects of protein kinase inhibitors. *Brain Res Brain Res Protoc 2*: 250-258, 1998.

[112] Williams EJ, Mittal B, Walsh FS, Doherty P. A Ca2+/calmodulin kinase inhibitor, KN-62, inhibits neurite outgrowth stimulated by CAMs and FGF. *Mol Cell Neurosci 6*: 69-79, 1995.

[113] Means AR. Regulatory cascades involving calmodulin-dependent protein kinases. *Mol Endocrinol 14*: 4-13, 2000.

[114] Choi J, Krushel LA, Crossin KL. NF-kappaB activation by N-CAM and cytokines in astrocytes is regulated by multiple protein kinases and redox modulation. *Glia 33*: 45-56, 2001.

[115] Hughes K, Edin S, Antonsson A, Grundstrom T. Calmodulin-dependent kinase II mediates T cell receptor/CD3- and phorbol ester-induced activation of IkappaB kinase. *J Biol Chem 276*: 36008-36013, 2001.

[116] Fink CC, Meyer T. Molecular mechanisms of CaMKII activation in neuronal plasticity. *Curr Opin Neurobiol 12*: 293-299, 2002.

[117] Benowitz LI, Routtenberg A. GAP-43: an intrinsic determinant of neuronal development and plasticity. *Trends Neurosci 20*: 84-91, 1997.

[118] Dent EW, Meiri KF. Distribution of phosphorylated GAP-43 (neuromodulin) in growth cones directly reflects growth cone behavior. *J Neurobiol 35*: 287-299, 1998.

[119] He Q, Meiri KF. Isolation and characterization of detergent-resistant microdomains responsive to NCAM-mediated signaling from growth cones. *Mol Cell Neurosci 19*: 18-31, 2002.

[120] Meiri KF, Saffell JL, Walsh FS, Doherty P. Neurite outgrowth stimulated by neural cell adhesion molecules requires growth-associated protein-43 (GAP-43) function and is associated with GAP-43 phosphorylation in growth cones. *J Neurosci 18*: 10429-10437, 1998.

[121] Schmid RS, Graff RD, Schaller MD, Chen S, Schachner M, Hemperly JJ, Maness PF. NCAM stimulates the Ras-MAPK pathway and CREB phosphorylation in neuronal cells. *J Neurobiol 38*: 542-558, 1999.

[122] Perron JC, Bixby JL. Distinct neurite outgrowth signaling pathways converge on ERK activation. *Mol Cell Neurosci 13*: 362-378, 1999.

[123] Klint P, Kanda S, Claesson-Welsh L. Shc and a novel 89-kDa component couple to the Grb2-Sos complex in fibroblast growth factor-2-stimulated cells. *J Biol Chem 270*: 23337-23344, 1995.

[124] Kouhara H, Hadari YR, Spivak-Kroizman T, Schilling J, Bar-Sagi D, Lax I, Schlessinger J. A lipid-anchored Grb2-binding protein that links FGF-receptor activation to the Ras/MAPK signaling pathway. *Cell 89*: 693-702, 1997.

[125] Ong SH, Guy GR, Hadari YR, Laks S, Gotoh N, Schlessinger J, Lax I. FRS2 proteins recruit intracellular signaling pathways by binding to diverse targets on fibroblast growth factor and nerve growth factor receptors. *Mol Cell Biol 20*: 979-989, 2000.

[126] Marais R, Light Y, Mason C, Paterson H, Olson MF, Marshall CJ. Requirement of Ras-GTP-Raf complexes for activation of Raf-1 by protein kinase C. *Science 280*: 109-112, 1998.

[127] Chen HC, Chan PC, Tang MJ, Cheng CH, Chang TJ. Tyrosine phosphorylation of focal adhesion kinase stimulated by hepatocyte growth factor leads to mitogen-activated protein kinase activation. *J Biol Chem 273*: 25777-25782, 1998.

[128] Kolkova K, Pedersen N, Berezin V, Bock E. Identification of an amino acid sequence motif in the cytoplasmic domain of the NCAM-140 kDa isoform essential for its neuritogenic activity. *J Neurochem 75*: 1274-1282, 2000.

[129] Niethammer P, Delling M, Sytnyk V, Dityatev A, Fukami K, Schachner M. Cosignaling of NCAM via lipid rafts and the FGF receptor is required for neuritogenesis. *J Cell Biol 157*: 521-532, 2002.

[130] Hakomori SI. The glycosynapse. *Proc Natl Acad Sci USA 99*: 225-232, 2002.

[131] Simons K, Ikonen E. Functional rafts in cell membranes. *Nature 387*: 569-572, 1997.

[132] Lisanti MP, Scherer PE, Vidugiriene J, Tang Z, Hermanowski-Vosatka A, Tu YH, Cook RF, Sargiacomo M. Characterization of caveolin-rich membrane domains isolated from an endothelial-rich source: implications for human disease. *J Cell Biol 126*: 111-126, 1994.

[133] Mineo C, James GL, Smart EJ, Anderson RGW. Localization of Epidermal Growth Factor-stimulated Ras/Raf-1 Interaction to Caveolae Membrane. *J Biol Chem 271*: 11930-11935, 1996.

[134] Iwabuchi K, Yamamura S, Prinetti A, Handa K, Hakomori SI. GM3-enriched Microdomain Involved in Cell Adhesion and Signal Transduction through Carbohydrate-Carbohydrate Interaction in Mouse Melanoma B16 Cells. *J Biol Chem 273*: 9130-9138, 1998.

[135] Arudchandran R, Brown MJ, Peirce MJ, Song JS, Zhang J, Siraganian RP, Blank U, Rivera J. The Src Homology 2 Domain of Vav Is Required for Its Compartmentation to the Plasma Membrane and Activation of c-Jun NH2-terminal Kinase 1. *J Exp Med 191*: 47-60, 2000.

[136] Maekawa S, Kumanogoh H, Funatsu N, Takei N, Inoue K, Endo Y, Hamada K, Sokawa Y. Identification of NAP-22 and GAP-43 (neuromodulin) as major protein components in a Triton insoluble low density fraction of rat brain. *Biochim Biophys Acta 1323* :1-5, 1997.

[137] Yuyama K, Sekino-Suzuki N, Sanai Y, Kasahara K. Lipid rafts in cellular signaling and disease. *Trend Glycosci Glycotechnol 15*: 139-151, 2003.

[138] Krämer EM, Klein C, Koch T, Boytinck M, Trotter J. Compartmentation of Fyn kinase with glycosylphosphatidylinositol-anchored molecules in oligodendrocytes facilitates kinase activation during myelination. *J Biol Chem 274*: 29042-29049, 1999.

[139] Tsui-Pierchala BA, Encinas M, Milbrandt J, Johnson EM Jr. Lipid rafts in neuronal signaling and function. *Trends Neurosci 25*: 412-417, 2002.

[140] Walsh FS, Doherty P. Neural cell adhesion molecules of the immunoglobulin superfamily: role in axon growth and guidance. *Annu Rev Cell Dev Biol 13*: 425-456, 1997.

[141] Langley K, Gratzl M. Neural cell adhesion molecule NCAM in neural and endocrine cells. In: Gratzl M, Langley K (eds) *Markers for Neural and Endocrine Cells: Molecular and Cell Biology, Diagnotic Applications*. John Wiley & Sons Inc., Weinheim, German, pp 133-178, 1991.

[142] Filiz S, Dalcik H, Yardimoglu M, Gonca S, Ceylan S. Localization of neural cell adhesion molecule (N-CAM) immunoreactivity in adult rat tissues. *Biotech Histochem 77*: 127-135, 2002.

[143] Lanier LL, Chang C, Azuma M, Ruitenberg JJ, Hemperly JJ, Phillips JH. Molecular and functional analysis of human natural killer cell-associated neural cell adhesion molecule (N-CAM/CD56). *J Immunol 146*: 4421-4426, 1991.

[144] Nakatani K, Kaneda K, Seki S, Nakajima Y. Pit cells as liver-associated natural killer cells: morphology and function. *Med Electron Microsc 37*: 29-36, 2004.

[145] Grumet M, Rutishauser U, Edelman GM. Neural cell adhesion molecule is on embryonic muscle cells and mediates adhesion to nerve cells in vitro. *Nature 295*: 693-695, 1982.

[146] Covault J, Sanes JR. Distribution of N-CAM in synaptic and extrasynaptic portions of developing and adult skeletal muscle. *J Cell Biol 102*: 716-730, 1986.

[147] Covault J, Merlie JP, Goridis C, Sanes JR. Molecular forms of N-CAM and its RNA in developing and denervated skeletal muscle. *J Cell Biol 102*: 731-739, 1986.

[148] Cashman NR, Covault J, Wollman RL, Sanes JR. Neural cell adhesion molecule in normal, denervated, and myopathic human muscle. *Ann Neurol 21*: 481-489, 1987.

[149] Gosztonyi G, Naschold U, Grozdanovic Z, Stoltenburg-Didinger G, Gossrau R. Expression of Leu-19 (CD56, N-CAM) and nitric oxide synthase (NOS) I in denervated and reinnervated human skeletal muscle. *Microsc Res Tech 55*: 187-97, 2001.

[150] Andersson AM, Olsen M, Zhernosekov D, Gaardsvoll H, Krog L, Linnemann D, Bock E. Age-related changes in expression of the neural cell adhesion molecule in skeletal muscle: a comparative study of newborn, adult and aged rats. *Biochem J 290:* 641-648, 1993.

[151] Wharton J, Gordon L, Walsh FS, Flanigan TP, Moore SE, Polak JM. Neural cell adhesion molecule (N-CAM) expression during cardiac development in the rat. *Brain Res 483*: 170-176, 1989.

[152] al-Mahdawi S, Shallal A, Wyse RK. Neural cell adhesion molecule (N-CAM) in fetal and mature human heart. *FEBS Lett 267*: 183-185, 1990.

[153] Gordon L, Wharton J, Moore SE, Walsh FS, Moscoso JG, Penketh R, Wallwork J, Taylor KM, Yacoub MH, Polak JM. Myocardial localization and isoforms of neural cell adhesion molecule (N-CAM) in the developing and transplanted human heart. *J Clin Invest 86*: 1293-1300, 1990.

[154] Gordon L, Wharton J, Moore SE, Flanigan TP, Gulbenkian S, Walsh FS, David-Ferreira JF, Winter RJ, Polak JM. Expression of neural cell adhesion molecule immunoreactivity in hypertrophic myocardium. *Life Sci 47*: 601-609, 1990.

[155] Gattenlohner S, Waller C, Ertl G, Bultmann BD, Muller-Hermelink HK, Marx A. NCAM(CD56) and RUNX1(AML1) are up-regulated in human ischemic cardiomyopathy and a rat model of chronic cardiac ischemia. *Am J Pathol 163*: 1081-1090, 2003.

[156] Akeson RA, Wujek JR, Roe S, Warren SL, Small SJ. Smooth muscle cells transiently express NCAM. *Brain Res 464*: 107-120, 1988.

[157] Gulbenkian S, Santos J, Gordon L, Wharton J, Polak JM, David-Ferreira JF. Neural cell adhesion molecule is expressed by smooth muscle cells during the development of the rat vascular system. *J Neurocytol 18*: 809-817, 1989.

[158] Probstmeier R, Martini R, Tacke R, Schachner M. Expression of the adhesion molecules L1, N-CAM and J1/tenascin during development of the murine small intestine. *Differentiation 44*: 42-55, 1990.

[159] Romanska HM, Bishop AE, Moscoso G, Walsh FS, Spitz L, Brereton RJ, Polak JM. Neural cell adhesion molecule (NCAM) expression in nerves and muscle of developing human large bowel. *J Pediatr Gastroenterol Nutr 22*: 351-358, 1996.

[160] Langley OK, Aletsee MC, Gratzl M. Endocrine cells share expression of N-CAM with neurones. *FEBS Lett 220*: 108-112, 1987.

[161] Lahr G, Mayerhofer A, Bucher S, Barthels D, Wille W, Gratzl M. Neural cell adhesion molecules in rat endocrine tissues and tumor cells: distribution and molecular analysis. *Endocrinology 132*: 1207-1217, 1993.

[162] Langley OK, Aletsee-Ufrecht MC, Grant NJ, Gratzl M. Expression of the neural cell adhesion molecule NCAM in endocrine cells. *J Histochem Cytochem 37*: 781-791, 1989.

[163] Jin L, Hemperly JJ, Lloyd RV. Expression of neural cell adhesion molecule in normal and neoplastic human neuroendocrine tissues. *Am J Pathol 138*: 961-969, 1991.

[164] Esni F, Taljedal IB, Perl AK, Cremer H, Christofori G, Semb H. Neural cell adhesion molecule (N-CAM) is required for cell type segregation and normal ultrastructure in pancreatic islets. *J Cell Biol 144*: 325-337, 1999.

[165] Mayerhofer A, Seidl K, Lahr G, Bitter-Suermann D, Christoph A, Barthels D, Wille W, Gratzl M. Leydig cells express neural cell adhesion molecules in vivo and in vitro. *Biol Reprod 47*: 656-664, 1992.

[166] Gatchalian CL, Schachner M, Sanes JR. Fibroblasts that proliferate near denervated synaptic sites in skeletal muscle synthesize the adhesive molecules tenascin(J1), N-CAM, fibronectin, and a heparan sulfate proteoglycan. *J Cell Biol 108*: 1873-1890, 1989.

[167] Chuong CM, Chen HM. Enhanced expression of neural cell adhesion molecules and tenascin (cytotactin) during wound healing. *Am J Pathol 138*: 427-440, 1991.

[168] Chakravortty D, Kumar KS. Interaction of lipopolysaccharide with human small intestinal lamina propria fibroblasts favors neutrophil migration and peripheral blood mononuclear cell adhesion by the production of proinflammatory mediators and adhesion molecules. *Biochim Biophys Acta 1453*: 261-272, 1999.

[169] Bioulac-Sage P, Lafon ME, Saric J, Balabaud C. Nerves and perisinusoidal cells in human liver. *J Hepatol 10*: 105-112, 1990.

[170] Whiteside TL, Heo DS, Takagi S, Johnson JT, Iwatsuki S, Herberman RB. Cytolytic antitumor effector cells in long-term cultures of human tumor-infiltrating lymphocytes in recombinant interleukin 2. *Cancer Immunol Immunother 26*: 1-10, 1988.

[171] Roskams T, van den Oord JJ, De Vos R, Desmet VJ. Neuroendocrine features of reactive bile ductules in cholestatic liver disease. *Am J Pathol 137*: 1019-1025, 1990.

[172] Roskams T, De Vos R, van den Oord JJ, Desmet V. Cells with neuroendocrine features in regenerating human liver. *APMIS Suppl 23*: 32-39, 1991.

[173] Libbrecht L, Cassiman D, Desmet V, Roskams T. Expression of neural cell adhesion molecule in human liver development and in congenital and acquired liver diseases. *Histochem Cell Biol 116*: 233-239, 2001.

[174] Torbenson M, Wang J, Abraham S, Maitra A, Boitnott J. Bile ducts and ductules are positive for CD56 (N-CAM) in most cases of extrahepatic biliary atresia. *Am J Surg Pathol 27*: 1454-1457, 2003.

[175] Fabris L, Strazzabosco M, Crosby HA, Ballardini G, Hubscher SG, Kelly DA, Neuberger JM, Strain AJ, Joplin R. Characterization and isolation of ductular cells coexpressing neural cell adhesion molecule and Bcl-2 from primary cholangiopathies and ductal plate malformations. *Am J Pathol 156*: 1599-15612, 2000.

[176] Tiniakos DG, Lee JA, Burt AD. Innervation of the liver: morphology and function. *Liver 16*: 151-160, 1996.

[177] Friedman MI. Hepatic nerve function. In: Arias IM, Jakoby WB, Popper H, Schachter D, Shafritz DA (eds), *The Liver: Biology and Pathobiology*, 2nd ed, Raven Press Ltd, NY, pp 949-959, 1988.

[178] Scoazec JY, Racine L, Couvelard A, Moreau A, Flejou JF, Bernuau D, Feldmann G. Parenchymal innervation of normal and cirrhotic human liver: a light and electron microscopic study using monoclonal antibodies against the neural cell-adhesion molecule. *J Histochem Cytochem 41*: 899-908, 1993.

[179] Jaskiewicz K, Voigt MD, Robson SC. Distribution of hepatic nerve fibers in liver diseases. *Digestion 55*: 247-252, 1994.

[180] Kanda N, Fukuda Y, Imoto M, Koyama Y, Nakano I, Urano F. Localization of synaptophysin immunoreactivity in the human liver. *Scand J Gastroenterol 29*: 275-279, 1994.

[181] Ueno T, Tanikawa K. Intralobular innervation and lipocyte contractility in the liver. *Nutrition 13*: 141-148, 1996.

[182] Ueno T, Sata M, Sakata R, Torimura T, Sakamoto M, Sugawara H, Tanikawa K. Hepatic stellate cells and intralobular innervation in human liver cirrhosis. *Hum Pathol 28*: 953-959, 1997.

[183] Kobayashi S, Seki S, Kawada N, Morikawa H, Nakatani K, Uyama N, Ikeda K, Nakajima Y, Arakawa T, Kaneda K. Apoptosis of T cells in the hepatic fibrotic tissue of the rat: a possible inducing role of hepatic myofibroblast-like cells. *Cell Tissue Res 311:* 353-364, 2003.

[184] Knittel T, Dinter C, Kobold D, Neubauer K, Mehde M, Eichhorst S, Ramadori G. Expression and regulation of cell adhesion molecules by hepatic stellate cells (HSC) of rat liver: involvement of HSC in recruitment of inflammatory cells during hepatic tissue repair. *Am J Pathol 154*: 153-167, 1999.

[185] Edvardsen K, Chen W, Rucklidge G, Walsh FS, Obrink B, Bock E. Transmembrane neural cell-adhesion molecule (NCAM), but not glycosyl-phosphatidylinositol-anchored NCAM, down-regulates secretion of matrix metalloproteinases. *Proc Natl Acad Sci USA 90*: 11463-11467, 1993.

[186] Akiyoshi H, Terada T. Mast cell, myofibroblast and nerve terminal complexes in carbon tetrachloride-induced cirrhotic rat livers. *J Hepatol 29*: 112-119, 1998.

[187] Pinzani M, Marra F. Cytokine receptors and signaling in hepatic stellate cells. *Semin Liver Dis 21*: 397-416, 2001.

[188] Okuyama H, Shimahara Y, Kawada N. The hepatic stellate cell in the post-genomic era. *Histol Histopathol 17*: 487-495, 2002.

[189] Yamada E, and Hirosawa K. The possible existence of a vitamin A-storing cell system. *Cell Struct Funct 1*: 201-204, 1976.

[190] Kusumoto Y, and Fujita T. Vitamin A uptake cells distributed in the liver and other organs of the rat. *Arch Histol Jap 40*: 121-136, 1977.

[191] Nagy NE, Holven KB, Roos N, Senoo H, Kojima N, Norum KR, and Blomhoff R. Storage of vitamin A in extrahepatic stellate cells in normal rats. *J Lipid Res 38*: 645-658, 1997.

[192] Wake K. The sternzellen of von Kupffer – after 106 years. In: Knook DL, and Wisse E, editors. Sinusoidal liver cells, the Netherlands: *Elsevier Biomedical Press, 1-12*, 1982.

[193] Wake K. One hundred years of sinusoidal cells in the liver. *Acta Anat Nippon 72*: 407-423, 1997.

[194] Apte MV, Haber PS, Applegate TL, Norton ID, McCaughan GW, Korster MA, Pirola RC, and Wilson JS. Periacinar stellate shaped cells in rat pancreas: identification, isolation, and culture. *Gut 43*: 128-133, 1998.

[195] Bachem MG, Schneider E, Gross H, Weidenbach H, Schmid RM, Menke A, Siech M, Beger H, Grunert A, and Adler G. Identification, culture, and characterization of pancreatic stellate cells in rats and humans. *Gastroenterology 115*: 421-432, 1998.

[196] Haber PS, Keogh GW, Apte MV, Moran CS, Stewart NL, Crawford DH, Pirola RC, McCaughan GW, Ramm GA, and Wilson JS. Activation of pancreatic stellate cells in human and experimental pancreatic fibrosis. *Am J Pathol 155*: 1087-1095, 1999.

[197] Termine JD, Kleinman HK, Whitson SW, Conn KM, McGarvey ML, Martin GR. Osteonectin, a bone-specific protein linking mineral to collagen. *Cell 26*: 99-105, 1981.

[198] Mason IJ, Taylor A, Williams JG, Sage H, Hogan BL. Evidence from molecular cloning that SPARC, a major product of mouse embryo parietal endoderm, is related to an endothelial cell 'culture shock' glycoprotein of Mr 43,000. *EMBO J 5*: 1465-1472, 1986.

[199] Mann K, Deutzmann R, Paulsson M, Timpl R. Solubilization of protein BM-40 from a basement membrane tumor with chelating agents and evidence for its identity with osteonectin and SPARC. *FEBS Lett 218*: 167-172, 1987.

[200] Bornstein P. Diversity of function is inherent in matricellular proteins: an appraisal of thrombospondin 1. *J Cell Biol 130*: 503-506, 1995.

[201] Yan Q, Sage EH. SPARC, a matricellular glycoprotein with important biological functions. *J Histochem Cytochem 47*: 1495-1506, 1999.

[202] Brekken RA, Sage EH. SPARC, a matricellular protein: at the crossroads of cell-matrix. *Matrix Biol 19*: 816-827, 2000.

[203] Sage H, Johnson C, Bornstein P. Characterization of a novel serum albumin-binding glycoprotein secreted by endothelial cells in culture. *J Biol Chem 259*: 3993-4007, 1984.

[204] Romberg RW, Werness PG, Lollar P, Riggs BL, Mann KG. Isolation and characterization of native adult osteonectin. *J Biol Chem 260*: 2728-2736, 1985.

[205] Maurer P, Mayer U, Bruch M, Jeno P, Mann K, Landwehr R, Engel J, Timpl R. High-affinity and low-affinity calcium binding and stability of the multidomain extracellular 40-kDa basement membrane glycoprotein (BM-40/SPARC/osteonectin). *Eur J Biochem 205*: 233-240, 1992.

[206] Vernon RB, Sage H. The calcium-binding protein SPARC is secreted by Leydig and Sertoli cells of the adult mouse testis. *Biol Reprod 40*: 1329-1340, 1989.

[207] Funk SE, Sage EH. Differential effects of SPARC and cationic SPARC peptides on DNA synthesis by endothelial cells and fibroblasts. *J Cell Physiol 154*: 53-63, 1993.

[208] Lane TF, Iruela-Arispe ML, Johnson RS, Sage EH. SPARC is a source of copper-binding peptides that stimulate angiogenesis. *J Cell Biol 125*: 929-943, 1994.

[209] Hohenester E, Maurer P, Hohenadl C, Timpl R, Jansonius JN, Engel J. Structure of a novel extracellular Ca(2+)-binding module in BM-40. *Nat Struct Biol 3*: 67-73, 1996.

[210] Mayer U, Aumailley M, Mann K, Timpl R, Engel J. Calcium-dependent binding of basement membrane protein BM-40 (osteonectin, SPARC) to basement membrane collagen type IV. *Eur J Biochem 198*: 141-150, 1991.

[211] Sasaki T, Hohenester E, Gohring W, Timpl R. Crystal structure and mapping by site-directed mutagenesis of the collagen-binding epitope of an activated form of BM-40/SPARC/osteonectin. *EMBO J 17*: 1625-1634, 1998.

[212] Nomura S, Hashmi S, McVey JH, Ham J, Parker M, Hogan BL. Evidence for positive and negative regulatory elements in the 5'-flanking sequence of the mouse sparc (osteonectin) gene. *J Biol Chem 264*: 12201-12207, 1989.

[213] Neri M, Descalzi-Cancedda F, Cancedda R. Heat-shock response in cultured chick embryo chondrocytes. Osteonectin is a secreted heat-shock protein. *Eur J Biochem 205*: 569-574, 1992.

[214] Lane TF, Sage EH. The biology of SPARC, a protein that modulates cell-matrix interactions. *FASEB J 8*: 163-173, 1994.

[215] Wrana JL, Overall CM, Sodek J. Regulation of the expression of a secreted acidic protein rich in cysteine (SPARC) in human fibroblasts by transforming growth factor beta. Comparison of transcriptional and post-transcriptional control with fibronectin and type I collagen. *Eur J Biochem 197*: 519-528, 1991.

[216] Nakamura S, Kamihagi K, Satakeda H, Katayama M, Pan H, Okamoto H, Noshiro M, Takahashi K, Yoshihara Y, Shimmei M, Okada Y, Kato Y. Enhancement of SPARC (osteonectin) synthesis in arthritic cartilage. Increased levels in synovial fluids from patients with rheumatoid arthritis and regulation by growth factors and cytokines in chondrocyte cultures. *Arthritis Rheum 39*: 539-551, 1996.

[217] Chandrasekhar S, Harvey AK, Johnson MG, Becker GW. Osteonectin/SPARC is a product of articular chondrocytes/cartilage and is regulated by cytokines and growth factors. *Biochim Biophys Acta 1221*: 7-14, 1994.

[218] Delany AM, Canalis E. Basic fibroblast growth factor destabilizes osteonectin mRNA in osteoblasts. *Am J Physiol 274*: C734-740, 1998.

[219] Shiba H, Fujita T, Doi N, Nakamura S, Nakanishi K, Takemoto T, Hino T, Noshiro M, Kawamoto T, Kurihara H, Kato Y. Differential effects of various growth factors and cytokines on the syntheses of DNA, type I collagen, laminin, fibronectin, osteonectin/secreted protein, acidic and rich in cysteine (SPARC), and alkaline phosphatase by human pulp cells in culture. *J Cell Physiol 174*: 194-205, 1998.

[220] Clezardin P, Malaval L, Ehrensperger AS, Delmas PD, Dechavanne M, McGregor JL. Complex formation of human thrombospondin with osteonectin. *Eur J Biochem 175*: 275-284, 1988.

[221] Rosenblatt S, Bassuk JA, Alpers CE, Sage EH, Timpl R, Preissner KT. Differential modulation of cell adhesion by interaction between adhesive and counter-adhesive proteins: characterization of the binding of vitronectin to osteonectin (BM40, SPARC). *Biochem J 324*: 311-319, 1997.

[222] Sage H, Vernon RB, Funk SE, Everitt EA, Angello J. SPARC, a secreted protein associated with cellular proliferation, inhibits cell spreading in vitro and exhibits Ca+2-dependent binding to the extracellular matrix. *J Cell Biol 109*: 341-356, 1989.

[223] Bradshaw AD, Sage EH. SPARC, a matricellular protein that functions in cellular differentiation and tissue response to injury. *J Clin Invest 107*: 1049-1054, 2001.

[224] Raines EW, Lane TF, Iruela-Arispe ML, Ross R, Sage EH. The extracellular glycoprotein SPARC interacts with platelet-derived growth factor (PDGF)-AB and -BB and inhibits the binding of PDGF to its receptors. *Proc Natl Acad Sci USA 89*: 1281-1285, 1992.

[225] Pichler RH, Bassuk JA, Hugo C, Reed MJ, Eng E, Gordon KL, Pippin J, Alpers CE, Couser WG, Sage EH, Johnson RJ. SPARC is expressed by mesangial cells in experimental mesangial proliferative nephritis and inhibits platelet-derived-growth-factor-medicated mesangial cell proliferation in vitro. *Am J Pathol 148*: 1153-1167, 1996.

[226] Thomas KA. Vascular endothelial growth factor, a potent and selective angiogenic agent. *J Biol Chem 271*: 603-606, 1996.

[227] Kupprion C, Motamed K, Sage EH. SPARC (BM-40, osteonectin) inhibits the mitogenic effect of vascular endothelial growth factor on microvascular endothelial cells. *J Biol Chem 273*: 29635-29640, 1998.

[228] Hasselaar P, Sage EH. SPARC antagonizes the effect of basic fibroblast growth factor on the migration of bovine aortic endothelial cells. *J Cell Biochem 49*: 272-283, 1992.

[229] Francki A, Bradshaw AD, Bassuk JA, Howe CC, Couser WG, Sage EH. SPARC regulates the expression of collagen type I and transforming growth factor-beta1 in mesangial cells. *J Biol Chem 274*: 32145-32152, 1999.

[230] Sodek J, Zhu B, Huynh MH, Brown TJ, Ringuette M. Novel functions of the matricellular proteins osteopontin and osteonectin/SPARC. *Connect Tissue Res 43*: 308-319, 2002.

[231] Gooden MD, Vernon RB, Bassuk JA, Sage EH. Cell cycle-dependent nuclear location of the matricellular protein SPARC: association with the nuclear matrix. *J Cell Biochem 74*: 152-167, 1999.

[232] Funk SE, Sage EH. The Ca2(+)-binding glycoprotein SPARC modulates cell cycle progression in bovine aortic endothelial cells. *Proc Natl Acad Sci USA 88*: 2648-2652, 1991.

[233] Funk SE, Sage EH. Differential effects of SPARC and cationic SPARC peptides on DNA synthesis by endothelial cells and fibroblasts. *J Cell Physiol 154*: 53-63, 1993.

[234] Bradshaw AD, Francki A, Motamed K, Howe C, Sage EH. Primary mesenchymal cells isolated from SPARC-null mice exhibit altered morphology and rates of proliferation. *Mol Biol Cell 10*: 1569-1579, 1999.

[235] Motamed K, Sage EH. SPARC inhibits endothelial cell adhesion but not proliferation through a tyrosine phosphorylation-dependent pathway. *J Cell Biochem 70*: 543-552, 1998.

[236] Murphy-Ullrich JE, Lane TF, Pallero MA, Sage EH. SPARC mediates focal adhesion disassembly in endothelial cells through a follistatin-like region and the Ca(2+)-binding EF-hand. *J Cell Biochem*. 1995 Feb;57(2):341-50, 1995.

[237] Basu A, Kligman LH, Samulewicz SJ, Howe CC. Impaired wound healing in mice deficient in a matricellular protein SPARC (osteonectin, BM-40*). BMC Cell Biol 2*: 15, 2001.

[238] Bradshaw AD, Reed MJ, Sage EH. SPARC-null mice exhibit accelerated cutaneous wound closure. J *Histochem Cytochem 50*: 1-10, 2002.

[239] Bradshaw AD, Puolakkainen P, Dasgupta J, Davidson JM, Wight TN, Helene Sage E. SPARC-null mice display abnormalities in the dermis characterized by decreased collagen fibril diameter and reduced tensile strength. *J Invest Dermatol 120*: 949-955, 2003.

[240] Puolakkainen P, Bradshaw AD, Kyriakides TR, Reed M, Brekken R, Wight T, Bornstein P, Ratner B, Sage EH. Compromised production of extracellular matrix in mice lacking secreted protein, acidic and rich in cysteine (SPARC) leads to a reduced foreign body reaction to implanted biomaterials. *Am J Pathol 162*: 627-635, 2003.

[241] Francki A, Sage EH. SPARC and the kidney glomerulus: matricellular proteins exhibit diverse functions under normal and pathological conditions. *Trends Cardiovasc Med 11*: 32-37, 2001.

[242] Nomura S, Wills AJ, Edwards DR, Heath JK, Hogan BL. Developmental expression of 2ar (osteopontin) and SPARC (osteonectin) RNA as revealed by in situ hybridization. *J Cell Biol 106*: 441-450, 1988.

[243] Floege J, Alpers CE, Sage EH, Pritzl P, Gordon K, Johnson RJ, Couser WG. Markers of complement-dependent and complement-independent glomerular visceral epithelial cell injury in vivo. Expression of antiadhesive proteins and cytoskeletal changes. *Lab Invest 67*: 486-497, 1992.

[244] Floege J, Johnson RJ, Alpers CE, Fatemi-Nainie S, Richardson CA, Gordon K, Couser WG. Visceral glomerular epithelial cells can proliferate in vivo and synthesize platelet-derived growth factor B-chain. *Am J Pathol 142:* 637-650, 1993.

[245] Kopp JB, Bianco P, Young MF, Termine JD, Robey PG. Renal tubular epithelial cells express osteonectin in vivo and in vitro. *Kidney Int 41*: 56-64, 1992.

[246] Pichler RH, Hugo C, Shankland SJ, Reed MJ, Bassuk JA, Andoh TF, Lombardi DM, Schwartz SM, Bennett WM, Alpers CE, Sage EH, Johnson RJ, Couser WG. SPARC is expressed in renal interstitial fibrosis and in renal vascular injury. *Kidney Int 50*: 1978-1989, 1996.

[247] Bassuk JA, Pichler R, Rothmier JD, Pippen J, Gordon K, Meek RL, Bradshaw AD, Lombardi D, Strandjord TP, Reed M, Sage EH, Couser WG, Johnson R. Induction of TGF-beta1 by the matricellular protein SPARC in a rat model of glomerulonephritis. *Kidney Int 57*: 117-128, 2000.

[248] Francki A, Motamed K, McClure TD, Kaya M, Murri C, Blake DJ, Carbon JG, Sage EH. SPARC regulates cell cycle progression in mesangial cells via its inhibition of IGF-dependent signaling. *J Cell Biochem 88*: 802-811, 2003.

[249] Walpen S, Beck KF, Eberhardt W, Apel M, Chatterjee PK, Wray GM, Thiemermann C, Pfeilschifter J. Downregulation of SPARC expression is mediated by nitric oxide in rat mesangial cells and during endotoxemia in the rat. *J Am Soc Nephrol 11*: 468-476, 2000.

[250] Frizell E, Liu SL, Abraham A, Ozaki I, Eghbali M, Sage EH, Zern MA. Expression of SPARC in normal and fibrotic livers. *Hepatology 21*: 847-854, 1995.

[251] Inagaki H, Lin KH, Maeda S, Saito T. Osteonectin gene expression in fibrotic liver. *Life Sci 58*: 927-934, 1996.

[252] Blazejewski S, Le Bail B, Boussarie L, Blanc JF, Malaval L, Okubo K, Saric J, Bioulac-Sage P, Rosenbaum J. Osteonectin (SPARC) expression in human liver and in cultured human liver myofibroblasts. *Am J Pathol 151*: 651-657, 1997.

[253] Stolz DB, Ross MA, Salem HM, Mars WM, Michalopoulos GK, Enomoto K. Cationic colloidal silica membrane perturbation as a means of examining changes at the sinusoidal surface during liver regeneration. *Am J Pathol 155*: 1487-1498, 1999.

[254] Lamireau T, Le Bail B, Boussarie L, Fabre M, Vergnes P, Bernard O, Gautier F, Bioulac-Sage P, Rosenbaum J. Expression of collagens type I and IV, osteonectin and transforming growth factor beta-1 (TGFbeta1) in biliary atresia and paucity of intrahepatic bile ducts during infancy. *J Hepatol 31*: 248-255, 1999.

[255] Grenard P, Bresson-Hadni S, El Alaoui S, Chevallier M, Vuitton DA, Ricard-Blum S. Transglutaminase-mediated cross-linking is involved in the stabilization of extracellular matrix in human liver fibrosis. *J Hepatol 35*: 367-375, 2001.

[256] Le Bail B, Faouzi S, Boussarie L, Guirouilh J, Blanc JF, Carles J, Bioulac-Sage P, Balabaud C, Rosenbaum J. Osteonectin/SPARC is overexpressed in human hepatocellular carcinoma. *J Pathol 189*: 46-52, 1999.

[257] Goldenberg D, Ayesh S, Schneider T, Pappo O, Jurim O, Eid A, Fellig Y, Dadon T, Ariel I, de Groot N, Hochberg A, Galun E. Analysis of differentially expressed genes in hepatocellular carcinoma using cDNA arrays. *Mol Carcinog 33*: 113-124, 2002.

[258] Rempel SA, Ge S, Gutierrez JA. SPARC: a potential diagnostic marker of invasive meningiomas. *Clin Cancer Res 5*: 237-241, 1999.

[259] Ledda MF, Adris S, Bravo AI, Kairiyama C, Bover L, Chernajovsky Y, Mordoh J, Podhajcer OL. Suppression of SPARC expression by antisense RNA abrogates the tumorigenicity of human melanoma cells. *Nat Med 3*: 171-176, 1997.

[260] Senoo H. Structure and function of hepatic stellate cells. *Med Electron Microsc 37*: 16-28, 2004.

[261] Carloni V, Romanelli RG, Pinzani M, Laffi G, Gentilini P. Expression and function of integrin receptors for collagen and laminin in cultured human hepatic stellate cells. *Gastroenterology 110*: 1127-1136, 1996.

[262] Ankoma-Sey V, Matli M, Chang KB, Lalazar A, Donner DB, Wong L, Warren RS, Friedman SL. Coordinated induction of VEGF receptors in mesenchymal cell types during rat hepatic wound healing. *Oncogene 17*: 115-121, 1998.

[263] Olaso E, Ikeda K, Eng FJ, Xu L, Wang LH, Lin HC, Friedman SL. DDR2 receptor promotes MMP-2-mediated proliferation and invasion by hepatic stellate cells. *J Clin Invest 108*: 1369-1378, 2001.

[264] Vogel W, Gish GD, Alves F, Pawson T. The discoidin domain receptor tyrosine kinases are activated by collagen. *Mol Cell 1:* 13-23, 1997.

[265] Ikeda K, Wang LH, Torres R, Zhao H, Olaso E, Eng FJ, Labrador P, Klein R, Lovett D, Yancopoulos GD, Friedman SL, Lin HC. Discoidin domain receptor 2 interacts with Src and Shc following its activation by type I collagen. *J Biol Chem 277*: 19206-19212, 2002.

[266] Mao TK, Kimura Y, Kenny TP, Branchi A, Gishi RG, Van de Water J, Kung HJ, Friedman SL, Gershwin ME. Elevated expression of tyrosine kinase DDR2 in primary biliary cirrhosis. *Autoimmunity 35*: 521-529, 2002.

In: New Cell Adhesion Research
Editors: Patrick Nott and Matthew Temple

ISBN 978-1-60692-378-8

Chapter X

EPITHELIAL CELL ADHESION MOLECULE EPCAM: PAST, PRESENT, AND FUTURE

Olivier Gires, Dorothea Maetzel and Markus Munz
The Clinical Cooperation Group Molecular Oncology, Department of Head and Neck Research, Ludwig-Maximilians-University Munich and GSF-National Research Center for Environment and Health, Germany

ABSTRACT

This chapter will focus on current findings on the epithelial cell adhesion molecule EpCAM, recapitulate past achievements, and eventually look into the future of research on EpCAM and therapeutic potentialities.

1. EPCAM ON ITS JOURNEY FROM DISCOVERY TO CLINICAL APPLICATION

Human EpCAM is a transmembrane glycoprotein that was amongst the first tumour-associated antigens discovered in the late 1970s with the emerging monoclonal antibody technology (Herlyn et al., 1979). *Koprowski* and colleagues fused murine myeloma cells with splenocytes from mice immunized with human colon carcinoma cells. The resulting hybridoma cell lines were selected on the basis of their capacity to secrete immunoglobulins exclusively specific for colon carcinoma cells (Herlyn et al., 1979). One antibody termed GA733 displayed anti-tumoural activity (Herlyn et al., 1984) and hence was of great interest to oncologists. In the 1980s both the protein and the cognate DNA sequence of the GA733 antigen were circumstantially described (Linnenbach et al., 1989; Ross et al., 1986). From thereon, GA733, which now is referred to as EpCAM (see section 3), gained increasing attention as a potential therapeutic target for the treatment of cancer. Various therapeutic strategies make use of a plethora of antibodies (Braun et al., 1999; Morecki et al., 2006; Riethmüller et al., 1994; Ruf and Lindhofer, 2001) and antibody-like molecules (Schanzer et al., 2006; Schlereth et al., 2005; Schlereth et al., 2006; Wolf et al., 2005) to target EpCAM since it is heavily overexpressed in a variety of human malignancies (Armstrong and Eck, 2003; Balzar et al., 1999b; Went et al., 2006; Went et al., 2004). Vaccination protocols with recombinant EpCAM protein have been successful and induced a Th1-biased immune response in colorectal carcinoma patients (Mellstedt et al., 2000; Mosolits et al., 2004; Ullenhag et al., 2003). Although being the first and largest entity to be targeted more than ten years ago, colon carcinoma is not the only human malignancy included in encouraging clinical trials related to EpCAM since then. Cervical, ovarian, breast, and head and neck are to be mentioned.

To summarise, EpCAM can be considered one of the most frequently overexpressed tumour-associated antigens in a great variety of cancer entities (Went et al., 2006), which is highly immunogenic, a prognostic factor, and moreover a valuable target for immunotherapy.

2. GENE STRUCTURE AND REGULATION

The *epcam* gene (GA733-2, PubMed entry NM_002354, TACSTD1) is composed of nine coding exons (see Figure 1) and has an estimated size of 14 kilobases (kb). Exon 1 encodes a signal peptide, which is indispensable for membrane targeting of the protein product. Exon 2-6 encode a large extracellular domain, exon 7 a single transmembrane domain, and exon 8-9 a small peptide facing the cytoplasm. In 2001 the genomic locus of the *epcam* gene was revisited, which led to its definite assignment to chromosome 2p21 (Calabrese et al., 2001).

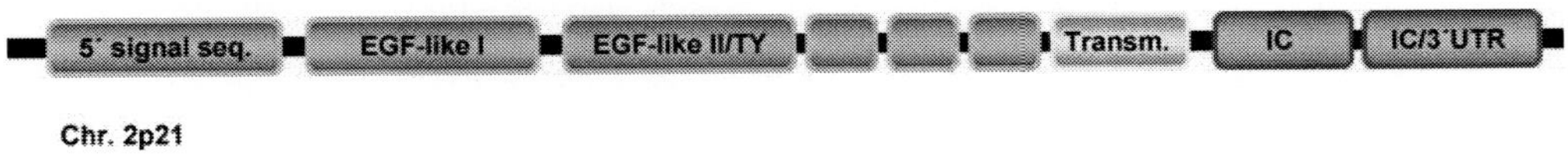

Figure 1. Schematic representation of the genomic structure of the *epcam* gene. *Epcam* maps to chromosome 2p21 and is composed of nine exons including 5′ and 3′ untranslated regions (UTR). The gene encodes two EGF-like repeats, a transmembrane domain (Transm.), and an intracellular domain (IC).

It is worth to notice that the localization of *epcam* to chromosome 4 has been reported (Balzar et al., 1999b; Linnenbach et al., 1993) but is nonetheless incorrect. A second gene termed GA733-1 shares a 50% homology with *epcam/GA733-2* and localizes on chromosome 1. *GA733-1* is an intronless gene whose promoter sequences share no significant homology with the promoter region of *epcam*. Supposedly, *GA733-1* arose from the retroposition of an *epcam* mRNA intermediate into chromosome 1, a process which was additionally accompanied by exon shuffling (Linnenbach et al., 1993). As no conclusive functional homology between EpCAM and GA733-1 was deciphered, it is a consensus to term GA733-2 EpCAM and GA733-1 Trop-2 (Baeuerle and Gires, 2006, *BJC manuscript in press*)(Fornaro et al., 1995).

Molecular mechanisms underlying the regulation of *epcam* transcription are still poorly understood, albeit they are of utmost interest in the light of the abovementioned overexpression of EpCAM in several human malignancies. Efforts toward the understanding of *epcam* gene regulation covered the cloning and characterization of its promoter region. DNA sequences ranging from 1.1kb to 3.4 kb upstream of the transcription start were isolated and revealed rich in GC residues, lacked a TATA or CAAT box, but comprised regulatory elements (Gires et al., 2003; Gires et al., 2001; Gires et al., 2004; McLaughlin et al., 2004). These elements confer a transcriptional specificity in EpCAM positive cells, which can even be of use for targeted expression of therapeutic genes (Gires et al., 2004; McLaughlin et al., 2004). Yet no insights into the identity of transcription factors which bind and regulate the *epcam* promoter were gained so far. Other approaches assessed the influence of promoter methylation at cytosine residues on the expression level of EpCAM in cell lines and in primary breast carcinoma. Promoter methylation at distinct CpG islands within the promoter was detectable in breast carcinoma cell lines and nicely correlated with a reduced presence of EpCAM at the plasma membrane (Spizzo et al., 2006a). Treatment of these cells with demethylating agents resulted in a heightened EpCAM expression and confirmed the latter data. Intriguingly, no correlation was observed between the level of EpCAM expression and the methylation status in primary breast cancer cells (Spizzo et al., 2006a). Hence, how EpCAM expression is regulated *in vivo* at the molecular level remains an unsolved question.

3. Protein Nomenclature, Structure/Function, and Expression Pattern

Human EpCAM has a multitude of names amongst them 17-1A, Ep-CAM, EGP-2, HEA125, GA733-2, KSA, EGP34, ESA, TROP-1, and KS1/4. All these different names are synonyms for one and the same protein and have been assigned based on the names of the monoclonal antibody that served to isolate the EpCAM antigen in each case. On the first symposium on EpCAM function and clinical application held in 2006 (*www.epcam-symposium.de*), scientists in the field unanimously agreed to use EpCAM as the standard name for the protein (Baeuerle and Gires, 2006).

Structurally, EpCAM is a type I transmembrane glycoprotein with its N-terminus facing the outer cellular space and the C-terminus directed to the cytoplasm. As is the case for several cell surface receptors, EpCAM displays a modular structure composed of a large extracellular domain of 265 amino acids (aa) including a signal peptide, a single transmembrane part of 23 aa, and a short intracellular module of 26 aa (see Figure 2). Screening of the extracellular domain for sequence homologies with other membrane proteins revealed the presence of two epidermal growth factor (EGF)-like repeats which are typically rich in cystein residues. These two EGF-like repeats reside at amino acids positions 27-59 and 66-135 (Balzar et al., 1999b). A recent review of both cysteine-rich repeats confirmed EGF-like repeat one but disputed the nature of the second repeat, suggesting it rather fits a thyroglobuline-type repeat (TY-repeat) (Baeuerle and Gires, 2006). This reassessment of an assumed EGF-like repeat II to actually represent a TY domain may have deep implications for our understanding of EpCAM`s three-dimensional folding. It clearly has to be envisaged that the most distal part of EpCAM will be in the vicinity of the plasma membrane according to the ever seen 180° protein turn characteristic for all TY-repeats.

EpCAM was described as a homophilic cell adhesion protein, a fact which was decisive for its nomenclature. Litvinov *et al* reported on this cell adhesive function in great details. Briefly, the group described a calcium-independent adhesion mediated by the full-length protein in epithelia (Litvinov et al., 1994a; Litvinov et al., 1994b). Adhesion mediated by EpCAM is rather weak and competes with cadherin-dependent cell-cell interactions and adhesions (Balzar et al., 1999a; Litvinov et al., 1997). Binding of the intracellular domain of EpCAM to the cytoskeleton via an interaction with □-actinin is mandatory for its full functionality (Balzar et al., 1998). Meanwhile it is well accepted that EpCAM transmits cell adhesion features via homophilic interactions, which depend on the EGF/TY-like motifs. EGF-like repeat I is responsible for an intercellular interaction of EpCAM molecules on opposing cells, while EGF-like/TY repeat II is necessary for interplay on the plasma membrane of the same cell (Balzar et al., 2001).

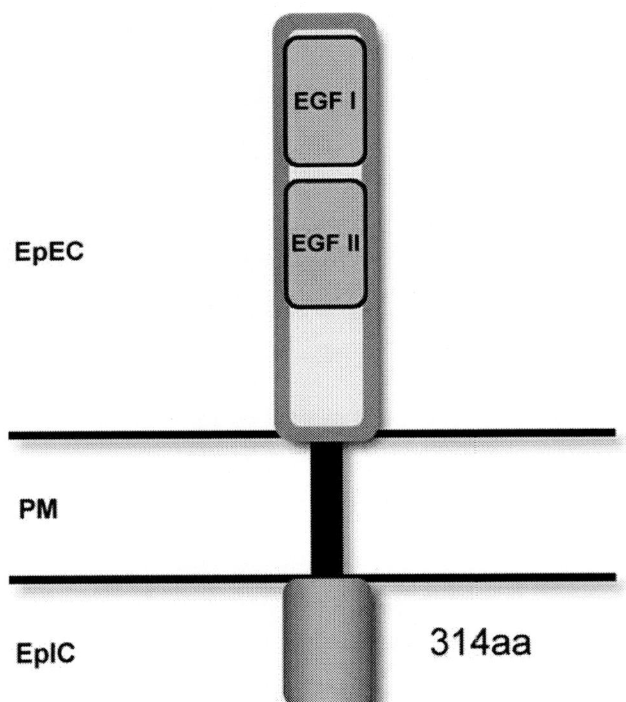

Figure 2. Schematic representation of the EpCAM protein in the plasma membrane. EpCAM is a transmembrane protein composed of a large extracellular domain with two EGF-like repeats EpEC, a single transmembrane domain, a short cytoplasmic tail (EpIC). Full-length EpCAM is 314 amino acids (aa) long. For the sake of simplicity, potential turns occurring due to the TY-repeat were omitted in this scheme.

The expression pattern of EpCAM is heterogeneous: EpCAM is expressed in a variety of simple, glandular, and pseudo-stratified epithelia (Balzar et al., 1999b). In normal adult tissue EpCAM levels may greatly vary. Nonetheless, a number of normal tissues express high levels of EpCAM including gastric and colonic epithelium, cells of the lower respiratory tract, ductal epithelium of the pancreas, and the kidney. It is expressed during embryonic development on various tissues including fetal lung, hepatocytes, kidney, pancreas, skin, germ cells and more, possibly including tumour stem cells (Al-Hajj et al., 2003; Anderson et al., 1999; Balzar et al., 1999b; Schmelzer et al., 2006). In a substantial selection of tissues, cells cease to express EpCAM when they are terminally differentiated, but tissues may re-express the protein if injury occurs and tissue repopulation from more undifferentiated or stem cell populations occurs (de Boer et al., 1999). Hence, EpCAM has characteristics of a differentiation marker, leading to the recent assignment of the cluster of differentiation number CD326. Re-expression of this differentiation marker becomes even more striking when appreciating EpCAM levels in normal squamous epithelia, which are basically devoid of EpCAM, versus squamous cell carcinomas where EpCAM is *de novo* expressed in a large diversity of entities (Quak et al., 1990; Went et al., 2006). The repartition on cells in ordered normal tissue is baso-lateral but may switch to a presence across the entire plasma membrane on culture cells and, most interestingly, on carcinoma cells (Litvinov et al., 1994a).

4. EpCAM as a Therapeutic Target Antigen and Prognostic Marker

Understandably, EpCAM attracted considerable attention as a target antigen for cancer therapy. Its strong overexpression or even *de novo* expression in some carcinoma entities predestinated EpCAM to serve as a target for immunotherapeutics (Armstrong and Eck, 2003). Ten years after discovery the potential of EpCAM-directed therapeutic antibodies (Ab) was pioneered by Riethmüller and coworkers in clinical trials enrolling colon carcinoma Dukes′ C staged patients. The EpCAM-specific monoclonal antibody m17-1A (Panorex®) proved to possess clinical efficacy and to ameliorate overall survival of those patients (n=189) (Riethmuller et al., 1998; Riethmüller et al., 1994). As to now, seven antibodies or antibody-like single chain recombinant molecules are to our knowledge in different clinical phases ranging from small, non-controlled trials to phase II/III and pivotal trials. At present, immunotherapeutics include *Catumaxomab* (bispec. Ab, TrionPharma GmbH), *Adecatumomab* and BiTE® (humanized mAb and engineered single chain bispecific construct, Micromet Inc./Serono), *ING-101* (vaccine for the elicitation of an anti-idiotype reaction, Aphton), *ING-1* (human engineered Ab, Xoma), *Proxinium/Vivendium* (immunotoxin conjugate, Viventia), *EMD273 066* (fusion of humanized KS1/4 to human IL2, Lexigen), and different vaccines developed in the academic field. Hence, forthcoming years are crucial for the conclusive implementation of EpCAM-directed therapies as an adjuvant in a multimodal concept of cancer treatment.

Clinical interventions aim at (*1*) the reduction of ascites in the case of cervical and ovarial carcinomas, and more generally of carcinomas of the abdomen, (*2*) the targeting of minimal residual disease cells as an adjuvant therapy, (*3*) the targeting of solid tumours, albeit being hardly conceivable as a single therapy, and (*4)* the delivery of toxins and radiolabels to $EpCAM^+$ cancer cells.

Besides its role as a therapeutic target, EpCAM advanced to the position of a valuable but negative prognostic marker for node-positive breast cancer, gallbladder cancer, ovarian cancer, and Barrett´s metaplasia (Spizzo et al., 2003; Spizzo et al., 2006b; Spizzo et al., 2004; Varga et al., 2004; Wong et al., 2006). Also, disease-specific variants of EpCAM have been reported for head and neck carcinomas (Pauli et al., 2003), although the significance for its function or repartition is unknown as yet.

5. EpCAM and Signal Transduction

More recent work concentrated on the link of EpCAM to the pathogenesis of carcinomas. This issue was of special interest in order to understand the meaning of EpCAM overexpression in carcinomas. Moreover, a role in proliferation of epithelia was assumed for a long time, although EpCAM was regarded as an "*artless adhesive molecule*". The notion of a growth-promoting capacity was mainly supported by the finding that EpCAM was co-expressed with Ki67, a marker for proliferation, in cervical intraepithelial neoplasia (Litvinov et al., 1996).

Few groups dedicated themselves to the elucidation of EpCAM`s role in carcinogenesis using a panoply of molecular and cell biology technologies. It could be demonstrated that EpCAM is a regulator of cell proliferation. *De novo* expression of EpCAM drives human epithelial cells and murine fibroblasts into proliferation most probably *via* the direct up-regulation of the proto-oncogenes c-Myc and cyclin A/E (Munz et al., 2004). Reversely, the inhibition of EpCAM expression with anti-sense constructs or siRNA oligonucleotides had a strong detrimental effect on cell growth, migration, and invasion of different carcinoma cell lines (Munz et al., 2004; Osta et al., 2004). Hence, EpCAM is not a trivial cell adhesion molecule but a potent oncogene, which signals to the nucleus to regulate target genes involved in cell cycle regulation. These striking features are in accordance with its ability to confer a reduced requirement for growth factors, anchorage-independent growth, and increased proliferation.

Hence, EpCAM raised itself from the status of a "*molecular glue*" present on carcinoma cells to a cell adhesion molecule with mitogenic signalling properties, which influence cell proliferation, migration, invasive capacity, and intracellular pathways in tumours.

Though, ever since a central question remained: *how does EpCAM achieve to transmit regulatory signal to the nucleus?* – an issue recently pronounced in News of the *Journal of the National Cancer Institute* (Beckman, 2006). The revelation of EpCAM`s molecular mechanism of action awaits further attention and should represent a major topic in the near future.

6. Future Perspectives

Certainly, at least three future perspectives are imaginable or yet underway:

- the routine usage of EpCAM as a diagnostic and prognostic factor for the purpose of determining individualised therapy modalities as is the case for breast cancer and Her2/neu status of the patient,

- the application of monoclonal, bispecific, or recombinant antibodies to target EpCAM *in vivo* in the context of an adjuvant or even monotherapy,
- and revealing EpCAM`s molecular background of mitogenic signal transduction.

Great progress was made in the field of diagnostics and prognostics due to solid improvement of the standardisation of staining and scoring procedures. As is the case for EpCAM, Gastl and co-workers adopted an elaborate scoring system which reflects both the intensity of staining and the percentage of positive cells (Gastl et al., 2000; Varga et al., 2004). Additionally, the development of large tissue arrays including thousands of specimens with a well-documented disease history beautifully demonstrated EpCAM`s prognostic value (Went et al., 2006). Future challenge will be the implementation of this knowledge into daily clinical applications in order to adapt potential therapies to the EpCAM-status of the patient. Eventually, adapting and improving cancer therapy guided by EpCAM`s diagnostic and prognostic value would be highly desirable.

However, in the vast majority of cases, EpCAM therapeutics still have to prove their efficacy in phase III clinical trials and obtain approvals by the authorities. It is to be expected that this will happen within the forthcoming two to five years, depending on the evaluated product. Here, the development of humanised antibodies is of great importance to reduce severe side-effects such as human anti-mouse antibody immune reactions (HAMA), which occur when administrating murine antibodies. The deployment of single chain antibody cloning strategies (Laffly and Sodoyer, 2005; Weiner, 2006) and the availability of mouse models for the generation of human antibodies *in vivo* (Lonberg, 2005) revealed decisive in this respect. Last but the not least, the impressive revenues to be expected for supplying companies will be a driving force toward the development of cancer therapeutics, including EpCAM-based strategies.

Finally, the youngest research area which investigates the molecular biology of EpCAM is not only highly relevant to basic science but also possesses a great potential for the development of novel therapeutics. The knowledge of an oncogenic potential of EpCAM opens the possibility to interfere with its growth-promoting signals, *e.g.* using specific small molecule inhibitors, *in vivo* applicable siRNAs, or inhibitory peptides. For this purpose a more capacious knowledge of EpCAM`s signalling features and molecules involved is indispensable.

In summary, a great body of evidence qualifies EpCAM as a major player in the ever-growing collection of useful cancer antigens. The adequacy of the molecule as a diagnostic, prognostic, and therapeutic target is accepted and well documented. Previous shortfalls on the actual function of the molecule in the process of carcinogenesis are now being addressed and should decisively complement the field of basic and clinical research on EpCAM.

REFERENCES

Al-Hajj, M., Wicha, M.S., Benito-Hernandez, A., Morrison, S.J. and Clarke, M.F. (2003) Prospective identification of tumorigenic breast cancer cells. *Proc. Natl. Acad. Sci. U S A*, *100*, 3983-3988.

Anderson, R., Schaible, K., Heasman, J. and Wylie, C. (1999) Expression of the homophilic adhesion molecule, Ep-CAM, in the mammalian germ line. *J. Reprod. Fertil*, *116*, 379-384.

Armstrong, A. and Eck, S.L. (2003) EpCAM: A new therapeutic target for an old cancer antigen. *Cancer Biol. Ther*, *2*, 320-326.

Baeuerle, P.A. and Gires, O. (2006) EpCAM finding its way in cancer. *British Journal of Cancer* (*manuscript accepted*).

Balzar, M., Bakker, H.A., Briaire-de-Bruijn, I.H., Fleuren, G.J., Warnaar, S.O. and Litvinov, S.V. (1998) Cytoplasmic tail regulates the intercellular adhesion function of the epithelial cell adhesion molecule. *Mol. Cell Biol*, *18*, 4833-4843.

Balzar, M., Briaire-de Bruijn, I.H., Rees-Bakker, H.A., Prins, F.A., Helfrich, W., de Leij, L., Riethmuller, G., Alberti, S., Warnaar, S.O., Fleuren, G.J. and Litvinov, S.V. (2001) Epidermal growth factor-like repeats mediate lateral and reciprocal interactions of Ep-CAM molecules in homophilic adhesions. *Mol. Cell Biol*, *21*, 2570-2580.

Balzar, M., Prins, F.A., Bakker, H.A., Fleuren, G.J., Warnaar, S.O. and Litvinov, S.V. (1999a) The structural analysis of adhesions mediated by Ep-CAM. *Exp. Cell Res*, *246*, 108-121.

Balzar, M., Winter, M.J., de Boer, C.J. and Litvinov, S.V. (1999b) The biology of the 17-1A antigen (Ep-CAM). *J. Mol. Med*, *77*, 699-712.

Beckman, M. (2006) CAMs are stopping cancer in its metastatic tracks. *J. Natl. Cancer Inst*, *98*, 576-577.

Braun, S., Hepp, F., Kentenich, C.R., Janni, W., Pantel, K., Riethmuller, G., Willgeroth, F. and Sommer, H.L. (1999) Monoclonal antibody therapy with edrecolomab in breast cancer patients: monitoring of elimination of disseminated cytokeratin-positive tumor cells in bone marrow. *Clin. Cancer Res*, *5*, 3999-4004.

Calabrese, G., Crescenzi, C., Morizio, E., Palka, G., Guerra, E. and Alberti, S. (2001) Assignment of TACSTD1 (alias TROP1, M4S1) to human chromosome 2p21 and refinement of mapping of TACSTD2 (alias TROP2, M1S1) to human chromosome 1p32 by in situ hybridization. *Cytogenet. Cell Genet*, *92*, 164-165.

de Boer, C.J., van Krieken, J.H., Janssen-van Rhijn, C.M. and Litvinov, S.V. (1999) Expression of Ep-CAM in normal, regenerating, metaplastic, and neoplastic liver. *J. Pathol*, *188*, 201-206.

Fornaro, M., Dell'Arciprete, R., Stella, M., Bucci, C., Nutini, M., Capri, M.G. and Alberti, S. (1995) Cloning of the gene encoding Trop-2, a cell-surface glycoprotein expressed by human carcinomas. *Int. J. Cancer*, *62*, 610-618.

Gastl, G., Spizzo, G., Obrist, P., Dunser, M. and Mikuz, G. (2000) Ep-CAM overexpression in breast cancer as a predictor of survival. *Lancet*, *356*, 1981-1982.

Gires, O., Eskofier, S., Lang, S., Zeidler, R. and Munz, M. (2003) Cloning and characterisation of a 1.1 kb fragment of the carcinoma-associated epithelial cell adhesion molecule promoter. *Anticancer Res*, *23*, 3255-3261.

Gires, O., Kieu, C., Fix, P., Schmitt, B., Munz, M., Wollenberg, B. and Zeidler, R. (2001) Tumor necrosis factor alpha negatively regulates the expression of the carcinoma-associated antigen epithelial cell adhesion molecule. *Cancer*, *92*, 620-628.

Gires, O., Pockl, S., Chapman, R.D. and Munz, M. (2004) Targeted gene expression using a 1.1 kilobase promoter fragment of the tumour-associated antigen EpCAM. *Anticancer Res*, *24*, 3715-3721.

Herlyn, D., Herlyn, M., Ross, A.H., Ernst, C., Atkinson, B. and Koprowski, H. (1984) Efficient selection of human tumor growth-inhibiting monoclonal antibodies. *J. Immunol. Methods*, *73*, 157-167.

Herlyn, M., Steplewski, Z., Herlyn, D. and Koprowski, H. (1979) Colorectal carcinoma-specific antigen: detection by means of monoclonal antibodies. *Proc. Natl. Acad. Sci. USA*, *76*, 1438-1442.

Laffly, E. and Sodoyer, R. (2005) Monoclonal and recombinant antibodies, 30 years after. *Hum. Antibodies*, *14*, 33-55.

Linnenbach, A.J., Seng, B.A., Wu, S., Robbins, S., Scollon, M., Pyrc, J.J., Druck, T. and Huebner, K. (1993) Retroposition in a family of carcinoma-associated antigen genes. *Mol. Cell Biol*, *13*, 1507-1515.

Linnenbach, A.J., Wojcierowski, J., Wu, S.A., Pyrc, J.J., Ross, A.H., Dietzschold, B., Speicher, D. and Koprowski, H. (1989) Sequence investigation of the major gastrointestinal tumor-associated antigen gene family, GA733. *Proc. Natl. Acad. Sci. USA*, *86*, 27-31.

Litvinov, S.V., Bakker, H.A., Gourevitch, M.M., Velders, M.P. and Warnaar, S.O. (1994a) Evidence for a role of the epithelial glycoprotein 40 (Ep-CAM) in epithelial cell-cell adhesion. *Cell Adhes. Commun*, *2*, 417-428.

Litvinov, S.V., Balzar, M., Winter, M.J., Bakker, H.A., Briaire-de Bruijn, I.H., Prins, F., Fleuren, G.J. and Warnaar, S.O. (1997) Epithelial cell adhesion molecule (Ep-CAM) modulates cell-cell interactions mediated by classic cadherins. *J. Cell Biol*, *139*, 1337-1348.

Litvinov, S.V., van Driel, W., van Rhijn, C.M., Bakker, H.A., van Krieken, H., Fleuren, G.J. and Warnaar, S.O. (1996) Expression of Ep-CAM in cervical squamous epithelia correlates with an increased proliferation and the disappearance of markers for terminal differentiation. *Am. J. Pathol*, *148*, 865-875.

Litvinov, S.V., Velders, M.P., Bakker, H.A., Fleuren, G.J. and Warnaar, S.O. (1994b) Ep-CAM: a human epithelial antigen is a homophilic cell-cell adhesion molecule. *J. Cell Biol*, *125*, 437-446.

Lonberg, N. (2005) Human antibodies from transgenic animals. *Nat. Biotechnol*, *23*, 1117-1125.

McLaughlin, P.M., Trzpis, M., Kroesen, B.J., Helfrich, W., Terpstra, P., Dokter, W.H., Ruiters, M.H., de Leij, L.F. and Harmsen, M.C. (2004) Use of the EGP-2/Ep-CAM promoter for targeted expression of heterologous genes in carcinoma derived cell lines. *Cancer Gene Ther*, *11*, 603-612.

Mellstedt, H., Fagerberg, J., Frodin, J.E., Hjelm-Skog, A.L., Liljefors, M., Markovic, K., Mosolits, S. and Ragnhammar, P. (2000) Ga733/EpCAM as a target for passive and active specific immunotherapy in patients with colorectal carcinoma. *Ann. N Y. Acad. Sci*, *910*, 254-261; discussion 261-252.

Morecki, S., Lindhofer, H., Yacovlev, E., Gelfand, Y. and Slavin, S. (2006) Use of trifunctional bispecific antibodies to prevent graft versus host disease induced by allogeneic lymphocytes. *Blood, 107*, 1564-1569.

Mosolits, S., Markovic, K., Frodin, J.E., Virving, L., Magnusson, C.G., Steinitz, M., Fagerberg, J. and Mellstedt, H. (2004) Vaccination with Ep-CAM protein or anti-idiotypic antibody induces Th1-biased response against MHC class I- and II-restricted Ep-CAM epitopes in colorectal carcinoma patients. *Clin. Cancer Res, 10*, 5391-5402.

Munz, M., Kieu, C., Mack, B., Schmitt, B., Zeidler, R. and Gires, O. (2004) The carcinoma-associated antigen EpCAM upregulates c-myc and induces cell proliferation. *Oncogene, 23*, 5748-5758.

Osta, W.A., Chen, Y., Mikhitarian, K., Mitas, M., Salem, M., Hannun, Y.A., Cole, D.J. and Gillanders, W.E. (2004) EpCAM is overexpressed in breast cancer and is a potential target for breast cancer gene therapy. *Cancer Res, 64*, 5818-5824.

Pauli, C., Munz, M., Kieu, C., Mack, B., Breinl, P., Wollenberg, B., Lang, S., Zeidler, R. and Gires, O. (2003) Tumor-specific glycosylation of the carcinoma-associated epithelial cell adhesion molecule EpCAM in head and neck carcinomas. *Cancer Lett, 193*, 25-32.

Quak, J.J., Van Dongen, G., Brakkee, J.G., Hayashida, D.J., Balm, A.J., Snow, G.B. and Meijer, C.J. (1990) Production of a monoclonal antibody (K 931) to a squamous cell carcinoma associated antigen identified as the 17-1A antigen. *Hybridoma, 9*, 377-387.

Riethmuller, G., Holz, E., Schlimok, G., Schmiegel, W., Raab, R., Hoffken, K., Gruber, R., Funke, I., Pichlmaier, H., Hirche, H., Buggisch, P., Witte, J. and Pichlmayr, R. (1998) Monoclonal antibody therapy for resected Dukes' C colorectal cancer: seven-year outcome of a multicenter randomized trial. *J. Clin. Oncol, 16*, 1788-1794.

Riethmüller, G., Schneider Gadicke, E., Schlimok, G., Schmiegel, W., Raab, R., Hoffken, K., Gruber, R., Pichlmaier, H., Hirche, H., Pichlmayr, R. and et al. (1994) Randomised trial of monoclonal antibody for adjuvant therapy of resected Dukes' C colorectal carcinoma. *Lancet, 343*, 1177-1183.

Ross, A.H., Lubeck, M., Steplewski, Z. and Koprowski, H. (1986) Identification and characterization of the CO17-1A carcinoma-associated antigen. *Hybridoma, 5* Suppl 1, S21-28.

Ruf, P. and Lindhofer, H. (2001) Induction of a long-lasting antitumor immunity by a trifunctional bispecific antibody. *Blood, 98*, 2526-2534.

Schanzer, J.M., Baeuerle, P.A., Dreier, T. and Kufer, P. (2006) A human cytokine/single-chain antibody fusion protein for simultaneous delivery of GM-CSF and IL-2 to Ep-CAM overexpressing tumor cells. *Cancer Immun, 6*, 4.

Schlereth, B., Fichtner, I., Lorenczewski, G., Kleindienst, P., Brischwein, K., da Silva, A., Kufer, P., Lutterbuese, R., Junghahn, I., Kasimir-Bauer, S., Wimberger, P., Kimmig, R. and Baeuerle, P.A. (2005) Eradication of tumors from a human colon cancer cell line and from ovarian cancer metastases in immunodeficient mice by a single-chain Ep-CAM-/CD3-bispecific antibody construct. *Cancer Res, 65*, 2882-2889.

Schlereth, B., Kleindienst, P., Fichtner, I., Lorenczewski, G., Brischwein, K., Lippold, S., Silva, A., Locher, M., Kischel, R., Lutterbuse, R., Kufer, P. and Baeuerle, P.A. (2006) Potent inhibition of local and disseminated tumor growth in immunocompetent mouse models by a bispecific antibody construct specific for Murine CD3. *Cancer Immunol. Immunother*, *55*, 785-796.

Schmelzer, E., Wauthier, E. and Reid, L.M. (2006) The Phenotypes of Pluripotent Human Hepatic Progenitors. *Stem Cells*.

Spizzo, G., Gastl, G., Obrist, P., Fong, D., Haun, M., Grunewald, K., Parson, W., Eichmann, C., Millinger, S., Fiegl, H., Margreiter, R. and Amberger, A. (2006a) Methylation status of the Ep-CAM promoter region in human breast cancer cell lines and breast cancer tissue. *Cancer Lett*.

Spizzo, G., Gastl, G., Wolf, D., Gunsilius, E., Steurer, M., Fong, D., Amberger, A., Margreiter, R. and Obrist, P. (2003) Correlation of COX-2 and Ep-CAM overexpression in human invasive breast cancer and its impact on survival. *Br. J. Cancer*, *88*, 574-578.

Spizzo, G., Went, P., Dirnhofer, S., Obrist, P., Moch, H., Baeuerle, P.A., Mueller-Holzner, E., Marth, C., Gastl, G. and Zeimet, A.G. (2006b) Overexpression of epithelial cell adhesion molecule (Ep-CAM) is an independent prognostic marker for reduced survival of patients with epithelial ovarian cancer. *Gynecol. Oncol*.

Spizzo, G., Went, P., Dirnhofer, S., Obrist, P., Simon, R., Spichtin, H., Maurer, R., Metzger, U., von Castelberg, B., Bart, R., Stopatschinskaya, S., Kochli, O.R., Haas, P., Mross, F., Zuber, M., Dietrich, H., Bischoff, S., Mirlacher, M., Sauter, G. and Gastl, G. (2004) High Ep-CAM expression is associated with poor prognosis in node-positive breast cancer. *Breast Cancer Res. Treat*, *86*, 207-213.

Ullenhag, G.J., Frodin, J.E., Mosolits, S., Kiaii, S., Hassan, M., Bonnet, M.C., Moingeon, P., Mellstedt, H. and Rabbani, H. (2003) Immunization of colorectal carcinoma patients with a recombinant canarypox virus expressing the tumor antigen Ep-CAM/KSA (ALVAC-KSA) and granulocyte macrophage colony- stimulating factor induced a tumor-specific cellular immune response. *Clin. Cancer Res*, *9*, 2447-2456.

Varga, M., Obrist, P., Schneeberger, S., Muhlmann, G., Felgel-Farnholz, C., Fong, D., Zitt, M., Brunhuber, T., Schafer, G., Gastl, G. and Spizzo, G. (2004) Overexpression of epithelial cell adhesion molecule antigen in gallbladder carcinoma is an independent marker for poor survival. *Clin. Cancer Res*, *10*, 3131-3136.

Weiner, L.M. (2006) Fully human therapeutic monoclonal antibodies. *J. Immunother*, 29, 1-9.

Went, P., Vasei, M., Bubendorf, L., Terracciano, L., Tornillo, L., Riede, U., Kononen, J., Simon, R., Sauter, G. and Baeuerle, P.A. (2006) Frequent high-level expression of the immunotherapeutic target Ep-CAM in colon, stomach, prostate and lung cancers. *Br. J. Cancer*, *94*, 128-135.

Went, P.T., Lugli, A., Meier, S., Bundi, M., Mirlacher, M., Sauter, G. and Dirnhofer, S. (2004) Frequent EpCam protein expression in human carcinomas. *Hum. Pathol*, *35*, 122-128.

Wolf, E., Hofmeister, R., Kufer, P., Schlereth, B. and Baeuerle, P.A. (2005) BiTEs: bispecific antibody constructs with unique anti-tumor activity. *Drug Discov. Today*, *10*, 1237-1244.

Wong, N.A., Warren, B.F., Piris, J., Maynard, N., Marshall, R. and Bodmer, W.F. (2006) EpCAM and gpA33 are markers of Barrett's metaplasia. *J. Clin. Pathol*, *59*, 260-263.

In: New Cell Adhesion Research
Editors: Patrick Nott and Matthew Temple

ISBN 978-1-60692-378-8

Short Communication

MUCIN COATINGS FOR CONTROLLED CELL-MATERIAL INTERACTION

Tomas Sandberg* and Anita Vitina#

Department of Physical and Analytical Chemistry, Division of Surface Biotechnology, Box 577, BMC, Uppsala University, SE-751 23 Uppsala, Sweden.

ABSTRACT

Recently, we showed that surface coatings of mucins, i.e. large-sized amphiphilic glycoproteins found at the mucosal surfaces of all vertebrates, are capable of reducing the uptake of human neutrophils to a polymeric model biomaterial. In the present investigation, human-derived epithelial HEK293, fibroblastic MRC-5 and osteoblastic MG63 cells were studied microscopically during contact with tissue culture-treated polystyrene (TCPS) substrates coated with bovine (BSM), porcine (PGM) and human (MG1) mucins, respectively. We found that, for an initial period of two days, all mucin-coated substrates exhibited strongly reduced cell uptakes compared to the non-coated controls. Over time, successively more cells established on the surfaces. However, excluding the BSM-coated substrate which probably was insufficiently coated, the in-growth of cells was significantly slower to the mucin-coated than to the non-coated substrates. Particularly, the PGM- and MG1 coated substrates showed a very limited in-growth of fibroblastic cells over a period of one week. Importantly, the cell viabilities remained at a high level (≥95 %) during all time of cell-material contact, indicating that the mucin-coated substrates were essentially non-toxic. A screening experiment suggested that mucin coatings also could be used to prevent the surface adhesion of neuronal cells. Altogether, our results demonstrate that mucin coatings could be used to control the in-growth of cells to an artificial material. Specifically, the strong inhibitory effect on fibroblastic cell adhesion suggested that mucin coatings could be used to reduce fibrous encapsulation of implanted materials.

---

* Correspondence concerning this article should be addressed to: Tomas Sandberg, phone: +46-18-4714682; fax: +46-18-555016; email: tomas.sandberg@ytbioteknik.uu.se.

Present address: GE Healthcare Biosciences AB, SE-751 84 Uppsala, Sweden.

Keywords: Mucin; BSM; PGM; MG1; Biomaterial coating; Tissue culture-treated polystyrene (TCPS); Cell adhesion; HEK293; MRC-5; MG63; SH-SY5Y.

INTRODUCTION

The fate of an implanted artificial material depends on its interactions with the surrounding tissues and fluids of the host [1]. Long-term implant survival normally calls for contact with cells from one or more of the four basic types of host tissue, i.e. epithelial, connective, muscle and nervous tissue. However, from an integrative point of view, the implant surface presents difficult challenges for the invading host cells. Specifically, while too strong cell-material interaction may inhibit the formation of higher-order (tissue-like) cell assemblies, too weak interaction disfavors integration and eventually leads to implant inactivation by fibrous encapsulation [1].

From studying the native mucosa that lines the epithelial surfaces of humans and other vertebrates, we proposed in the past [2-4] that mucins, a class of large-sized amphiphilic glycoproteins [5,6], could be favorably used to coat biomedical materials. Particularly, mucin coatings of a bovine salivary mucin fraction (BSM) were shown to reduce the uptake of proteins, cells and bacteria to various polymeric substrates [2-4]. Recently we expanded our mucin research platform by preparing two more defined BSM fractions as well as three mucin fractions from porcine (PGM) and human (MG1) sources [7]. In two subsequent studies [8,9], surface coatings of these mucins were shown capable of suppressing the immuno-response (adhesion and activation) of human neutrophils towards a polymeric model biomaterial. In particular, the performance of the mucin coatings was proposed to depend on their ability to shield the material surface.

Clearly, cell interaction data is crucial for understanding the *in vivo* compatibility of mucin-coated materials. Thus, the present study evaluated the behavior of human-derived epithelial, fibroblastic, osteoblastic and neuronal cells during contact with tissue culture-treated polystyrene substrates coated with the BSM, PGM and MG1 mucins, respectively. Based on the results, a discussion is made on the potential use of mucins as biomaterial coatings for controlled cell-material interaction.

MATERIALS AND METHODS

Mucins from bovine salivary glands (BSM), porcine stomach scrapings (PGM) and human whole saliva (MG1) were prepared according to recently developed protocols [7]. In brief, partially purified bovine submaxillary gland mucin (Sigma M3895), partially purified pig gastric mucin (Sigma M1778) and human whole saliva were dissolved, equilibrated and then gel-filtered, before being desalted and finally freeze-dried. The bovine material was pre-fractionated using anion exchange chromatography. In the present study, only the low molar mass variants of BSM and PGM were used. The prepared mucins differed in chemical composition and biophysical properties [7]. Specifically, while the BSM and PGM mucins were essentially pure, the MG1 mucin contained significant amounts of bound non-mucin material, including albumin, which constituted 1.6 wt-% of the MG1 preparation. Furthermore, while the BSM and PGM mucins adopted random coil conformations in solution, the MG1 mucin appeared compact. The predominant mucin species in the BSM, PGM and MG1 preparations had molar masses of 0.8, 1.2 and 4.2 MDa, respectively. The mucins were dissolved and equilibrated in 20 mM phosphate-buffered saline of pH 7.4 (PBS20) for at least 12 h before any experiment.

All mucins were adsorbed for 20 h at room temperature onto tissue culture-treated polystyrene (TCPS) 24-well multidish substrates (Costar 3524; Corning). The surface of this material, which is hydrophilic and negatively charged, is considered a relevant model for surface-oxidized polymeric biomaterials [10]. The MG1 mucin was adsorbed at 0.125 mg/mL and the BSM and PGM mucins were adsorbed at 0.25 mg/mL. These concentrations were based on the near-plateau concentrations recently found for a hydrophobic model surface [7]. The non-coated material (PBS20 buffer) served as a reference. In one experimental setup, the center of the substrate surface was spot-coated with BSM at 1 mg/mL under the conditions given above. The coated substrates were rinsed three times with PBS20 and stored in the same buffer until used.

The cellular response towards the mucin-coated substrates was evaluated using epithelial-like HEK293 (human embryonic kidney; ECACC 85120602), fibroblast-like MRC-5 (human embryonic lung; ECACC 84101801), osteoblast-like MG63 (human osteosarcoma; ECACC 86051601) and neuronal-like SH-SY5Y (human peripheral neuroblastoma) cells derived from human sources. To increase readability, these cells are in the following referred to as "epithelial", "fibroblastic", "osteoblastic" and "neuronal" cells, respectively. Cells were cultured in sterile-filtered Dulbecco's Modified Eagle Medium/Nutrient Mixture F-12 Ham (DMEM/F-12 Ham; Sigma D8437) supplemented with 10 % fetal bovine serum (HyClone CH30160003) and gentamycin (Sigma G1264) at a concentration of 10 μg/mL. Before cell seeding, the mucin-coated substrates were washed two times with serum-free cell culture medium and one last time with cell culture medium. Cells, treated with trypsin/EDTA (Sigma T4049) and washed in cell culture medium, were then diluted to 50,000 cells/mL in fresh cell culture medium and added in 1 mL aliquots to the mucin-coated substrates (26,300 cells/cm^2). The seeded samples were incubated at 37°C in a humidified atmosphere of 95 % air and 5 % CO_2. At 4 days of incubation fresh cell culture medium was added to prevent any effects of starvation. The total number of live and dead cells was counted after trypsin/EDTA treatment and staining with 0.4 % Trypan Blue (Sigma T8154), using a hemocytometer in combination with a light microscope (Nikon Eclipse TE 300). Cell viability was calculated as the fraction of live cells in relation to the total number of cells. Cell morphology was documented using a standard light microscope (Zeiss) equipped with a CoolSNAP Pro/Cf CCD camera (Photometrics). The cell surface coverage was analyzed using the ImageJ software (Version 1.40g; NIH). Data are generally based on 3 measurements (n=3). The images shown are representative of 2-3 separate experiments.

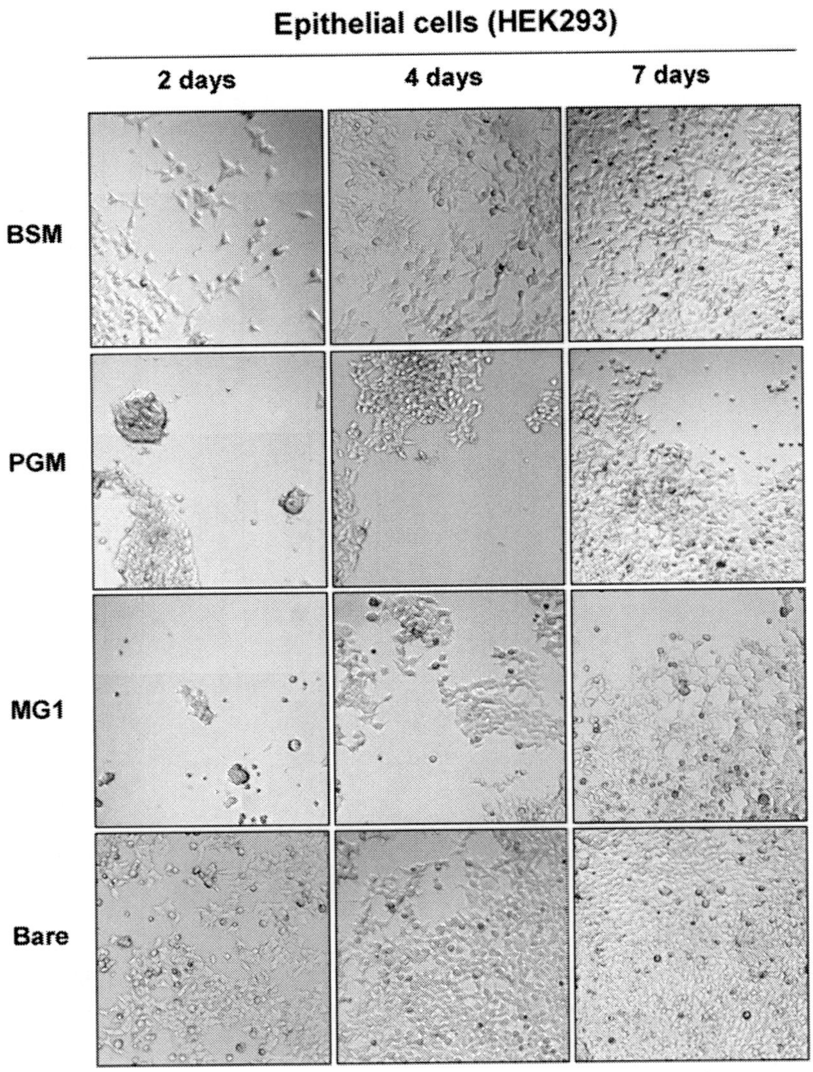

Figure 1. Epithelial HEK293 cells after 2, 4 and 7 days of incubation with tissue culture-treated polystyrene (TCPS) coated with BSM, PGM and MG1 mucins, respectively. Cells were seeded at 50,000 cells/mL in cell culture medium and grown at 37°C in humidified atmosphere of 95 % air and 5 % CO_2. Bare TCPS (PBS20) substrates served as positive controls. Images were taken using a standard light microscope setup. See Materials and Methods section for further details.

RESULTS AND DISCUSSION

Figures 1 and 2 present images of differently mucin-coated (BSM, PGM or MG1) tissue culture-treated polystyrene (TCPS) substrates after 2, 4 and 7 days of incubation with epithelial HEK293 and fibroblastic MRC-5 cells, respectively. For comparison, each figure includes images of non-coated (bare) substrates incubated the same way. Corresponding cell surface coverages are presented in Figure 3. After 2 days of incubation, the epithelial cells covered significantly lower surface fractions of the mucin-coated substrates than of the non-coated control. In addition, substantial cell aggregation was observed on the mucin-coated substrates, indicating that cell-cell contact was favored and cell-material contact disfavored [11]. Over time, successively more epithelial cells established on the surfaces and fewer aggregates were observed. However, still after 7 days of incubation, all mucin-coated substrates were associated with lower cell surface coverages than the non-coated control. Specifically, while the cell coverage of the BSM-coated substrate (83 %) was only slightly lower than that of the non-coated control (100 %), the cell coverages of the PGM- and MG1 coated substrates were relatively low (on average 55 %).

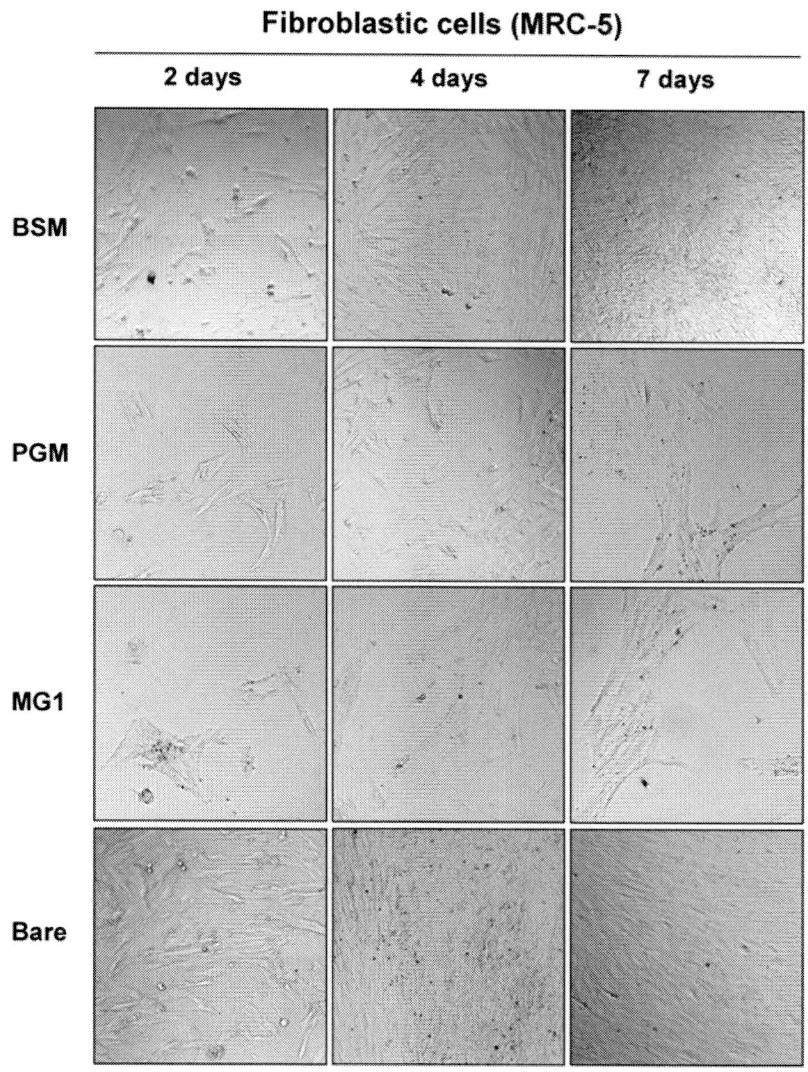

Figure 2. Fibroblastic MRC-5 cells after 2, 4 and 7 days of incubation with tissue culture-treated polystyrene (TCPS) coated with BSM, PGM and MG1 mucins, respectively. The experimental conditions were the same as those described in Figure 1.

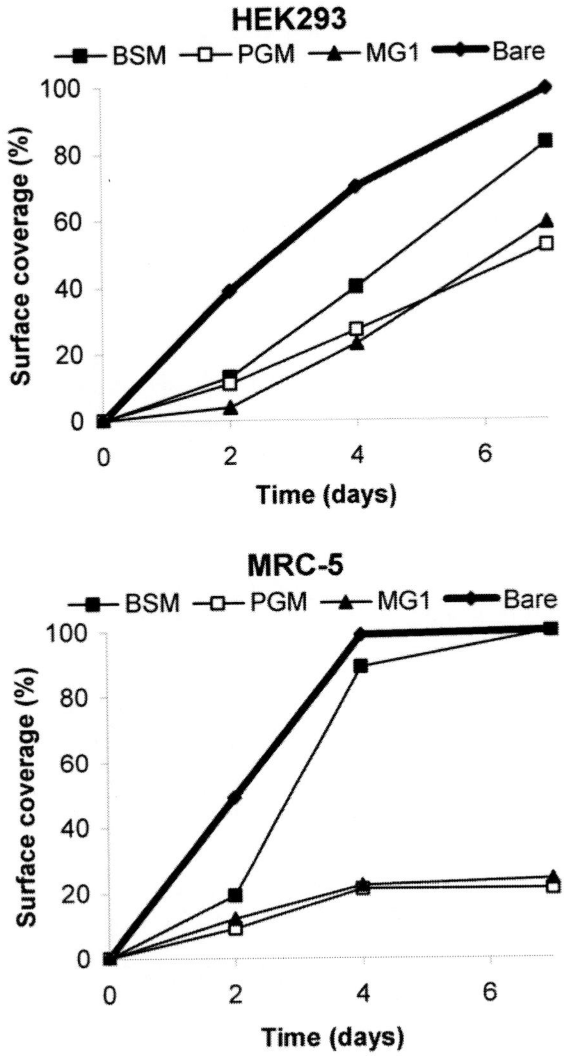

Figure 3. Diagrams showing the surface coverages of HEK293 and MRC-5 cells, respectively, on mucin-coated TCPS substrates over 7 days of incubation. Bare TCPS (PBS20) substrates served as positive controls and are marked with bold lines.

Initially, also the fibroblastic cells adhered poorly to all mucin-coated substrates. Although cell aggregation was observed, this effect was not as pronounced as with the epithelial cells. While the BSM-coated substrate over time became associated with essentially the same high cell coverages as the non-coated control, the PGM- and MG1 coated substrates showed essentially no uptake of fibroblastic cells between 2 and 7 days of incubation.

In a parallel experiment (Figure 4), the uptake of osteoblastic MG63 cells was shown to be significantly lower to the BSM- and PGM coated substrates than to the non-coated control after 7 days of incubation. Moreover, in resemblance with our observations on the epithelial cells, a substantial fraction of the osteoblastic cells grew in clusters on the mucin-coated surfaces.

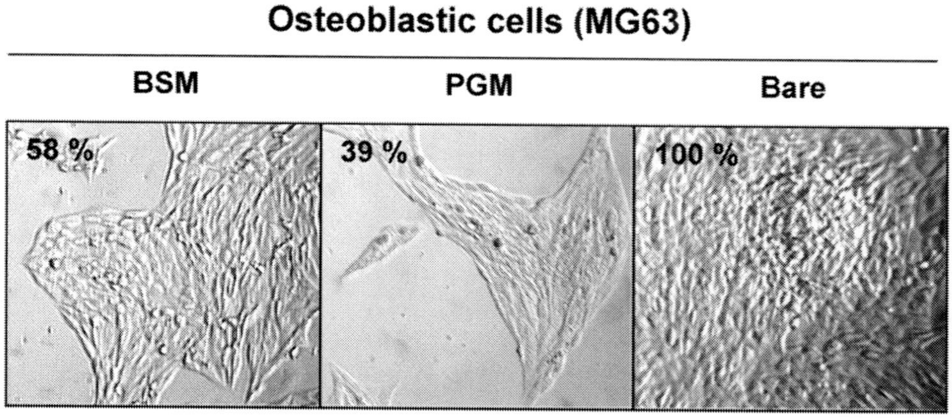

Figure 4. Osteoblastic cells (MG63) after 7 days of incubation with tissue culture-treated polystyrene (TCPS) coated with BSM and PGM mucins, respectively. The experimental conditions were the same as those described in Figure 1. Numbers indicate cell surface coverages compared to the bare control.

Although not shown, the cell viability remained at a high level (≥95 %) for the epithelial, fibroblastic and osteoblastic cells during one week of contact with the mucin-coated substrates. This finding indicated that the mucin coatings exerted essentially no toxic effect on the cells.

Figure 5 summarizes the total number of viable cells (adherent and non-adherent) as a function of time for each cell-type, during cell contact with the differently treated substrates. In good correlation with our microscopic observations, the fibroblastic cells contacting the PGM- and MG1 coated substrates showed essentially no growth over one week. During an initial period of 4 days, the epithelial and osteoblastic cells showed the same cell growth on the mucin-coated substrates as on the non-coated control. This was somewhat surprising, considering that the mucin-coated substrates were associated with considerably lower cell coverages than the non-coated substrates. Our explanation to this was that a large fraction of the cell population initially grew in three-dimensional clusters on the mucin-coated surfaces and therefore "disappeared" in the cell coverage data. As cell aggregation became less pronounced, the growth curves leveled off and a better correlation could be found between the trends in cell count and cell surface coverage.

In a 24-hour screening experiment, the BSM coating was also shown capable of preventing the uptake of neuronal SH-SY5Y cells to the used model substrate (Figure 6). As in the case of the epithelial and osteoblastic cells, the neuronal cells showed an increased propensity to aggregate on the mucin-coated surfaces.

Normal cells in the human body experience a three-dimensional environment, surrounded by other cells, membranes, connective layers, and adhesion proteins. Ryan and co-workers demonstrated that tissue spreading on implantable substrates is a complex process, which depends on the outcome of competitive cell-cell versus cell-substrate adhesivity [11]. Thus, the fact that a large fraction of the epithelial, osteoblastic and neuronal cells initially preferred to grow in clusters on the mucin-coated surfaces could possibly have implications for the field of tissue engineering.

Recent data [8] demonstrated that the uptake of human neutrophils to a mucin-coated polymeric material relates to the mucin concentration used during the coating process. Specifically, coatings formulated at successively higher mucin concentrations were successively more effective at preventing cell adhesion. Thus, the high uptakes of epithelial and fibroblastic cells, respectively, to the BSM-coated substrate were probably a consequence of insufficient coating and could be reduced by coating the material at a higher BSM concentration.

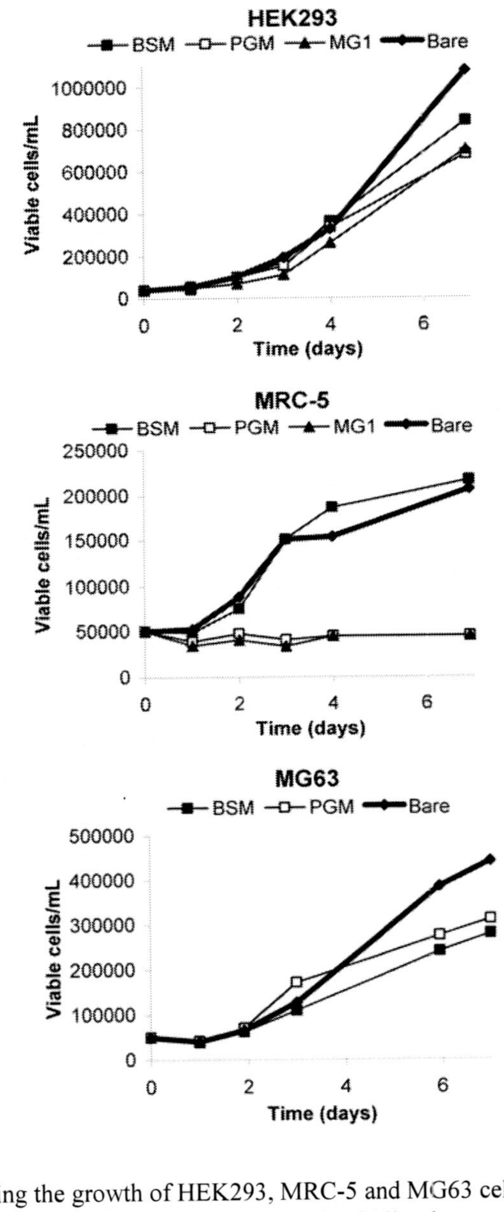

Figure 5. Diagrams showing the growth of HEK293, MRC-5 and MG63 cells during one week of contact with mucin-coated TCPS substrates. Bare TCPS (PBS20) substrates served as positive controls. See Materials and Methods section for further details.

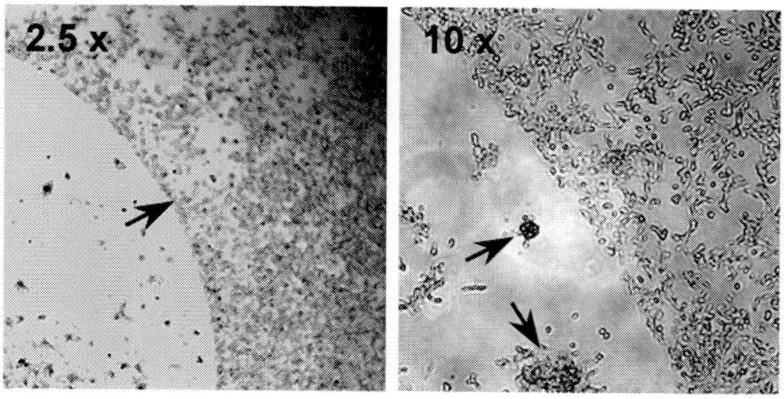

Figure 6. Human neuronal cells (SH-SY5Y) incubated for 24 h on a TCPS substrate spot-coated overnight with BSM at 1 mg/mL. Images shown at 2.5 and 10 times magnification. The arrow in the left image marks the interface between the mucin-coated and non-coated surface. The arrows in the right image indicate cell aggregates.

To mimic the *in vivo* situation the cell medium contained serum. It is well known that the uptake of serum proteins to an implant surface promotes cell adhesion [11]. Mucins are amphiphilic molecules and have been suggested to interact by non-specific repulsive as well as specific adhesive mechanisms [12]. A plausible explanation to the cell-repelling effect of the mucin coatings is that these could prevent cell-binding proteins, such as fibronectin, vitronectin, laminin and collagen IV from adsorbing to the surface. Alternatively or additionally, the mucin coatings could possibly affect the cell-binding properties of the adsorbing serum proteins. In support of such synergistic mechanism, recent studies on the sequential adsorption of BSM and bovine serum albumin to a polystyrene substrate have shown that several biologically relevant parameters, such as the layer viscoelasticity [13] and the interfacial friction and layer stability (to be published) are significantly different for the BSM-albumin composite layer and the layers of its separately adsorbed constitutents.

The present data demonstrate that mucin coatings could be used to control the in-growth of cells to an artificial material. Of particular interest, fibroblastic cells were strongly and durably prevented from establishing on the mucin-coated surfaces, suggesting that mucin coatings could be used to reduce fibrous encapsulation of implanted materials [14].

CONCLUSION

In conclusion, bovine (BSM), porcine (PGM) and human (MG1) mucin coatings were shown capable of reducing the in-growth of human-derived epithelial, fibroblastic, osteoblastic and neuronal cells to a polymeric material. Cell viabilities remained at a high level during contact with the mucin-coated substrates, indicating that the mucin coatings were essentially non-toxic. The fibroblastic cells showed very limited uptake to the PGM- and

MG1 coated substrates. Altogether, our results demonstrate that mucin coatings could be used to control the in-growth of cells to an artificial material.

Acknowledgments

The authors greatly acknowledge Karin Caldwell, Uppsala University for introducing us to the mucin field and for providing working space and facilities. We also thank Finn Hallböök, Uppsala University for kindly donating the SH-SY5Y cell line and Klara Sandberg for providing T.S. with valuable time during manuscript preparation. Finally, T.S is grateful for financial support from the Lennander foundation.

References

[1] Anderson, JM. Inflammation, wound healing and the foreign-body response. In: Ratner BD, Hoffman AS, Schoen FJ, Lemons JE, editors. *Biomaterials Science*. Oxford: Elsevier; 2004; 296-304.

[2] Shi, L; Caldwell, KD. Mucin adsorption to hydrophobic surfaces. *J Colloid Interf Sci*, 2000 224, 372-381.

[3] Shi, L. Biomimetic surfaces of biomaterials using mucin-type glycoproteins. *Trends Glycosci Glycotechnol,* 2000 12, 229-239.

[4] Shi, L; Ardehali, R; Caldwell KD; Valint P. Mucin coating on polymeric material surfaces to suppress bacterial adhesion. *Colloids Surf B,* 2000 17, 229-239.

[5] Perez-Vilar, J; Hill, RL. Mucin family of glycoproteins. In: Lennarz WJ, Lane MD, editors. *Encyclopedia of Biological Chemistry.* Oxford: Elsevier; 2004; 758-764.

[6] Bansil, R; Turner, BS. Mucin structure, aggregation, physiological functions and biomedical applications. *Curr Opinion Coll Interface Sci*, 2006 11, 164-170.

[7] Sandberg, T; Blom, H; Caldwell, KD. Potential use of mucins as biomaterial coatings. I. Fractionation, characterization, and model adsorption of bovine, porcine, and human mucins. *J Biomed Mater Res A,* 2008. Published online, DOI: 10.1002/jbm.a.32266.

[8] Sandberg, T; Carlsson, J; Karlsson Ott, M. Mucin coatings suppress neutrophil adhesion to a polymeric model biomaterial. *Microsc Res Tech*, 2007 70, 864-868.

[9] Sandberg, T; Carlsson, J; Karlsson Ott, M. Interactions between human neutrophils and mucin-coated surfaces. *J Mater Sci Mater Med*, 2008. Published online, DOI: 10.1007/s10856-008-3595-y.

[10] Ratner, BD; Hoffman, AS. Physicochemical surface modification of materials used in medicine. In: Ratner BD, Hoffman AS, Schoen FJ, Lemons JE, editors. *Biomaterials Science*. Oxford: Elsevier; 2004; 67-79.

[11] Ryan, PL; Foty, RA; Kohn, J; Steinberg, MS. Tissue spreading on implantable substrates is a competitive outcome of cell-cell vs. cell-substratum adhesivity. *PNAS*, 2001 98, 4323-4327.

[12] O'Boyle, KP; Chen, T; Kozlowski, S. Mucin inhibition of lymphocyte function does not require specific mucin-ligand interactions. *Scand J Immunol*, 2000 52, 46-52.

[13]Feiler, AA; Sahlholm, A; Sandberg, T; Caldwell KD. Adsorption and viscoelastic properties of fractionated mucin (BSM) and bovine serum albumin (BSA) studied with QCM-D. *J Colloid Interf Sci,* 2007 315, 475-481

[14]Thomsen, P; Gretzer, C. Macrophage interactions with modified material surfaces. *Curr Opin Sol State Mater Sci*, 2001 5, 163-176.

INDEX

A

B

C

D

E

F

J

K

L

M

N

O

P

Q

R

S

U

V

W

X

Y

Z